Synthetic Biology

Synthetic Biology

Nana Addo
Amina Bello

Kruger Brentt
Publishers
2026

Kruger Brentt Publishers UK. LTD.
Company Number 9728962

Regd. Office: 68 St Margarets Road, Edgware, Middlesex HA8 9UU

ISBN: 978-1-78715-658-6

For information on all our publications visit our website at http://krugerbrentt.com/

Preface

Synthetic Biology represents one of the most dynamic, interdisciplinary, and impactful scientific frontiers of the twenty-first century. By integrating principles from biology, engineering, genetics, chemistry, computer science, and biotechnology, synthetic biology aims to design, construct, and optimize biological systems with unprecedented precision and creativity. This emerging field transcends traditional boundaries, offering transformative solutions in healthcare, agriculture, environmental remediation, energy production, and industrial biotechnology.

This book introduces readers to the foundational concepts and methodologies of synthetic biology, beginning with the engineering mindset applied to living systems—standardized biological parts, modular genetic circuits, chassis organisms, and computational design tools. It explains how synthetic biologists use DNA synthesis, genome editing, metabolic engineering, and systems biology to build customized organisms and metabolic pathways capable of performing novel functions. The text highlights landmark achievements such as synthetic genomes, designer microbes, gene circuits, biosensors, and bio-based manufacturing platforms.

Beyond the technical frameworks, the book examines the real-world applications that are reshaping industries—from therapeutic gene networks and engineered cells for cancer treatment to synthetic pathways for drug production, biodegradable materials, biofuels, and carbon capture. Ethical considerations, biosafety, biosecurity, and regulatory concerns are explored to provide a balanced understanding of the power and responsibility inherent in engineering living systems.

Written for students, researchers, innovators, policymakers, and multidisciplinary learners, Synthetic Biology offers an accessible yet rigorous guide to a rapidly evolving discipline. It aims to inspire inquiry, support academic learning, and empower the next generation of thinkers and practitioners to advance biological innovation responsibly and creatively.

Nana Addo

Amina Bello

Contents

CHAPTER 1

Introduction to Synthetic Biology

1.1 WHAT IS SYNTHETIC BIOLOGY?

Synthetic biology represents a paradigm shift in how humans interact with living systems. By applying engineering principles to biology, it allows for the design of organisms with novel capabilities, offering solutions to pressing challenges in health, agriculture, energy, and the environment. However, the field is accompanied by ethical, ecological, and societal considerations that require careful governance and public engagement. As technology advances, synthetic biology has the potential to redefine the boundaries of life and transform multiple industries. Its future will depend not only on scientific breakthroughs but also on responsible innovation, collaboration, and informed societal discourse. The integration of biology, engineering, and computer science in this field continues to inspire both awe and critical reflection, positioning synthetic biology as a key driver of 21st-century science and technology.

1.1.1 Historical Background

The roots of synthetic biology can be traced back to the early experiments in genetic engineering during the 1970s when scientists began inserting foreign DNA into bacteria to study gene function. However, the modern concept of synthetic biology emerged in the early 2000s when researchers sought to design biological systems with engineering principles, including standardization, modularity, and abstraction. The first synthetic genome was created in 2010 by the J. Craig Venter Institute, which synthesized a complete bacterial genome and transplanted it into a host cell, effectively creating the world's first synthetic organism. This milestone demonstrated that life could be programmed, setting the stage for a new era in biology.

1.1.2 Core Principles of Synthetic Biology

Synthetic biology is built on several core principles. First, standardization involves creating a library of genetic parts that can be reliably combined to produce desired functions. Second, modularity emphasizes designing biological systems as interchangeable units, allowing for easier construction and troubleshooting.

Third, abstraction simplifies complex biological networks by representing them as components with defined inputs and outputs. Fourth, modeling and simulation enable scientists to predict how engineered organisms will behave before physically creating them. These principles collectively transform biology from a largely descriptive science into a predictive and design-oriented discipline, bridging the gap between engineering and life sciences.

1.1.3 Techniques and Tools

The practice of synthetic biology relies on cutting-edge techniques. DNA synthesis and assembly technologies allow researchers to build long stretches of genetic material with precision. CRISPR-Cas systems enable targeted editing of genomes, making it possible to switch genes on or off or insert new functions. Computational tools are increasingly used to model genetic circuits, metabolic pathways, and cellular behavior, reducing trial-and-error experiments. Microfluidics and lab-on-a-chip devices facilitate high-throughput testing and screening of synthetic constructs. Synthetic biologists also utilize directed evolution, a process in which genes are deliberately mutated and selected for desired traits, to optimize engineered systems.

1.1.4 Applications of Synthetic Biology

The applications of synthetic biology are vast and transformative. In medicine, engineered bacteria can produce pharmaceuticals, vaccines, or therapeutic molecules with high efficiency. Synthetic biology enables the development of biosensors that detect diseases, toxins, or pathogens rapidly. In agriculture, synthetic organisms can promote plant growth, increase crop resistance to pests, and improve nutrient absorption. Environmental applications include microbes engineered to degrade pollutants, capture carbon dioxide, or produce sustainable biofuels. Additionally, synthetic biology plays a crucial role in industrial biotechnology by optimizing the production of chemicals, enzymes, and materials.

1.1.5 Ethical and Safety Considerations

The rapid growth of synthetic biology raises significant ethical, legal, and safety concerns. The creation of novel organisms could have unintended ecological consequences if they escape into natural ecosystems. Dual-use research, where technology could be misused for harmful purposes, presents a biosecurity risk. There are also philosophical and ethical debates about "playing God" and the moral implications of creating life forms for human purposes. Regulatory frameworks are still evolving, with international organizations and governments attempting to balance innovation with biosafety and ethical responsibility. Public engagement and transparent communication are essential to ensure societal acceptance and informed decision-making.

1.1.6 Challenges in Synthetic Biology

Despite its promise, synthetic biology faces several challenges. The complexity of biological systems makes predictability difficult, and unintended interactions between synthetic and natural components can arise. Scaling up laboratory experiments to industrial or environmental applications remains a hurdle. Ethical and regulatory uncertainties can slow progress, especially in countries with stringent biosafety laws. The high cost of synthetic DNA and sophisticated lab infrastructure limits accessibility for smaller research groups. Additionally, public skepticism or fear of genetically modified organisms could impede widespread adoption, highlighting the importance of education and outreach.

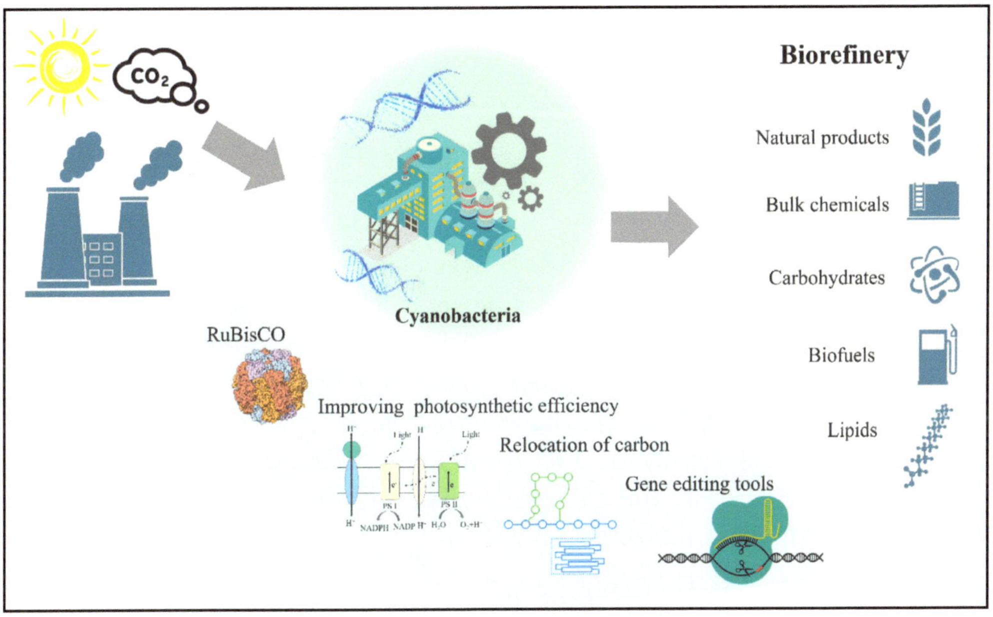

Complexities in the Practice of Synthetic Biology

1.1.7 Concepts and Components

Synthetic biology relies on foundational concepts and components that enable precise design and control of biological systems:

- **Biobricks:** Standardized DNA sequences that perform specific functions and can be combined modularly.
- **Genetic Circuits:** Engineered networks of genes designed to produce predictable responses, akin to electronic circuits.
- **Metabolic Pathway Engineering:** Modification of cellular pathways to optimize the production of desired chemicals.
- **Minimal Genomes:** Stripped-down versions of organisms containing only essential genes for survival and function.

- **Chassis Organisms:** Host organisms chosen as platforms for synthetic biology experiments.
- **Biosensors:** Engineered systems that detect environmental or biological signals.
- **Directed Evolution:** Artificial selection of genetic variants to improve specific traits.
- **CRISPR-Cas Gene Editing:** Targeted modification of DNA sequences for precise control.
- **Computational Modeling:** Simulation of biological systems to predict behavior before experimentation.
- **Biofoundries:** Facilities equipped with automated systems for high-throughput synthetic biology research.

1.2 HISTORICAL DEVELOPMENT OF SYNTHETIC BIOLOGY

1.2.1 Conceptual Forerunners and Theoretical Foundations

Long before the term synthetic biology was coined, the mid-twentieth century provided the essential conceptual framework. The landmark discovery of the double helix structure of DNA by Watson and Crick in 1953 unveiled the molecular alphabet of life, suggesting a mechanism for heredity and information storage. This was followed by the deciphering of the genetic code, which revealed how sequences of DNA nucleotides are translated into the amino acid sequences of proteins. These breakthroughs established the central dogma of molecular biology, outlining the flow of genetic information from DNA to RNA to protein.

The 1970s witnessed the birth of genetic engineering with the development of recombinant DNA technology, pioneered by scientists like Paul Berg, Herbert Boyer, and Stanley Cohen. This technique, which allowed for the cutting and pasting of DNA fragments from different organisms, was a revolutionary step. It demonstrated that genetic material could be manipulated, and biological components could be functionally transferred across species boundaries, providing the first tangible proof that biology could be engineered.

1.2.2 Dawn of a New Discipline and the Engineering Ethos

The term synthetic biology first appeared in scientific literature in the 1970s and 1980s, but it was at the turn of the millennium that it began to coalesce into a defined discipline. A pivotal moment came in 2000 with the publication of two seminal papers that established the core engineering principles of the field. Researchers from Princeton University engineered a synthetic genetic circuit in E. coli bacteria, known as the toggle switch, which could flip between two stable states like a genetic light switch. Simultaneously, scientists at the University of Boston created a rhythmic oscillator, dubbed the repressilator, which caused cells to produce fluorescent proteins in periodic, oscillating waves. These achievements were monumental not merely for their technical prowess but for their conceptual impact; they demonstrated that biological components could be rewired to create predictable, digital-like logic functions, treating the cell as a programmable substrate. This period also saw the establishment of the International Genetically Engineered Machine competition, which has played an instrumental role in popularizing the field, promoting standardization, and inspiring generations of students to approach biology with an engineer's mindset.

1.2.3 Genome Synthesis Era and the Rise of Foundational Tools

Building on the success of genetic circuits, the ambition of synthetic biologists expanded from editing genes to writing entire genomes. This era was defined by a series of high-profile, proof-of-concept projects that tested the limits of DNA synthesis and assembly. In 2002, researchers at the State University of New York at Stony Brook achieved the first synthesis of a complete viral genome, creating poliovirus from scratch using mail-order DNA sequences.

A more significant milestone was reached in 2010 by the J. Craig Venter Institute, which announced the creation of the first self-replicating synthetic bacterial cell. The team synthesized the entire genome of the bacterium Mycoplasma mycoides from chemical building blocks and transplanted it into the cytoplasm of a closely related species, effectively booting up a cell controlled by synthetic DNA. This project was a landmark demonstration of synthetic genomics, proving that a chemical recipe could produce a functioning chromosome capable of sustaining life. Concurrently, the cost of DNA synthesis began to plummet, driven by technological advances, making the reading and writing of DNA more accessible and scalable than ever before.

1.2.4 CRISPR Revolution and the Acceleration of Precision Editing

While the ability to synthesize DNA was advancing rapidly, the field lacked a truly precise and facile tool for editing genomes within living cells. This changed dramatically with the advent of CRISPR-Cas9 gene-editing technology, adapted from a bacterial immune system into a programmable genomic scalpel. Its development, for which Emmanuelle Charpentier and Jennifer Doudna were awarded the Nobel Prize in Chemistry in 2020, marked a watershed moment for synthetic biology.

CRISPR-Cas9 provided an unprecedented ability to make targeted cuts in DNA at virtually any location in the genome with relative ease and low cost. This technology removed a major bottleneck, shifting the primary challenge of synthetic biology from the technical difficulty of genetic modification to the intellectual challenge of design. Researchers could now iteratively engineer complex traits with high precision, accelerating applications in medicine, such as CAR-T cell therapies, and in agriculture, leading to the development of crops with enhanced nutritional profiles and disease resistance.

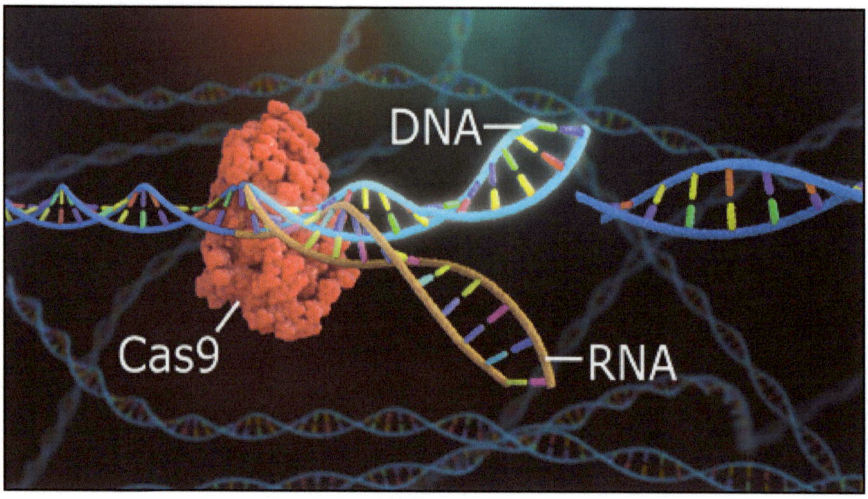

CRISPR Technological Revolution

Milestone	Year	Significance
Structure of DNA	1953	Provided the fundamental understanding of life's information molecule, the essential alphabet for all future engineering.
Recombinant DNA Technology	1970s	Enabled the cutting and pasting of genes from different organisms, proving genetic material was malleable and interchangeable.
The Toggle Switch & Repressilator	2000	Demonstrated that genetic components could be assembled into predictable, engineered circuits, establishing the field's core engineering ethos.
Synthesis of Poliovirus	2002	First chemical synthesis of a complete viral genome, proving that a pathogen could be recreated from written instructions.
iGEM Foundation	2003	Catalyzed a global community and culture of standardization and open-source innovation in biological engineering.
Synthetic M. mycoides Cell	2010	Creation of the first self-replicating bacterial cell controlled by a chemically synthesized genome, a landmark in synthetic genomics.
CRISPR-Cas9 Adaptation	2012	Provided a highly precise, programmable, and accessible tool for genome editing, dramatically accelerating the pace of biological design.

1.3 KEY CONCEPTS AND PRINCIPLES IN SYNTHETIC BIOLOGY

1.3.1 Engineering Biology: From Evolution to Design

The future of engineering biology points toward a world where biological design becomes an indispensable part of human creativity and infrastructure. The integration of AI, robotics, and bioinformatics will lead to fully automated biological design platforms capable of producing customizable organisms. From on-demand biofabrication of materials to self-healing living buildings, engineered biology will extend beyond laboratories into daily life and industrial ecosystems. As technology matures, it will also interface with evolutionary biology in new ways, allowing researchers to accelerate adaptive processes in silico and implement them in vivo. The long-term vision of this discipline is not merely to imitate nature but to expand upon its possibilities, creating new forms of life optimized for human and planetary well-being. Engineering biology, standing on the shoulders of evolution, transforms the randomness of natural selection into the precision of design, ushering in an era where biology itself becomes a programmable medium for innovation.

Foundations in Evolutionary Processes

The evolutionary foundation of life is essential to appreciate the principles of engineering biology. Evolution serves as nature's most powerful engineer, using random mutations and environmental selection pressures to generate biological diversity and efficiency. Over billions of years, evolution has optimized complex cellular mechanisms such as metabolism, replication, and signaling pathways, effectively producing systems that scientists now study for inspiration. By decoding evolutionary logic, researchers can identify which genetic designs yield successful functionalities under specific conditions. Evolutionary biology thus provides both the map and the experimental archive for synthetic biologists seeking to replicate or enhance nature's designs. Unlike traditional evolutionary adaptation, biological engineering is directed and goal-oriented. It does not rely on random variation but on rational design and controlled experimentation guided by computational and molecular tools.

Emergence of Synthetic Biology

The field of synthetic biology emerged from molecular genetics advances that made it possible to edit and assemble genetic sequences with incredible precision. Where early genetic engineering focused on modifying single genes, synthetic biology aims to program entire cellular systems. This discipline draws analogies from computer science and electrical engineering, incorporating concepts like modularity, standardization, and reusability. Scientists design genetic circuits that perform logical operations inside cells, turning genes on or off in response to environmental signals. The culmination of this work enables the creation of cells with novel metabolic capabilities, such as bacteria that produce pharmaceuticals or yeast strains that synthesize biofuels. By shifting from random mutagenesis to rational coding, synthetic biology turns cells into living factories governed by human intention rather than natural selection.

Design-Driven Paradigm

Engineering biology reframes life as a design space defined by rules that can be modeled, predicted, and optimized. The core philosophy lies in applying engineering methodologies—system design, modeling, modularization, and testing—to biological systems. This design-driven approach introduces a cycle of "design-build-test-learn" that continuously refines genetic and metabolic constructs.

Computer-aided design tools and AI-driven algorithms now allow biologists to simulate the behavior of engineered pathways before they are built experimentally. Through this paradigm, biological components such as promoters, ribosome-binding sites, and enzymes can be quantitatively tuned for desired outputs. The result is a rapidly accelerating capability to design synthetic organisms that outperform their natural analogs in efficiency, yield, or stability.

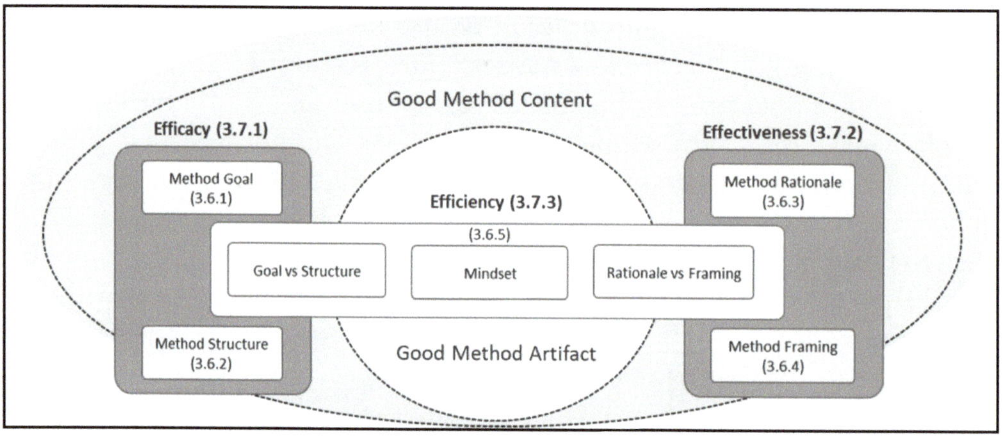

Framework Guided by Design Principles

Tools and Technologies Enabling Engineering Biology

The advancement of engineering biology relies heavily on powerful enabling technologies. CRISPR-Cas systems have revolutionized genome editing by allowing high-precision, site-specific DNA modifications. Advances in sequencing technologies provide comprehensive insights into genomes, transcriptomes, and proteomes, enabling the identification of patterns needed for design. DNA synthesis platforms now make it possible to assemble large genomes in laboratories, reducing the time required to prototype living systems. Artificial intelligence accelerates this process by predicting protein folding, gene regulatory effects, and metabolic outcomes of engineered constructs. Computational modeling helps simulate entire cellular environments, offering predictive insights that reduce costly experimental iterations. These technologies collectively create an ecosystem where designing biology is akin to writing a program—ordered, purposeful, and iterative.

Metabolic Engineering and Biomanufacturing

One of the most practical applications of engineering biology lies in metabolic engineering and biomanufacturing. By redesigning metabolic pathways, scientists can redirect cellular resources toward producing valuable compounds such as biofuels, bioplastics, and pharmaceuticals. Microbes can be optimized to synthesize complex molecules traditionally derived from petroleum or plant extraction, thereby reducing environmental impact and improving sustainability. For instance, engineered yeast has been used to produce precursors for antimalarial drugs, while bacteria have been modified to generate biodegradable polymers. This convergence of biology and industry redefines manufacturing by replacing chemical synthesis with biological assembly lines. The scalable and renewable nature of engineered organisms offers enormous ecological and economic advantages for global production systems.

Biomedical and Therapeutic Innovations

In medicine, engineering biology has opened unprecedented opportunities for disease treatment, diagnostics, and personalized health solutions. Engineered cells now serve as "living drugs," capable of detecting and destroying cancer cells or delivering precise doses of therapeutic molecules. Synthetic circuits can reprogram immune cells to enhance their targeting efficiency in conditions like leukemia or autoimmune disorders. Meanwhile, gene therapies derived from synthetic biology principles enable genetic correction at the source of inherited diseases. Researchers are also designing programmable probiotics that detect and treat gut inflammation or metabolic imbalances in situ. Beyond therapy, engineered biology supports diagnostics through biosensors that detect pathogens and biochemical markers in bodily fluids, allowing rapid and accurate disease detection. The long-term potential includes on-demand synthesis of vaccines and drugs within compact biological reactors.

Environmental and Agricultural Applications

Engineering biology also plays a transformative role in addressing ecological and agricultural challenges. Synthetic microorganisms are being developed for bioremediation, capable of degrading plastic waste, capturing heavy metals, or detoxifying pollutants from soil and water. In agriculture, engineered crops exhibit enhanced resistance to drought, pests, and diseases while maintaining nutritional quality. Precision design allows for nitrogen-fixing capabilities to be introduced into non-leguminous plants, reducing dependence on synthetic fertilizers. Engineered microbiomes can improve soil fertility and plant growth, offering a biologically sustainable alternative to conventional agrochemicals. Importantly, these innovations support the larger goal of transitioning to a circular bioeconomy, where waste materials become feedstocks for new biological processes.

Ethical and Safety Considerations

The design and manipulation of living systems inevitably raise profound ethical and safety challenges. Concerns revolve around biosafety, biosecurity, and the moral boundaries of constructing artificial life. There is an ongoing debate about how to prevent misuse of synthetic biology technologies in creating harmful pathogens or ecological disruptions. Regulatory frameworks are being developed to ensure that engineered organisms are contained and that their environmental interactions remain predictable. Ethical discussions also address questions about the ownership of genetic constructs, the modification of human embryos, and the definition of life itself. Scientists and policymakers advocate for open, responsible innovation that balances progress with precaution. Public engagement and global collaboration are essential to govern this rapidly evolving field and to ensure that the benefits outweigh the risks.

1.3.2 DNA Synthesis and Biological Circuitry

DNA synthesis and biological circuitry are inextricably linked technologies that are redefining our relationship with the living world. DNA synthesis provides the essential physical substrate–the genetic code–that allows scientists to move from observation to creation. Biological circuitry provides the conceptual framework and design principles to imbue cells with novel, programmable functions. Together, they empower the engineering of life with unprecedented precision. From creating pulsating cellular clocks and decision-making microbes to sensitive biosensors and sustainable production platforms, the applications are vast and transformative. While challenges in fidelity, predictability, delivery, and ethics persist, rapid technological advancements fueled by innovation in synthesis, assembly, and computational design are steadily overcoming these hurdles. The ongoing convergence of these fields holds the potential to unlock solutions to some of humanity's most pressing challenges, paving the way for a future where biological systems are not just studied, but rationally designed and constructed to serve human and environmental needs. The blueprint of life, once a mystery, is now increasingly a canvas for human ingenuity.

Engine of Creation: Understanding DNA Synthesis

DNA synthesis is the chemical or enzymatic process of creating deoxyribonucleic acid (DNA) strands with a specified nucleotide sequence. Unlike the natural replication of DNA within cells, which relies on existing templates, de novo DNA synthesis builds sequences from scratch, nucleotide by nucleotide. Initially, chemical synthesis dominated, employing phosphoramidite chemistry on solid supports, enabling the rapid and automated production of short oligonucleotides (typically up to 200 bases). These oligonucleotides serve as primers, probes, and crucially, as building blocks for assembling larger constructs. More recently, enzymatic synthesis has emerged as a powerful complementary approach.

Techniques like Terminal deoxynucleotidyl Transferase (TdT) mediated synthesis or engineered polymerase systems allow for the template-independent addition of nucleotides, offering potential advantages in speed, cost, and the ability to incorporate modified bases. The significance of DNA synthesis cannot be overstated; it provides the essential raw material–the genetic code–those synthetic biologists use to write new programs for life. It allows for the creation of genes that never existed in nature, the modification of existing genomes, and the construction of complex genetic assemblies from the ground up. This capability is fundamental to realizing the vision of biological circuitry.

Wiring Life: The Concept and Components of Biological Circuitry

Biological circuitry draws a powerful analogy from electrical engineering, treating genetic components within a cell as analogous to resistors, capacitors, transistors, and wires. Instead of electrons flowing, however, biological circuits manipulate the flow of information via molecules: DNA, RNA, and proteins. The core idea is to design and implement networks of interacting genetic parts that process inputs (like chemical signals or light) to produce desired outputs (such as the production of a protein, a change in cell growth, or a specific metabolic activity). Key components include:

- **Promoters:** DNA sequences that act as switches, initiating the transcription of adjacent genes. They can be constitutive (always on), inducible (turned on/off by specific molecules like inducers or repressors), or responsive to environmental cues.
- **Genes:** The coding sequences that specify the structure of proteins or functional RNAs. These are the primary output devices.
- **Regulators:** Proteins (transcription factors) or RNAs (riboswitches, small RNAs) that bind to promoters or other regulatory elements to modulate gene expression. They act as the circuit's control logic.
- **Terminators:** Sequences that signal the end of transcription.
- **Ribosome Binding Sites (RBS):** Sequences that facilitate the translation of mRNA into protein.

- **Reporters:** Genes encoding easily detectable proteins (e.g., fluorescent proteins like GFP, enzymes like luciferase) used to monitor circuit activity.

By connecting these components in specific configurations–for example, placing a gene under the control of a promoter that is repressed by a protein which itself is encoded by another gene–one can create genetic devices that perform functions like signal amplification, filtering, oscillation, or Boolean logic operations (AND, OR, NOT gates).

Constructing the Code: DNA Synthesis in Building Biological Circuits

The practical construction of biological circuits heavily relies on DNA synthesis. It provides the means to physically create the DNA sequences encoding the designed circuit components and their interconnections. This involves several key steps:

- **Gene Synthesis:** The artificial synthesis of complete genes or genetic elements based on a digital sequence design. This allows for the creation of custom parts with optimized or novel functions, free from the constraints of natural DNA.
- **Assembly of Genetic Constructs:** Synthesized oligonucleotides and gene fragments are assembled into larger DNA constructs using various techniques. Traditional methods like restriction enzyme digestion and ligation are being largely superseded by more efficient and flexible methods like Gibson Assembly, Golden Gate Assembly, and Ligation Independent Cloning (LIC). These methods allow for the seamless joining of multiple DNA fragments in a specific order and orientation.
- **Standardization and Abstraction:** DNA synthesis facilitates the creation of standardized biological parts (like BioBricks, defined by specific prefix and suffix sequences) that can be easily combined like Lego bricks. This abstraction allows designers to focus on the circuit function rather than the low-level assembly details, accelerating the design-build-test-learn cycle.
- **Genome Editing and Integration:** Synthesized DNA constructs are used to edit genomes (e.g., via CRISPR-Cas systems) or to integrate entire synthetic circuits into the chromosomes of host cells, ensuring stable inheritance and function.

Without the ability to synthesize DNA to order, the construction of complex, custom-designed biological circuits would be prohibitively slow, expensive, and limited to what could be found in nature or assembled from existing fragments.

Design to Function: Exemplars of Biological Circuits

The power of combining DNA synthesis with circuit design is demonstrated by a growing array of functional biological systems. One landmark example is the repressilator, an oscillatory circuit. Designed by Michael Elowitz and Stanislas Leibler, it consists of three genes arranged in a loop, where each gene represses the expression of the next.

Synthesized DNA encoding the repressors and their cognate promoters was assembled and introduced into bacteria. The result was a population of cells where the levels of the three fluorescent reporter proteins cycled periodically, creating a biological clock–a function not naturally present in those cells. Another critical class of circuits involves genetic logic gates. By synthesizing DNA that places genes under the control of specific combinations of regulators (e.g., an AND gate requiring two different activators to be present simultaneously), researchers have created cells that perform complex decision-making.

For instance, engineered bacteria can be programmed to detect the presence of multiple cancer biomarkers and only produce a therapeutic drug when all are detected. Biosensors represent another vital application. Circuits are built using synthesized DNA that links a promoter responsive to a specific environmental signal (e.g., a heavy metal, a toxin, or a specific pathogen) to a reporter gene. When the target molecule is present, the circuit activates, producing a visible signal (like fluorescence) or triggering a response, enabling rapid and sensitive detection. These examples underscore the transformative potential of engineered biological circuitry.

1.3.3 Standardization of Biological Parts

Standardization of biological parts is a transformative concept in synthetic biology, enabling the systematic design, construction, and analysis of complex genetic systems. By providing modular, well-characterized, and interoperable components, it allows researchers to predictably engineer organisms for a wide range of applications. While challenges remain, including context-dependence, ethical considerations, and biosafety concerns, ongoing advances in DNA synthesis, computational modeling, and high-throughput characterization continue to enhance the reliability and scope of standardized parts. The integration of standardized biological parts into synthetic biology projects promises to accelerate innovation, improve reproducibility, and unlock new possibilities for solving some of the most pressing problems in medicine, industry, and the environment.

Historical Context

The idea of standardizing biological parts emerged in the early 2000s alongside the rise of synthetic biology. Traditional genetic engineering focused on modifying single genes or pathways in a specific organism, but the complexity and variability of biological systems limited reproducibility and scalability. The iGEM (International

Genetically Engineered Machine) competition, launched in 2003, popularized the concept of standardized genetic parts and provided a framework for students and researchers to share and reuse genetic components. This initiative emphasized modularity, interoperability, and documentation, laying the groundwork for a global community dedicated to biological standardization.

Core Principles of Standardization

Standardization in biology rests on several key principles. First, parts must have clearly defined functions that are reproducible across different organisms or contexts. Second, parts should be modular, allowing them to be combined or rearranged without altering their core behavior. Third, documentation and characterization are essential to ensure predictable performance. Fourth, interoperability ensures that parts from different sources or labs can function together seamlessly. Finally, abstraction allows scientists to conceptualize complex genetic systems as networks of standard components, similar to electronic circuits, making design, testing, and optimization more efficient.

Types of Biological Parts

Biological parts can be categorized based on their function. Promoters regulate the transcription of genes, controlling when and how much of a protein is produced. Ribosome binding sites influence translation efficiency, determining how effectively a protein is synthesized.

Coding sequences define the amino acid sequence of proteins or functional RNAs. Terminators signal the end of transcription, ensuring proper gene expression. Regulatory elements, including repressors and activators, control the activity of other genes. Additionally, sensors and reporters enable detection of environmental conditions or internal cellular states, while scaffolds and connectors facilitate spatial organization of proteins within a cell.

Tools and Techniques for Standardization

Several tools and techniques support the standardization of biological parts. DNA synthesis and assembly technologies allow precise construction of parts with defined sequences. CRISPR-Cas systems enable targeted insertion or modification of parts in genomes. Computational modeling predicts how parts will behave individually and in combination, improving reliability before physical construction.

High-throughput screening and characterization provide quantitative data on part performance, including promoter strength, protein expression levels, and response to environmental signals. Databases such as the Registry of Standard Biological Parts store validated sequences and experimental data, making them accessible to the global research community.

Applications of Standardized Parts

Standardized biological parts are applied across medicine, industry, and environmental science. In medicine, engineered bacteria or viruses can deliver therapeutic molecules or detect disease biomarkers. In industrial biotechnology, standardized metabolic pathways enhance production of chemicals, biofuels, or enzymes. Agricultural applications include engineered microbes that promote plant growth, protect crops, or improve soil quality.

Environmental uses involve organisms capable of degrading pollutants, sequestering carbon, or monitoring ecosystems. By providing a predictable and modular toolkit, standardized parts accelerate research, reduce development costs, and improve reproducibility in synthetic biology projects.

Challenges and Limitations

Despite its advantages, standardization of biological parts faces several challenges. Biological systems are inherently complex, and a part may behave differently depending on the host organism, environmental conditions, or interactions with other components. Context-dependence limits predictability, requiring extensive testing and characterization. Intellectual property and patent issues can restrict access to certain standardized parts. Moreover, ethical and biosafety concerns arise when engineered organisms are released into the environment or used in medical applications. Continuous refinement of standards, improved characterization, and international collaboration are necessary to address these limitations.

Principles for Effective Implementation

Effective implementation of standardized parts requires careful planning and adherence to best practices. Documentation should include sequence information, functional characterization, and experimental context. Modular design allows parts to be combined flexibly without unintended interference. Validation under multiple conditions ensures robustness across applications.

Sharing parts and data through open-access repositories promotes collaboration and accelerates innovation. Integration with computational tools and automation streamlines the design-build-test cycle, enabling rapid iteration and optimization of engineered systems. Collaboration among interdisciplinary teams ensures that biological, engineering, and computational aspects are harmonized.

Key Features and Examples

The standardization of biological parts relies on well-defined characteristics and common examples:

- **Promoters:** DNA sequences controlling gene transcription, such as constitutive or inducible promoters.

- **Ribosome Binding Sites:** Elements influencing protein translation efficiency.
- **Coding Sequences:** Genes encoding proteins or functional RNAs.
- **Terminators:** Signals that end transcription to prevent unwanted read-through.
- **Regulatory Elements:** Activators, repressors, and enhancers that modulate gene activity.
- **Sensors:** Parts that detect environmental or intracellular signals.
- **Reporters:** Genes that produce measurable outputs, like fluorescent proteins.
- **Scaffolds:** Structural elements organizing proteins or enzymes spatially within cells.
- **Connectors:** Sequences that link parts while maintaining function.
- **Chassis Compatibility:** Testing parts in different host organisms for predictable performance.

1.4 THE RELATIONSHIP BETWEEN SYNTHETIC BIOLOGY AND GENETIC ENGINEERING

Genetic engineering and synthetic biology are inextricably linked in a hierarchical and progressive relationship. Genetic engineering is the essential, foundational discipline that unlocked the ability to manipulate DNA, providing the initial tools and proving that such manipulation was possible. Synthetic biology is the evolutionary successor that builds upon this foundation, introducing a rigorous engineering mindset to make biological design more predictable, reliable, and complex.

It is the difference between being a skilled mechanic who can expertly repair and modify an existing engine and being a mechanical engineer who designs a completely new engine from the ground up using standardized parts and computer models. One is not superior to the other; rather, they represent sequential stages in humanity's growing mastery over the code of life, with synthetic biology representing our most advanced and ambitious effort to date to write that code ourselves.

1.4.1 Genetic Engineering: The Foundational Precursor

Genetic engineering emerged in the 1970s as a revolutionary breakthrough, granting scientists the ability to directly alter an organism's DNA. Its core methodology is one of modification and augmentation. Using a suite of molecular tools, notably restriction enzymes which act as precise molecular scissors and ligases which function as genetic glue, researchers learned to cut genes from one organism and splice them into the genome of another. This process allowed for the transfer of specific, desirable traits across the biological kingdoms.

A classic and transformative example is the production of human insulin in genetically engineered E. coli bacteria. The human gene for insulin was isolated, inserted into the bacterial plasmid, and thereby commandeered the bacterial cell's machinery to produce a vast, pure, and life-saving human protein. This approach is inherently reductive and analog; it identifies a useful part from nature's vast library and transplants it into a new host to confer a new function. The focus is on the individual gene and its immediate product, and while powerful, it often requires significant trial and error, working with and around the complexities of the host's existing biological systems rather than fully redesigning them.

1.4.2 Synthetic Biology: The Paradigm Shift towards Engineering

Synthetic biology represents a paradigm shift from modification to comprehensive design. It adopts the core principles of modern engineering—standardization, abstraction, and decoupling—and applies them systematically to biology. While genetic engineering copies and pastes natural genes, synthetic biology often writes entirely new genetic code from scratch. Its practitioners aim to create standardized biological parts, known as BioBricks, which have predictable and reliable functions independent of their context.

These parts can then be assembled into larger, more complex systems called genetic circuits, which are designed to perform logical operations, such as oscillating like a clock or sensing an environmental condition and responding with a programmed behaviour. The goal is not merely to add a single trait but to install entirely new metabolic pathways, computational logic, or behaviours that do not exist in the natural world. For instance, while genetic engineering might move a gene for a red pigment into a flower, synthetic biology would design a circuit that causes the flower to change from red to blue when the soil temperature drops, a novel function engineered from the ground up. This represents a move from a craft-based approach to a truly design-driven one.

1.4.3 A Symbiotic Progression, not a Replacement

It is a critical mischaracterization to view synthetic biology as merely a rebranding of genetic engineering or as a field that renders its predecessor obsolete. The relationship is fundamentally symbiotic and cumulative. Synthetic biology is entirely dependent on the tools and techniques pioneered by genetic engineering. The molecular scissors of CRISPR-Cas9, a product of advanced genetic engineering research, are an indispensable tool in the synthetic biologist's toolkit for making precise edits. The polymerase chain reaction, a cornerstone of genetic manipulation, is equally vital for amplifying DNA sequences in synthetic workflows. Without the foundational capabilities provided by genetic engineering, the ambitious designs of synthetic biology would remain theoretical constructs trapped in computer models. Conversely, synthetic biology elevates and expands the potential of genetic engineering by providing a more robust conceptual framework and more powerful design methodologies. It brings a higher level of predictability, scalability, and ambition to the process of biological design. In this sense, genetic engineering provided the alphabet and basic grammar for writing biological code, while synthetic biology is using that foundation to write poetry, complex software, and entirely new languages.

1.4.4 Contrasting Methodologies through Application

The distinction between the two fields becomes starkly clear when comparing their approaches to a similar problem. Consider the challenge of producing a valuable biochemical, such as an anti-malarial drug precursor. A genetic engineering approach would likely identify the key gene or genes from the plant that naturally produces the compound and insert them into a microbial host like yeast. This is a transfer of an existing, natural pathway. A synthetic biology approach, however, would deconstruct the problem.

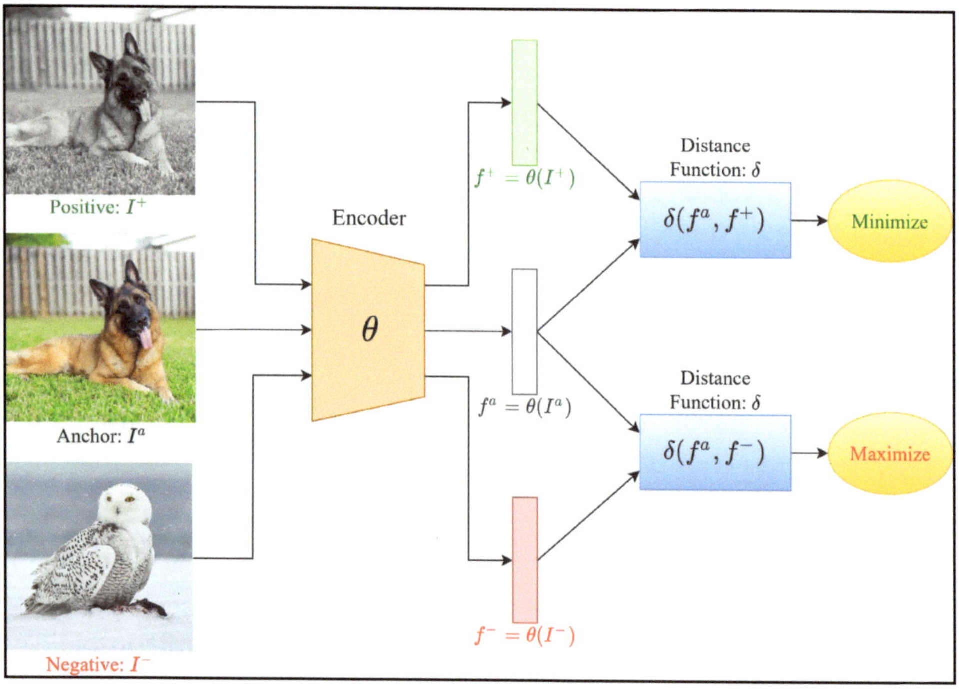

Contrastive Learning

Researchers would analyze the desired biochemical and design a completely novel metabolic pathway from first principles, potentially using enzyme components from multiple different organisms or even designing entirely novel enzymes in silico. They would then synthesize the DNA for this optimized pathway from scratch and assemble it into the host, often using standardised parts to allow for easy debugging and optimisation. The synthetic approach seeks to create a more efficient, streamlined, and predictable system that is not beholden to the evolutionary constraints of the original plant pathway. This exemplifies the shift from transferring natural systems to designing optimized ones.

1.4.5 Converging Tools and the Blurring Frontier

As both fields advance, the technological frontier between them is increasingly blurred, yet their philosophical cores remain distinct. The advent of powerful gene-editing tools like CRISPR and the plummeting cost of DNA synthesis have created a convergence point. A modern research project will fluidly employ techniques from both disciplines: using CRISPR for precise genomic edits (a genetic engineering technique) while also installing a completely synthetic gene circuit (a synthetic biology technique) into the same organism.

Table: Comparative Overview of Core Characteristics

Characteristic	Genetic Engineering	Synthetic Biology
Primary Objective	To modify or augment an existing biological system by introducing or altering genes.	To design and construct novel biological parts, devices, and systems from first principles.
Core Methodology	Cutting and pasting pre-existing genetic material from nature using enzymes like restriction enzymes and ligases.	Often involves the de novo synthesis of DNA sequences and the assembly of standardized, characterized biological parts.
Underlying Philosophy	A reductive approach focused on individual genes and their functions, often working within the constraints of host biology.	An engineering approach emphasizing standardization, abstraction, and predictable system-level design.
Analogy	Editing a sentence in a pre-existing book by rearranging or replacing a few words or paragraphs.	Writing an entirely new book, potentially with a new alphabet, to tell a story that has never been told before.

The toolkits have merged, but the intentions differentiate the work. A project motivated by correcting a single faulty gene in a human patient to cure a genetic disorder is rooted in the paradigm of genetic engineering—it is a corrective modification. A project aimed at designing a universal immune cell that can be programmed with synthetic receptors to target any cancer type is rooted in the paradigm of synthetic biology—it is the creation of a novel, platform technology. The relationship is thus best understood as a spectrum, with genetic modification at one end and the de novo synthesis of biological systems at the other, with most contemporary research existing somewhere in between.

1.5 INTERDISCIPLINARY NATURE OF SYNTHETIC BIOLOGY: BIOLOGY, ENGINEERING, AND COMPUTER SCIENCE

1.5.1 Biological Foundations of Synthetic Biology

At its core, synthetic biology emerges from molecular and cellular biology, which provide the necessary understanding of genetic and biochemical processes. Biology offers the fundamental language—the DNA code—that enables scientists to reprogram life at its most basic level. Decades of research into gene expression, protein synthesis, and metabolic regulation provide the blueprint for manipulating cellular functions. Advances in genomics and proteomics have revealed how living systems maintain homeostasis, adapt to changes, and evolve complexity. This understanding is critical because successful engineering of biological systems requires deep insight into how genes, enzymes, and pathways interact dynamically. Biological research also contributes vast libraries of genetic parts—promoters, regulatory sequences, enzymes, and structural elements—that serve as the modular components for synthetic designs. Ultimately, biology supplies both the materials and principles upon which synthetic systems are built, ensuring that every intervention remains grounded in the realities of life itself.

1.5.2 Role of Engineering in Synthetic Biology

Engineering provides synthetic biology with structure, discipline, and precision. It transforms biological curiosity into practical design by imposing systems-level methodologies. Engineers approach living cells as assemblable entities that can be rationally constructed and optimized for function. The engineering mindset brings concepts such as standardization, modularity, abstraction, and predictability into the life sciences. Standard biological parts can be characterized, catalogued, and recombined much like mechanical or electronic components. The process of "design-build-test-learn" forms the engineering backbone of synthetic biology, ensuring iterative refinement through data-driven feedback. Bioprocess engineering extends this logic to the industrial scale by optimizing conditions for microbial growth, metabolite production, and process efficiency. Through principles of control systems, fluid dynamics, and thermodynamics, engineers translate biological potential into manufacturable reality. By applying mechanical reasoning to living systems, engineering transforms biology from a descriptive to a prescriptive science capable of producing repeatable and scalable outcomes.

1.5.3 Computational Power and the Role of Computer Science

Computer science acts as the nervous system of synthetic biology, integrating and interpreting massive biological datasets while enabling predictive modeling. DNA sequencing technologies generate enormous volumes of data, and computational methods are required to extract meaning from them. Bioinformatics tools analyze gene patterns, protein structures, and regulatory networks, revealing functional relationships that would otherwise remain hidden. Beyond analysis, computer science contributes to the design phase through algorithmic modeling, simulation, and artificial intelligence.

Sophisticated computational platforms can predict how genetic circuits will behave under different environmental and cellular contexts, significantly reducing experimental failures. Machine learning algorithms further refine these predictions by identifying correlations between design parameters and outcomes. Computational modeling also assists in protein folding predictions, metabolic flux analysis, and optimization of synthetic pathways. In essence, computer science is the intelligence that links design and implementation, allowing researchers to visualize and model complex living systems before they exist in the laboratory.

Computational Advancement and the Importance of Computer Science

1.5.4 Integration of Biology, Engineering, and Computer Science

The true strength of synthetic biology lies not in the contribution of each discipline individually, but in their seamless integration. Biology provides the raw materials and phenomena; engineering delivers strategies for design and optimization; computer science supplies algorithms and simulations that drive decision-making. This triad creates a unified framework capable of designing living systems from the ground up. For example, when developing a synthetic pathway to produce a biofuel, computer models first predict optimal gene combinations. Engineers then construct the genetic circuit using standardized biological parts, while biologists test and refine the organism's performance. This cyclical process merges computational reasoning, experimental biology, and engineering problem-solving into a continuous feedback loop. Such integration exemplifies how synthetic biology embodies the essence of interdisciplinarity—each field amplifies the capabilities of the others, producing outcomes that no discipline could achieve alone.

1.5.5 Systems Biology and the Emergence of Design Principles

The understanding of whole-system behavior, rooted in systems biology, enhances the interdisciplinary nature of synthetic biology. Systems biology studies the interconnections among genes, proteins, and metabolites to reveal emergent properties of living networks. This systems-level thinking aligns closely with engineering, which relies on modeling and control of complex systems. Computer simulations can map out feedback loops and emergent behaviors arising from synthetic gene networks, enabling designers to anticipate unintended effects. These insights are essential when constructing genetic circuits with precise responses, such as biosensors that activate in the presence of pollutants or pathogens. Systems biology also helps establish quantitative models, linking changes in component parameters to measurable outcomes. The merging of systems-level analysis with design strategies transforms synthetic biology into a predictive science rather than an empirical one.

1.5.6 Applications Demonstrating Interdisciplinary Collaboration

The practical applications of synthetic biology vividly demonstrate its multidisciplinary foundation. In medicine, engineered immune cells designed through computational modeling and bioengineering respond selectively to cancer antigens, forming the basis of advanced immunotherapies. In energy, microbes have been programmed to convert carbon dioxide into fuels using pathways constructed through genetic and systems design principles. Agricultural systems benefit from engineered crops and microbes that enhance nutrient uptake, driven by computational models predicting optimal gene configurations. Environmental applications rely on genetically programmed organisms that sense and neutralize toxins, integrating molecular biology with circuit engineering and environmental modeling. In each case, success arises from the coordinated contribution of biological understanding, engineering design, and computational guidance. These collaborations exemplify how synthetic biology converts interdisciplinary complexity into tangible innovation with global impact.

1.5.7 Challenges and Ethical Dimensions of Interdisciplinary Work

While the interdisciplinary collaboration between biology, engineering, and computer science has expanded innovation, it also introduces significant challenges and ethical responsibilities. Communication barriers can arise as each discipline uses distinct conceptual frameworks and terminologies. Effective teamwork requires a shared language that bridges molecular details, mathematical abstraction, and algorithmic logic. Ethical considerations further complicate the landscape, particularly regarding the release of synthetic organisms into natural ecosystems or the modification of human genetic material. The computational power that enables design also increases the risk of dual-use misuse, where biological knowledge could be applied for harmful purposes. Establishing common regulatory and safety frameworks requires input from all participating disciplines to ensure that technological advancement aligns with societal values. Cultivating responsible innovation demands both technical proficiency and ethical awareness, emphasizing the need for comprehensive interdisciplinary education and governance.

CHAPTER 2

Core Technologies in Synthetic Biology

2.1 DNA SYNTHESIS AND ASSEMBLY TECHNOLOGIES

2.1.1 Oligonucleotide Synthesis

Chemical Crucible: Phosphoramidite Chemistry and the Synthesis Cycle

The dominant method for oligonucleotide synthesis is solid-phase phosphoramidite chemistry, a highly efficient and automated process. This method builds the oligonucleotide chain step-by-step, nucleotide by nucleotide, attached to a solid support (usually controlled pore glass or polystyrene beads). The process involves a repeating cycle of four key chemical reactions:

- **Deprotection:** The first nucleotide attached to the solid support has a protective group (typically a dimethoxytrityl group, DMT) on its 5' end. This group is removed using an acid (like trichloroacetic acid or dichloroacetic acid) to expose a reactive hydroxyl group, ready for the next coupling step.
- **Coupling:** The next nucleotide, in the form of a phosphoramidite monomer (activated with tetrazole or a similar activator), is added to the reaction column. Its 3' phosphoramidite group reacts with the free 5' hydroxyl group of the growing chain, forming a phosphite triester linkage. This coupling step is rapid but requires careful optimization to maximize efficiency and minimize side reactions.
- **Capping:** Any unreacted 5' hydroxyl groups on the chain are blocked ("capped") using acetic anhydride and a base. This prevents these chains from reacting in subsequent cycles, ensuring that only the full-length, correctly growing chains are extended further. Capping significantly improves the purity of the final product by reducing the formation of deletion sequences.
- **Oxidation:** The newly formed phosphite triester linkage is unstable and must be converted to a more robust phosphate triester (phosphodiester bond). This is achieved by treating the chain with an oxidizing agent,

typically iodine in a solution of water and pyridine, which stabilizes the backbone for the next cycle.

This cycle–deprotection, coupling, capping, oxidation–is repeated for each nucleotide added to the chain. After the desired sequence is assembled, a final deprotection step removes all remaining protecting groups from the nucleobases and the phosphate backbone. The completed oligonucleotide is then cleaved from the solid support using a strong base, such as ammonium hydroxide or methylamine, yielding the crude synthetic product.

Beyond Chemistry: Enzymatic Synthesis Pathways

While phosphoramidite chemistry dominates, enzymatic methods for oligonucleotide synthesis are gaining traction as complementary or alternative approaches. The most prominent is Terminal deoxynucleotidyl Transferase (TdT) mediated synthesis. TdT is an enzyme that naturally adds nucleotides to the 3' end of DNA strands in a template-independent manner. Leveraging this, scientists can use TdT in controlled reactions with specific nucleotide triphosphates (dNTPs) to synthesize DNA oligonucleotides.

This method offers potential advantages: it can incorporate a wide range of modified nucleotides (including those with bulky groups or unnatural bases) that might be challenging for phosphoramidite chemistry, and it can potentially achieve faster synthesis speeds for certain sequences. However, enzymatic synthesis currently faces challenges with controlling sequence specificity (TdT can add multiple nucleotides per cycle, leading to length heterogeneity) and achieving the same level of length and purity as solid-phase phosphoramidite for most applications. Research into engineered TdT variants and other enzymatic systems aims to overcome these limitations, potentially opening doors for novel applications.

Platform Matters: Solid-Phase vs. Solution-Phase Synthesis

Oligonucleotide synthesis primarily occurs on solid-phase supports. This approach offers significant advantages:

- **Automation:** The solid support allows the use of automated synthesizers, which perform the repetitive coupling cycles efficiently and precisely.
- **Excess Reagent Use:** Large excesses of reagents can be used to drive reactions to completion, as they can be easily washed away after each step, leaving only the desired product attached to the support.
- **Simplified Purification:** After synthesis, the oligonucleotide is cleaved from the support, and impurities (like salts, reagents, and truncated sequences) are removed through washing and purification steps, which are more straightforward than separating products from a complex mixture in solution.

- **Scalability:** Solid-phase synthesis is readily scalable for high-throughput production.

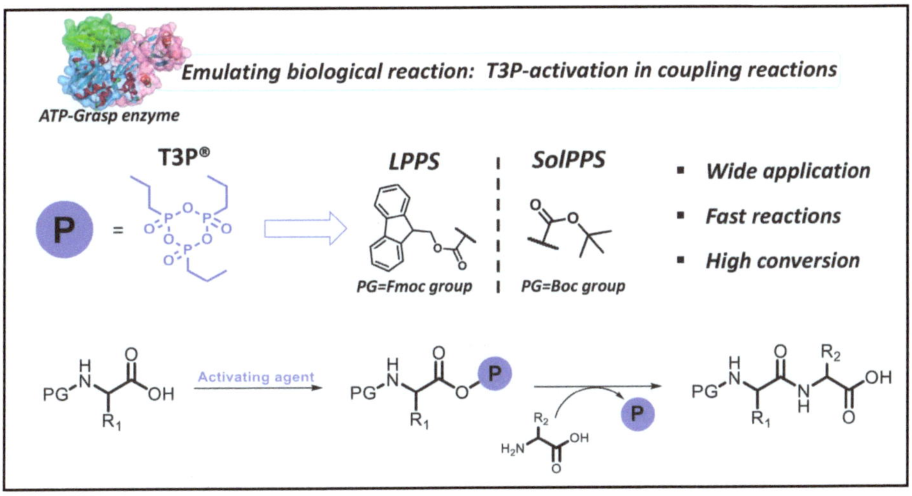

Solid-Phase Versus Solution-Phase Approaches in Chemical Synthesis

In contrast, solution-phase synthesis involves building the oligonucleotide chain in a homogeneous liquid solution. While it avoids the need for a solid support and may offer advantages for synthesizing very short oligomers or specific modifications, it suffers from significant drawbacks. Purification between each step is complex and time-consuming, as all components (including the growing chain and excess reagents) are in solution. This makes the process inefficient, expensive, and unsuitable for automated, high-volume production. Consequently, solid-phase synthesis remains the workhorse for the vast majority of oligonucleotide manufacturing.

Ensuring Perfection: Purification and Quality Control

The crude oligonucleotide mixture obtained after cleavage from the solid support contains not only the desired full-length product but also numerous impurities. These include:

- **Failure Sequences:** Chains that terminated early due to incomplete coupling (e.g., n-1, n-2, etc., missing one or more nucleotides).
- **Byproducts:** Chemicals from the synthesis or cleavage reactions.
- **Modified Oligonucleotides:** Unintended side products from incomplete deprotection or side reactions.

Rigorous purification is therefore essential to isolate the pure, full-length oligonucleotide. The two most common techniques are:

- **Reverse-Phase High-Performance Liquid Chromatography (RP-HPLC):** This separates molecules based on their hydrophobicity. The

DMT group, still attached to the full-length oligonucleotide after synthesis (if not removed in the final deprotection), increases its hydrophobicity, allowing separation from shorter, less hydrophobic failure sequences. After purification, the DMT group is removed.

- **Polyacrylamide Gel Electrophoresis (PAGE):** This separates molecules based on size and charge. The oligonucleotides are loaded onto a gel and subjected to an electric field. The full-length product, being larger, migrates slower than the failure sequences, allowing for excision and extraction from the gel. PAGE often provides higher purity but is more labor-intensive and lower throughput than HPLC.
- Quality control is paramount. Techniques like Capillary Electrophoresis (CE) and Mass Spectrometry (MS) are used to verify the purity, length, sequence, and molecular weight of the synthesized oligonucleotide. High-quality synthesis is critical because impurities can lead to failed experiments, inaccurate results, or adverse effects in therapeutic applications.

Ubiquitous Tool: Diverse Applications of Synthetic Oligonucleotides

The impact of oligonucleotide synthesis is felt across the life sciences and medicine due to the incredible versatility of these molecules:

- **PCR and Sequencing:** Synthetic oligonucleotides are essential primers for the Polymerase Chain Reaction (PCR), enabling the exponential amplification of specific DNA sequences. They are also used as probes and primers in various DNA sequencing technologies (Sanger, NGS).
- **Gene Editing (CRISPR-Cas9):** Guide RNAs (gRNAs), which are synthetic RNA oligonucleotides, are the critical components that direct the Cas9 enzyme to the specific DNA sequence to be edited. The ability to synthesize custom gRNAs rapidly is fundamental to CRISPR technology.
- **Antisense Therapy and RNA Interference (RNAi):** Synthetic oligonucleotides are the active ingredients in a growing class of therapeutics. Antisense oligonucleotides (ASOs) bind to complementary mRNA sequences, blocking their translation or triggering their degradation. Small interfering RNAs (siRNAs) are synthetic RNA duplexes that guide the RNA-induced silencing complex (RISC) to target mRNAs for degradation, silencing specific genes.
- **Diagnostics:** Oligonucleotides are used as probes in techniques like Fluorescence *In Situ* Hybridization (FISH) for visualizing specific DNA/RNA sequences in cells and tissues, and in microarrays for high-

throughput gene expression profiling. They are also key components in rapid diagnostic tests for pathogens (e.g., COVID-19 tests).

- **Research Tools:** From studying gene function (knockdown via siRNA, overexpression via synthetic genes) and protein-DNA interactions (EMSA, ChIP) to investigating RNA biology and epigenetics, synthetic oligonucleotides are indispensable reagents in molecular biology labs worldwide.
- **Vaccines:** Synthetic mRNA oligonucleotides, encoding viral antigens and stabilized by chemical modifications, form the basis of highly effective mRNA vaccines (like those for COVID-19).

Navigating the Hurdles: Challenges in Oligonucleotide Synthesis

Despite its maturity and widespread use, oligonucleotide synthesis faces ongoing challenges:

- **Error Rates and Purity:** Chemical synthesis is not perfect. The stepwise coupling efficiency is typically 99% or higher, but this still results in a small percentage of failure sequences for each coupling step. For a 20-mer, this could mean 10-20% impurities. While purification removes most, achieving 100% purity for long oligonucleotides is difficult and costly. Enzymatic synthesis also struggles with controlling sequence fidelity.
- **Length Limitations:** Solid-phase phosphoramidite synthesis becomes significantly less efficient beyond 100-150 nucleotides due to accumulating coupling failures and side reactions. Synthesizing very long oligonucleotides or genes requires alternative strategies like enzymatic assembly (e.g., using PCR) of shorter synthesized fragments.
- **Cost and Scalability:** While costs have plummeted over the decades, synthesizing large quantities of long, modified oligonucleotides, especially for therapeutic use, remains expensive. Scaling up production while maintaining high quality and purity is a constant challenge.
- **Chemical Modifications:** Incorporating non-standard nucleotides (e.g., locked nucleic acids - LNAs, phosphorothioates) or other chemical modifications is crucial for many applications (especially therapeutics for stability and efficacy), but it can complicate the synthesis chemistry, reduce coupling efficiency, and increase costs.
- **Delivery:** For therapeutic applications, efficiently and safely delivering synthetic oligonucleotides *into* specific cells and tissues in the body remains a major hurdle, limiting efficacy and causing potential off-target effects.

Horizon Beckons: Future Trends in Oligonucleotide Synthesis

The field is continuously evolving to address these challenges and unlock new possibilities:

- **Advanced Enzymatic Synthesis:** Engineered polymerases and TdT variants are being developed to enable faster, more accurate, and more flexible enzymatic synthesis, particularly for long and heavily modified oligonucleotides. This includes template-independent and template-dependent approaches.
- **Microfluidics and Automation:** Lab-on-a-chip technologies and highly integrated automated synthesizers are increasing throughput, reducing reagent consumption, minimizing errors, and accelerating the production of complex oligonucleotide libraries.
- **AI and Machine Learning:** These tools are being applied to optimize synthesis protocols, predict optimal sequences for specific functions (e.g., gRNA design), design novel modified nucleotides, and improve purification strategies.
- **Novel Chemistries:** Research into alternative phosphoramidite chemistries, new protecting groups, and more efficient coupling agents aims to further improve stepwise yields, reduce side reactions, and enable the synthesis of even longer and more complex molecules.
- **Synthetic Biology Integration:** Oligonucleotide synthesis is fundamental to synthetic biology, providing the parts for assembling genetic circuits, pathways, and even entire genomes. Future advancements will deepen this synergy.
- **Expanded Therapeutic Modalities:** Beyond ASOs and siRNAs, research is exploring synthetic oligonucleotides for targeted protein degradation (e.g., PROTACs), as aptamers (nucleic acid-based drugs that bind proteins), and in gene therapy.

2.1.2 PCR and Cloning Techniques

PCR and cloning techniques have transformed molecular biology, providing researchers with the ability to amplify, analyze, and manipulate DNA with unprecedented precision and efficiency. PCR allows rapid amplification of specific sequences, enabling detection, quantification, and study of genetic material even from minimal samples.

Cloning techniques facilitate the insertion, expression, and functional analysis of DNA fragments, supporting applications from recombinant protein production to gene therapy. Together, these tools have become indispensable in medicine, biotechnology, forensic science, and fundamental research. Despite certain challenges, ongoing innovations in enzyme design, automation, and genome editing continue to expand the capabilities and applications of PCR and cloning techniques, solidifying their role as cornerstones of modern molecular biology.

Historical Background

The concept of cloning DNA emerged in the 1970s with the advent of recombinant DNA technology, when scientists first inserted foreign DNA into plasmids and expressed them in bacterial hosts. This breakthrough laid the groundwork for modern molecular biology. PCR was introduced in 1983 by Kary Mullis, who developed a method to selectively amplify DNA sequences using a thermostable DNA polymerase.

This innovation addressed the limitation of working with tiny amounts of DNA, enabling rapid amplification without the need for complex enzymatic reactions. Since then, PCR has evolved into numerous variants, including quantitative PCR (qPCR), reverse transcription PCR (RT-PCR), and digital PCR, while cloning techniques have diversified to include traditional restriction enzyme-based cloning, gateway cloning, and seamless cloning methods.

Principles of PCR

PCR relies on the ability of DNA polymerase to synthesize a complementary DNA strand using a template and primers. The process involves repeated cycles of denaturation, annealing, and extension. During denaturation, the double-stranded DNA is heated to separate the strands. In the annealing step, short synthetic primers bind to complementary sequences on the target DNA.

The extension step, mediated by a thermostable DNA polymerase, synthesizes new DNA strands starting from the primers. Repeating these cycles exponentially amplifies the target sequence, producing millions of copies in a few hours. Key factors influencing PCR efficiency include primer design, template quality, enzyme fidelity, and reaction conditions.

Variants and Applications of PCR

PCR has diversified into several specialized forms to meet diverse research needs. Reverse transcription PCR (RT-PCR) converts RNA into complementary DNA (cDNA) for amplification, enabling the study of gene expression. Quantitative PCR (qPCR) allows real-time monitoring of DNA amplification, providing quantitative data on initial template concentration. Digital PCR partitions the reaction into thousands of micro-reactions, enabling highly sensitive detection of rare sequences. Multiplex PCR allows simultaneous amplification of multiple targets in a single reaction. PCR is applied in numerous areas, including genetic diagnostics, pathogen detection, forensic analysis, prenatal testing, and environmental monitoring.

Cloning Techniques

Cloning techniques are used to insert DNA fragments into vectors that can replicate independently within a host organism. Traditional cloning involves the use of restriction enzymes to cut both the DNA fragment and vector at specific sequences, followed by ligation to join the fragments. The recombinant DNA is then introduced into a host cell, such as bacteria, yeast, or mammalian cells, for propagation. Modern cloning methods, including gateway cloning, Gibson assembly, and ligase-independent cloning, enable seamless assembly of multiple DNA fragments without relying on restriction sites. Cloning allows the production of recombinant proteins, functional studies of genes, gene knockouts, and the creation of genetically modified organisms for research and therapeutic applications.

PCR in Cloning

PCR plays a critical role in cloning by enabling the amplification of DNA fragments before insertion into vectors. PCR-amplified DNA can be modified to include restriction enzyme recognition sites, facilitating subsequent cloning steps. High-fidelity polymerases reduce the risk of introducing mutations during amplification, which is especially important for functional studies or protein expression. PCR also allows the rapid generation of variants through site-directed mutagenesis, enabling detailed structure-function analysis of genes or proteins. By combining PCR with advanced cloning strategies, researchers can construct complex genetic constructs efficiently and accurately.

Tools and Enzymes

Several tools and enzymes are essential for PCR and cloning techniques. Thermostable DNA polymerases, such as Taq polymerase, enable repeated cycling without enzyme degradation. High-fidelity polymerases provide accurate DNA synthesis with minimal errors. Restriction enzymes cut DNA at specific sequences, while DNA ligases join DNA fragments. Competent host cells, such as Escherichia coli, facilitate DNA replication and propagation. Vectors, including plasmids, bacterial artificial chromosomes (BACs), and viral vectors, serve as carriers for DNA fragments. Additional tools include primers, nucleotides, buffers, and thermal cyclers for PCR, as well as electroporation and transformation systems for introducing recombinant DNA into host cells.

Applications in Medicine and Biotechnology

PCR and cloning techniques have widespread applications in medicine and biotechnology. In diagnostics, PCR enables rapid detection of infectious agents, genetic mutations, and disease biomarkers. Cloning techniques allow the production of therapeutic proteins, such as insulin, growth factors, and monoclonal antibodies, in large quantities. Gene therapy relies on vectors to deliver functional genes into patient cells, while synthetic biology uses PCR and cloning to construct novel genetic circuits. These techniques are also critical in vaccine development, including the creation of recombinant viral vaccines and mRNA-based vaccines. Additionally, forensic science uses PCR to analyze minute DNA samples for criminal investigations and paternity testing.

Research and Experimental Applications

PCR and cloning techniques facilitate fundamental research in molecular biology, genetics, and cell biology. PCR enables rapid genotyping, detection of gene variants, and analysis of gene expression patterns. Cloning allows detailed study of protein function, regulation of gene expression, and creation of model organisms. These methods are used in structural biology to express and purify proteins for crystallography or cryo-electron microscopy. In synthetic biology, researchers combine PCR and cloning to build genetic circuits, engineer metabolic pathways, and design organisms with novel traits. The versatility and precision of these techniques make them indispensable in both basic and applied research.

Challenges and Limitations

Despite their advantages, PCR and cloning techniques have limitations. PCR can introduce errors due to polymerase infidelity, and amplification bias may affect quantitative analysis. Contamination can lead to false-positive results, particularly in clinical diagnostics. Cloning procedures may be time-consuming, and certain DNA sequences, such as those with high GC content, repetitive elements, or secondary structures, can be difficult to clone. Ethical and biosafety concerns arise when cloning genes from pathogenic organisms or creating genetically modified organisms. Continuous improvements in enzyme fidelity, automation, and vector design aim to overcome these challenges and enhance the efficiency and reliability of these techniques.

Key Steps and Considerations

Successful PCR and cloning experiments require careful planning and execution. Critical steps and considerations include:

- **Primer Design:** Selecting primers with optimal length, melting temperature, and specificity.
- **Template Preparation:** Ensuring high-quality DNA or RNA for amplification.
- **Polymerase Selection:** Using high-fidelity enzymes for accurate DNA synthesis.
- **Cycling Conditions:** Optimizing denaturation, annealing, and extension parameters.

- **Restriction Enzyme Digestion:** Cutting DNA and vector at compatible sites.
- **Ligation:** Joining DNA fragments using DNA ligase.
- **Transformation:** Introducing recombinant DNA into competent host cells.
- **Screening and Selection:** Identifying successfully cloned constructs using markers or reporter genes.
- **Sequence Verification:** Confirming the integrity of cloned DNA through sequencing.
- **Documentation and Reproducibility:** Recording experimental conditions and results for reproducibility and future reference.

2.2 GENOME EDITING AND MODIFICATION TOOLS

2.2.1 CRISPR-Cas9 and other Gene Editing Tools

Predecessors: The Foundations of Targeted Gene Editing

Before the rise of CRISPR, scientists developed several sophisticated protein-based technologies that first demonstrated the feasibility of targeted genome engineering. The earliest of these were Meganucleases, which are naturally occurring enzymes that recognize and cut long DNA sequences, typically between 12 to 40 base pairs. Their high specificity was their main advantage, but their utility was limited because they were difficult to re-engineer to target new DNA sequences; each new target required the laborious discovery or design of a new meganuclease. This challenge was addressed by the subsequent development of Zinc Finger Nucleases. ZFNs were engineered proteins created by fusing a zinc finger DNA-binding domain, which could be designed to recognize specific three-base-pair sequences, to the DNA-cleaving domain of the FokI restriction enzyme.

By assembling multiple zinc fingers in a row, researchers could theoretically target a wide range of unique genomic sites. ZFNs were a monumental step forward, proving that customizable gene editing was possible and being used in the first-in-human clinical trials for gene therapy. However, they were expensive, complex to design, and often prone to off-target effects due to the difficulty of ensuring that each zinc finger module bound only to its intended triplet. The next iteration, Transcription Activator-Like Effector Nucleases, or TALENs, offered a significant improvement. TALENs also used the FokI nuclease domain, but they were fused to a DNA-binding domain derived from bacterial plant pathogens where each module recognized a single DNA base pair. This one-to-one correspondence made TALENs much easier and more reliable to design for a specific sequence than ZFNs. While TALENs were highly specific and effective, their construction was still a time-consuming and costly process, as it required the protein engineering of large, repetitive sequences for each new target. These tools were powerful and paved the way, but they lacked the simplicity and scalability needed to democratize gene editing.

CRISPR-Cas9 Revolution: Simplicity and Programmability

The limitations of these earlier systems were swept away by the emergence of Clustered Regularly Interspaced Short Palindromic Repeats and CRISPR-associated protein 9, known universally as CRISPR-Cas9. Its revolutionary power lies not in complex protein engineering, but in a simple and elegant guidance mechanism. The system is derived from a natural adaptive immune system in bacteria, which they use to recognize and destroy the DNA of invading viruses. The key insight was the realization that the Cas9 protein, a DNA-cutting enzyme, could be programmed to cut any DNA sequence of choice simply by providing it with a synthetic guide RNA molecule. This guide RNA is a short, custom-designed strand of RNA whose sequence is complementary to the target DNA site. The Cas9 protein, bound to this guide RNA, scans the genome until it finds a matching sequence adjacent to a short DNA motif known as a Protospacer Adjacent Motif, where it unleashes its enzymatic activity to make a double-strand break in the DNA. The true genius of this system is its programmability; to change the target, a researcher only needs to synthesize a new, inexpensive guide RNA sequence, a process that is vastly simpler and faster than engineering entirely new proteins as required by ZFNs and TALENs. This simplicity has democratized gene editing, placing a powerful and highly accessible tool into the hands of thousands of laboratories worldwide, dramatically accelerating the pace of biological discovery and therapeutic development across countless fields.

Cellular Repair Mechanisms: Enabling Precise Edits

The ability to make a precise cut in DNA is only half the story of gene editing. The true power of these technologies is realized through the cell's innate DNA repair machinery, which is activated in response to the double-strand break created by Cas9 or other nucleases. There are two primary repair pathways that can be harnessed, each leading to a different genetic outcome. The first is Non-Homologous End Joining, which is an error-prone process where the cell hastily re-joins the broken ends of the DNA. NHEJ often results in the insertion or deletion of a few base pairs at the cut site, which can disrupt or knockout the function of a target gene. This is highly useful for inactivating harmful genes or creating loss-of-function mutations for research. The second, more precise pathway is Homology-Directed Repair. HDR is a more accurate mechanism that uses a template to repair the break. Scientists can co-deliver a customized DNA repair template along with the gene-editing machinery. The cell then uses this template, which contains the desired new genetic sequence flanked by regions homologous to the target site, to copy the new information into the genome as it repairs the break. This allows for precise gene corrections, the insertion of new genes, or the introduction of specific single-nucleotide changes. While HDR is less efficient than NHEJ in most cell types, it represents the pathway to truly corrective gene therapy for a vast array of genetic disorders, from sickle cell anemia to cystic fibrosis.

Beyond Classic CRISPR: Next-Generation Editing Technologies

While CRISPR-Cas9 is a formidable tool, it is not without limitations, including the potential for off-target cuts at similar DNA sequences and the reliance on the cell's sometimes unpredictable repair mechanisms. In response, scientists have developed a suite of advanced, next-generation editors that refine and expand its capabilities. One major innovation is the creation of high-fidelity Cas9 variants, which have been engineered through mutations to drastically reduce off-target activity while maintaining high on-target efficiency. Another revolutionary development is the advent of Base Editing.

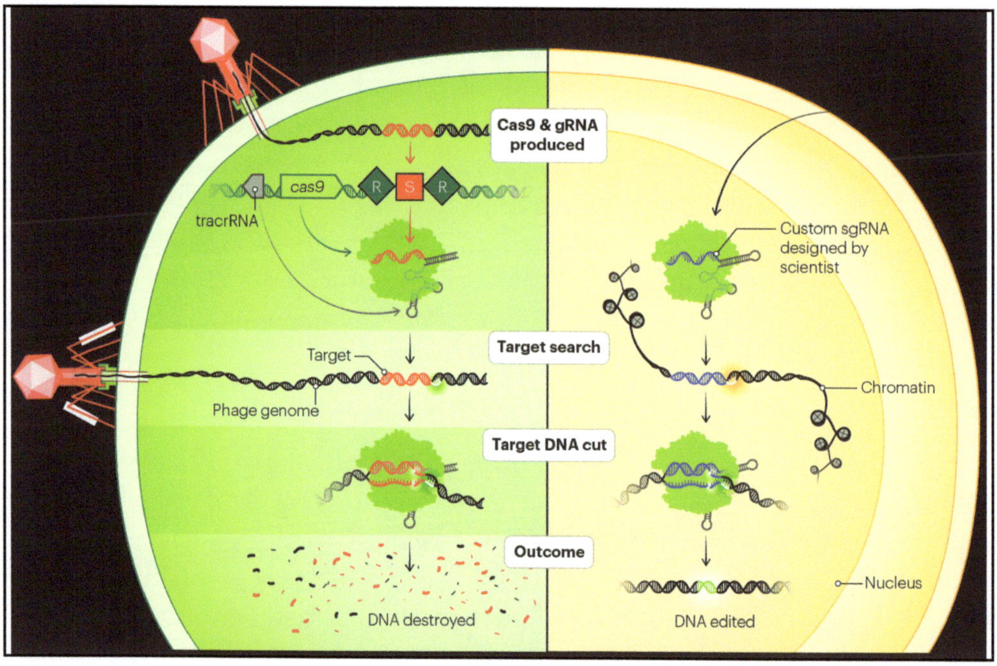

Moving Beyond Traditional CRISPR Techniques

This technology chemically converts a deactivated Cas9 protein, which can still target DNA but cannot cut it, into a precision molecular machine that directly converts one DNA base into another without causing a double-strand break. For example, a cytosine base editor can change a C-G base pair to a T-A pair, correcting a class of mutations that account for a significant portion of human genetic diseases. This bypasses the inefficiency and unpredictability of HDR. An even more precise technology, Prime Editing, functions like a genomic word processor. It uses an engineered Cas9 protein fused to a reverse transcriptase enzyme and is guided by a prime editing guide RNA that both specifies the target site and encodes the desired new genetic information. The system nicks a single strand of DNA and then directly writes the new, corrected sequence into the genome. Prime editing offers unprecedented versatility, capable of making all twelve possible base-to-base conversions, as well as small insertions and deletions, all with minimal off-target effects and without requiring a double-strand break. These advanced tools are transforming the therapeutic landscape, offering increasingly safer and more precise ways to correct genetic errors.

Broader Ecosystem and Ethical Considerations

The impact of these gene-editing tools extends far beyond the laboratory, sparking a global conversation about their ethical application. In medicine, clinical trials are already underway using CRISPR-based therapies to treat genetic blood disorders, certain cancers, and hereditary blindness, offering hope for previously untreatable conditions. In agriculture, gene editing is being used to develop crops with enhanced yield, nutritional value, and resistance to drought and disease, promising to bolster food security. However, these applications are accompanied by significant ethical and societal questions. The editing of somatic cells, which affects only the individual patient, is widely supported. In contrast, the editing of germline cells, which would result in heritable changes passed to future generations, raises profound ethical concerns about consent, long-term consequences, and the potential for a new form of eugenics. The scientific community has largely called for a moratorium on heritable human germline editing until the safety, efficacy, and societal implications are fully debated and a broad consensus is reached. The power to rewrite the code of life carries an immense responsibility, necessitating ongoing dialogue among scientists, ethicists, policymakers, and the public to ensure these powerful technologies are developed and deployed wisely, equitably, and for the benefit of all humanity.

2.2.2 Zinc Finger Nucleases (ZFNs) and TALENs

Zinc Finger Nucleases and TALENs have forged a new era in genome editing, empowering scientists to sculpt genomes with previously unimaginable precision and control. While both possess unique advantages and limitations, their development laid the foundation for transformative advances in biomedical research, agriculture, and therapeutics. Despite the rapid ascendancy of CRISPR/Cas9 systems, ZFNs and TALENs remain vital tools where reliability and specificity are essential, and their contributions to synthetic biology continue to shape progress in the field. The challenges and opportunities ahead underscore the enduring significance of these technologies as science and medicine strive for ever more targeted and effective genetic interventions.

Historical Development and Structural Principles

Zinc Finger Nucleases were the first programmable nucleases introduced to the field of gene editing, originating in the early 2000s as bioengineers learned to fuse zinc finger DNA-binding domains with the FokI DNA-cleaving enzyme. Zinc fingers are modular protein motifs, each typically recognizing a triplet of DNA nucleotides. By assembling these motifs in tandem, scientists could design ZFNs to bind specifically to any desired DNA sequence with high affinity and specificity.

When two ZFNs, recognizing adjacent DNA regions on opposite strands, bind their targets, the FokI domains dimerize and cleave the DNA, generating a double-stranded break that triggers endogenous repair mechanisms. This modular architecture delivered powerful genome editing capabilities, but the complexity of designing high-specificity, functional zinc finger arrays posed significant challenges. In contrast, TALENs were developed slightly later, inspired by the transcription activator-like effector proteins found in the plant-pathogenic bacterium Xanthomonas. Each TALEN motif recognizes a single nucleotide, allowing easier and more predictable targeting of DNA sequences.

By fusing multiple TALE repeats in series, researchers could generate nucleases capable of recognizing long stretches of DNA with high specificity. The activity of each TALE repeat is independent, simplifying design and enhancing targeting accuracy across the genome. Like ZFNs, TALENs use FokI as the cleavage domain, requiring dimerization for DNA cutting and minimizing off-target effects by necessitating dual binding events in close proximity.

Mechanism of Action and Gene Editing Applications

Both ZFNs and TALENs function by introducing targeted double-stranded breaks in DNA. The cell subsequently attempts to repair these breaks using either error-prone non-homologous end joining or homology-directed repair, the latter pathway allowing precise incorporation of engineered DNA templates. These approaches have enabled a broad spectrum of genetic modifications such as gene knockouts, insertions, deletions, and replacements in a variety of model organisms and cell types.

ZFNs and TALENs have been deployed successfully in mammalian cells, plants, zebrafish, fruit flies, nematodes, livestock, and even monarch butterflies, expanding the reach of genome engineering beyond classical systems. The precision and versatility of these nucleases make them ideal for investigating gene function, developing animal models for disease, and facilitating agricultural improvements. Their programmable nature unlocks the possibility to tailor genetic modifications for therapeutic interventions, fostering hope for treatments of inherited disorders and enabling ex vivo gene editing for cell-based therapies.

Comparative Advantages and Technical Challenges

While ZFNs and TALENs share a common conceptual framework, their practical differences are noteworthy. The design and synthesis of ZFNs is complex, time-consuming, and costly primarily due to the interdependent nature of zinc finger motifs. The specificity of one motif can influence neighboring motifs, and as a result, predicting and validating successful binding domains is often laborious and unpredictable.

This convoluted design process leads to frequent failure rates for new ZFN constructs and can result in substantial off-target activity and unwanted mutations. TALENs, on the other hand, offer a straightforward one-repeat-to-one-nucleotide binding code, making assembly and prediction of specificity much easier. TALENs are generally less expensive to produce, faster to design, and more flexible, and their performance is less dependent on context or sequence complexity. This difference has led to TALENs being widely adopted for many genome engineering applications, particularly where accuracy and simplicity are paramount.

Specificity, Efficiency, and Off-Target Effects

Both ZFNs and TALENs achieve high specificity through tailored protein-DNA interactions, reducing but not eliminating the risk of off-target mutations. ZFNs tend to have more unpredictable specificity profiles; interactions between zinc finger motifs can generate unanticipated binding and cleavage events, leading to potential genotoxicity. TALENs, with their individual nucleotide recognition, show reduced complexity and generally exhibit fewer off-target effects, with improved design reliability and mutation rates. Studies comparing the two technologies, especially against newer tools like CRISPR/Cas9, confirm that while ZFNs may cause a larger number of off-target cuts, TALENs achieve similar editing efficiency with fewer unwanted genomic alterations. Ultimately, both systems rely on careful design and thorough validation to optimize performance and safety, especially in applications with therapeutic potential.

Biomedical and Clinical Applications

The first therapeutic uses of ZFNs and TALENs have focused on ex vivo engineering of human cells for disease treatment. ZFNs have enabled targeted gene disruption to create HIV-resistant T cells by knocking out the CCR5 co-receptor gene. They have also been explored for correcting mutations underlying hemophilia, sickle cell disease, and lysosomal storage disorders. TALENs have supported the development of allogeneic CAR-T cells for cancer immunotherapy, using precise gene edits to prevent immune rejection and enhance efficacy. The broad applicability of both nucleases in stem cell editing, gene knockout, and gene correction underscores their pivotal role in translational medicine. Additionally, their use in animal models for disease and in agricultural biotechnology highlights the profound impact on research, drug development, and food security.

Evolution in the Era of CRISPR/Cas Technologies

The arrival of CRISPR/Cas9 systems, which use RNA-guided DNA endonuclease activity for gene editing, has revolutionized the field by offering even simpler, more efficient, and cost-effective genome engineering. CRISPR/Cas9 surpasses ZFNs and TALENs in ease of use and scalability, rapidly replacing earlier platforms for a majority of genome engineering projects. Nevertheless, ZFNs and TALENs retain significance, particularly in cases where CRISPR may not be optimal, such as in sequence contexts unfavorable for guide RNA design or where protein-DNA recognition is preferred. Continued comparison of these platforms informs the choice of editing tool for clinical and experimental needs, emphasizing specificity, efficiency, and safety.

2.2.3 Prime Editing and Base Editing

Base editing and prime editing are transformative technologies that have redefined the landscape of genome engineering. By enabling precise, targeted modifications without the collateral damage of double-strand breaks, they overcome critical limitations of earlier methods and open doors to previously unimaginable applications. Base editing provides a highly efficient solution for correcting specific point mutations, while prime editing offers unparalleled versatility for correcting virtually any small genetic alteration.

From curing monogenic disorders and combating cancer to engineering resilient crops and advancing fundamental biological research, the potential impact of these tools is profound. While challenges in delivery, specificity, and efficiency persist, the pace of innovation is rapid. As researchers continue to refine these molecular scalpels, base editing and prime editing are poised to revolutionize medicine and biology, moving us closer to the era of truly precise and safe genetic surgery. The future of genetic medicine is being written, one precise edit at a time.

Base Editing: The Art of Single-Base Conversion

Base editing is a technique designed to directly convert one base pair to another without requiring a DNA double-strand break or a donor DNA template. It achieves this by fusing a catalytically impaired Cas protein (nickase or dead Cas) to a base-modifying enzyme, specifically a deaminase. The most common base editors are:

- **Cytosine Base Editors (CBEs):** These consist of a catalytically dead Cas9 (dCas9) or a Cas9 nickase (nCas9) fused to a cytidine deaminase enzyme. The guide RNA (gRNA) directs the complex to the target DNA sequence. Once bound, the deaminase catalyzes the conversion of cytosine (C) to uracil (U) within a small, defined "editing window" (typically 4-5 nucleotides wide). Uracil is subsequently recognized by the cellular machinery as thymine (T) during DNA replication or repair. To increase efficiency and product purity, CBEs often include a uracil glycosylase inhibitor (UGI) domain, which prevents the immediate excision and repair of the U base, favoring its stable conversion to T. Thus, CBEs enable C•G to T•A base pair conversions.

- **Adenine Base Editors (ABEs):** Developed to address the limitation of CBEs (which cannot convert A•T to G•C), ABEs utilize a laboratory-evolved adenine deaminase enzyme fused to nCas9 or dCas9. The adenine deaminase converts adenine (A) to inosine (I). Inosine is treated as guanine (G) by the cellular replication and repair machinery, leading to an A•T to G•C base pair conversion.

The key advantages of base editing are its high efficiency for specific point mutations and its minimal creation of DSBs, which greatly reduces the risk of unwanted indels (insertions or deletions). This makes it particularly suitable for correcting pathogenic point mutations, such as those found in sickle cell disease, cystic fibrosis, or certain cancers.

Base editors have been successfully applied in a wide range of cell types and organisms, from plants to mammals, for research and therapeutic development. However, limitations exist. The editing window is constrained, restricting the targetable sites. Unintended "bystander" edits can occur within the window. Base editors cannot perform all possible base conversions (e.g., C•G to G•C or T•A) or make insertions or deletions larger than a single base. Furthermore, delivering the relatively large base editor fusion proteins efficiently in vivo remains a challenge.

Prime Editing: The Versatile Search-and-Replace System

Prime editing represents the next evolutionary step in precision genome editing, offering even greater versatility than base editing. Developed by David Liu and colleagues, prime editing can perform all 12 possible base-to-base conversions, as well as small insertions (up to ~40 bp) and deletions (up to ~80 bp), without requiring DSBs or double-stranded DNA donor templates. This is achieved through a sophisticated molecular machine consisting of three core components:

- **A Cas9 Nickase (nCas9):** This enzyme nicks one strand of the DNA double helix, creating a single-strand break. It is catalytically impaired so it cannot make a DSB.
- **A Reverse Transcriptase (RT) Enzyme:** Fused to the nCas9, this enzyme synthesizes DNA using an RNA template.

- **A Prime Editing Guide RNA (pegRNA):** This is a specially engineered guide RNA that has two key functions: (1) It contains the standard sequence for guiding the nCas9-RT complex to the target DNA site via complementary base pairing. (2) Crucially, it includes an additional "primer binding site" (PBS) and an "edit template" (RT template) at its 3' end. The PBS binds to the nicked DNA strand, and the RT template encodes the desired edit.
- **The Mechanism Involves Several Steps:** The pegRNA directs the nCas9-RT complex to the target DNA site. The nCas9 nicks the non-complementary (non-target) DNA strand. The PBS sequence of the pegRNA anneals to the exposed 3' end of the nicked DNA strand. Using the RT template sequence in the pegRNA as a blueprint, the reverse transcriptase enzyme synthesizes a new DNA flap containing the desired edit. Cellular DNA repair machinery then resolves the resulting DNA structures, preferentially incorporating the edited flap and repairing the nick, leading to a permanent change in the genome. This process is highly precise because the edit is directly copied from the pegRNA template.

Prime editing's ability to perform a wide range of edits–correcting all point mutations, making small insertions and deletions–without DSBs makes it exceptionally powerful. It has shown high efficiency and precision in various cell types and early therapeutic applications are being explored for diseases like sickle cell disease, progeria, and Tay-Sachs disease. However, the system is complex. Designing effective pegRNAs requires careful optimization of the PBS and RT template lengths. The efficiency can sometimes be lower than base editing for specific point mutations. The large size of the prime editor fusion (nCas9 + RT) poses significant delivery challenges, particularly for in vivo therapeutic applications. Off-target effects, while generally lower than CRISPR-Cas9, still require careful assessment.

Comparing the Editors: Strengths, Weaknesses, and Ideal Applications

Base editing and prime editing are complementary technologies, each with distinct advantages and ideal use cases.

Base Editing Excels At:

- Converting specific single base pairs (C•G to T•A or A•T to G•C) with very high efficiency.
- Applications where DSBs must be absolutely avoided.
- Situations where the target site falls within the optimal editing window and bystander edits are acceptable or can be minimized.
- High-throughput screening or applications in plants where simplicity is key.

Prime Editing Excels At:

- Performing any base-to-base conversion (all 12 types).
- Making precise small insertions or deletions.
- Correcting mutations where base editing is not possible (e.g., C•G to G•C) or where bystander edits are a major concern.
- Applications requiring versatile editing without DSBs, especially for complex mutations.

Shared Advantages:

- **No DSBs:** Both avoid the major pitfalls of CRISPR-Cas9-induced DSBs (indels, rearrangements).
- **No Donor Template:** Both eliminate the need for co-delivering a donor DNA template, simplifying the process.
- **High Precision:** Both offer superior precision for targeted edits compared to standard CRISPR-Cas9.

Shared Challenges:

- **Delivery:** Efficiently delivering the relatively large editor proteins (especially prime editors) and guide RNAs into target cells *in vivo* remains a significant hurdle for therapeutic translation. Viral vectors (like AAV) have limitations in cargo size and immunogenicity.
- **Off-Target Effects:** While generally lower than CRISPR-Cas9, both systems can potentially induce off-target edits. Careful design and assessment are crucial.
- **Efficiency Variability:** Editing efficiency can vary significantly depending on the target site, cell type, and delivery method.
- **Size Constraints:** The large size of the editor fusion proteins can be problematic for viral delivery.

Navigating the Roadblocks

Despite their immense promise, base editing and prime editing face several challenges that researchers are actively working to overcome:

- **Delivery:** This is arguably the biggest bottleneck for therapeutic applications. Strategies include optimizing viral vectors (e.g., using split-intein systems to deliver prime editors in smaller parts via AAV), developing non-viral delivery methods (e.g., lipid nanoparticles, polymer nanoparticles), and improving tissue-specific targeting.

- **Off-Target Effects:** Enhancing specificity is critical. This involves engineering high-fidelity versions of the Cas proteins, optimizing guide RNA/pegRNA design using computational tools, and developing improved methods for detecting and quantifying off-target edits.
- **Editing Efficiency and Window:** For base editing, expanding the targeting scope and reducing bystander edits are priorities. For prime editing, improving efficiency, especially for certain edits and cell types, and simplifying pegRNA design are key goals.
- **Understanding and Controlling Edits:** While precise, the cellular repair processes involved after the initial edit (especially for prime editing) are complex and can sometimes lead to unintended outcomes. Gaining a deeper understanding of these pathways will allow for better control.
- **Ethical and Safety Considerations:** As these technologies advance towards clinical use, careful consideration of ethical implications (e.g., germline editing, enhancement) and rigorous long-term safety assessment in relevant models are essential.

The future of base and prime editing is incredibly bright. Ongoing research focuses on:

- **Next-Generation Editors:** Engineering smaller, more efficient, and more specific editors. Developing base editors with expanded capabilities (e.g., C-to-G editors) and prime editors with higher efficiency.
- **Improved Delivery Systems:** Creating safer, more efficient, and tissue-targeted delivery platforms, particularly for *in vivo* therapies.
- **AI-Driven Design:** Utilizing artificial intelligence and machine learning to predict optimal gRNA/pegRNA designs, target sites, and editor variants for maximum efficiency and specificity.
- **Therapeutic Applications:** Moving promising candidates for genetic diseases into clinical trials. Exploring applications in complex diseases like cancer and neurodegeneration. Using these tools for *ex vivo* therapies (e.g., editing patient cells outside the body before transplantation).
- **Synthetic Biology:** Integrating base and prime editing into synthetic biology workflows for precise genome writing and the construction of complex genetic circuits.

2.3 SYNTHETIC BIOLOGY DESIGN TOOLS

2.3.1 BioCAD (Computer-Aided Design for Biology)

BioCAD represents a transformative approach in modern biology, combining computational design, modeling, and simulation with experimental implementation.

By providing tools for modular design, simulation of genetic circuits, metabolic pathway optimization, and protein engineering, BioCAD enables more predictable and efficient construction of biological systems. Its integration with laboratory automation and iterative design cycles enhances throughput, accuracy, and innovation in synthetic biology. While challenges related to biological complexity, data integration, and ethical considerations remain, the future of BioCAD is promising, driven by advances in AI, multi-omics integration, and global collaboration. As the field matures, BioCAD is poised to play a central role in biotechnology, medicine, environmental applications, and the engineering of life itself.

Historical Background

The concept of using computational tools to design biological systems emerged in the early 2000s alongside the rise of synthetic biology. Initial efforts focused on standardizing biological parts and developing frameworks to assemble them modularly. Early BioCAD tools allowed users to design simple genetic circuits and simulate their behavior using mathematical models. As the complexity of synthetic systems increased, more advanced platforms were developed that could integrate data from genomics, transcriptomics, and proteomics. Today, BioCAD encompasses a range of software for gene circuit design, metabolic pathway engineering, protein design, and whole-cell modeling, reflecting the growing computational sophistication in biology.

Principles of BioCAD

BioCAD operates on several foundational principles that parallel traditional engineering CAD systems. The first principle is modularity, which allows individual biological parts, such as promoters, coding sequences, and regulatory elements, to be designed and tested independently before integration. The second principle is predictability, achieved by simulating biological circuits to anticipate their dynamic behavior under varying conditions. Standardization is also critical, ensuring that parts have well-characterized properties that facilitate reproducible design. Abstraction simplifies complex biological systems into models that are easier to manipulate computationally. Finally, iteration allows rapid design-test cycles, where computational predictions are validated experimentally and refined to improve accuracy.

Tools and Platforms

A wide variety of tools and platforms are used in BioCAD. Software such as GenoCAD and CellDesigner enables the graphical assembly of genetic circuits using standardized parts. COPASI and MATLAB SimBiology facilitate the simulation of biochemical networks and metabolic pathways. Tools like Rosetta and Foldit assist in protein design, allowing users to model folding and stability computationally.

High-throughput BioCAD platforms can integrate genome-scale data to optimize entire metabolic pathways, predict flux distributions, and simulate cellular responses to environmental perturbations. Cloud-based platforms and collaborative repositories further enhance accessibility, allowing researchers worldwide to share designs, models, and data.

Applications in Synthetic Biology

BioCAD has numerous applications in synthetic biology. It is used to design genetic circuits that perform logical operations, control gene expression, or regulate metabolic flux. Metabolic pathway optimization through BioCAD allows for the efficient production of biofuels, pharmaceuticals, and fine chemicals.

Protein engineering benefits from BioCAD by enabling the design of enzymes with improved activity, specificity, or stability. In synthetic ecosystems, BioCAD can simulate interactions between engineered organisms and natural communities. Overall, BioCAD accelerates the translation of biological ideas into practical, functioning systems with reduced experimental uncertainty.

Integration with Laboratory Automation

BioCAD is increasingly integrated with laboratory automation, forming the basis for automated design-build-test-learn (DBTL) cycles in synthetic biology. Automated liquid handling systems can execute BioCAD-designed experiments with high precision, while high-throughput screening provides feedback for iterative optimization. Robotic platforms can construct large libraries of genetic variants based on BioCAD specifications, dramatically accelerating experimental throughput. This integration enables the construction of complex synthetic systems that would be challenging or impossible to achieve manually, combining computational precision with robotic efficiency.

Role in Genome and Metabolic Engineering

In genome and metabolic engineering, BioCAD provides a platform to integrate multiple layers of biological information. Whole-genome design and editing benefit from computational modeling that predicts gene interactions and essentiality. Metabolic engineering relies on BioCAD to simulate flux through pathways, identify bottlenecks, and propose interventions to enhance yield or efficiency.

By predicting the effects of genetic modifications before experimental implementation, BioCAD reduces time, cost, and resource expenditure. Moreover, BioCAD allows for iterative refinement, where computational predictions inform experimental design and experimental data are fed back into the model to enhance accuracy.

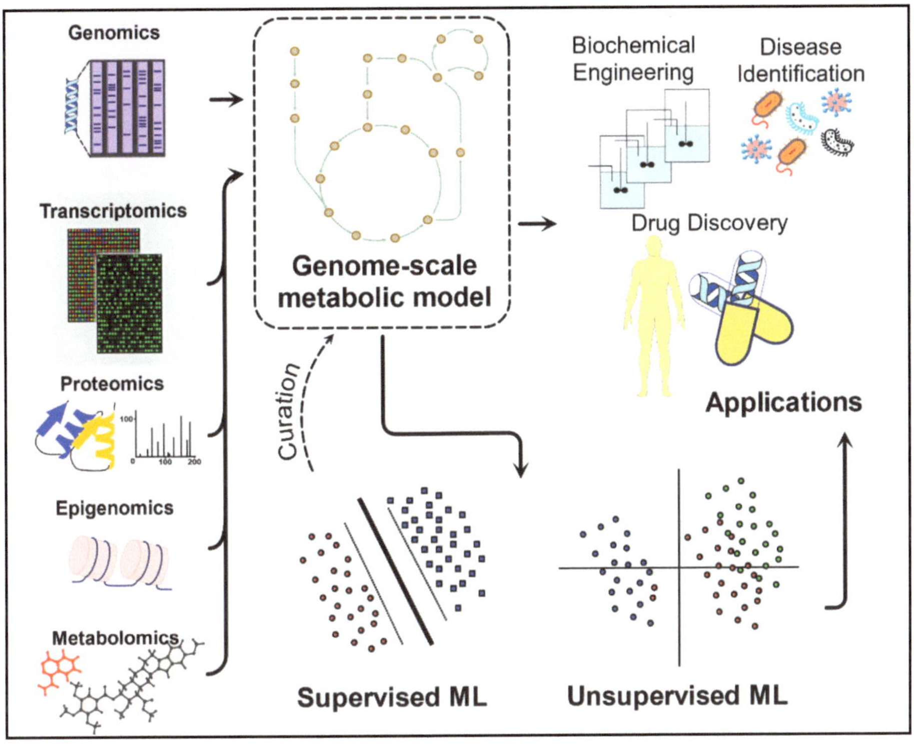

Genome-Scale Metabolic Modeling Enables in-Depth

Challenges in BioCAD

Despite its potential, BioCAD faces several challenges. Biological systems are inherently stochastic and context-dependent, making predictions less precise than in traditional engineering. Limited knowledge of regulatory networks and epigenetic effects can result in discrepancies between computational models and experimental outcomes. Data integration is challenging due to heterogeneous sources, varying formats, and incomplete datasets. Additionally, designing complex systems requires specialized expertise in both computational modeling and molecular biology, creating a barrier for widespread adoption. Computational limitations, including simulation speed and accuracy, also pose constraints on large-scale modeling efforts.

Effective use of BioCAD relies on understanding its key features and considerations:

- **Modularity:** Ability to design individual parts independently and assemble them predictably.
- **Standardized Biological Parts:** Use of well-characterized sequences with defined functions.
- **Simulation of Genetic Circuits:** Predicting dynamic behavior of genes and regulatory networks.

- **Metabolic Pathway Modeling:** Optimizing flux, yield, and efficiency in engineered pathways.
- **Protein Structure Prediction:** Designing enzymes or regulatory proteins computationally.
- **Whole-Cell Modeling:** Integrating genome-scale data to simulate cellular behavior.
- **Design Automation:** Generating multiple variants and optimizing design parameters computationally.
- **Integration with Laboratory Automation:** Translating in silico designs to wet-lab experiments.
- **Iterative Design Cycles:** Refining models and designs using experimental feedback.
- **Data Management and Sharing:** Maintaining repositories for reproducible and collaborative work.

Ethical and Safety Considerations

As BioCAD allows for the precise engineering of biological systems, ethical and safety considerations are paramount. Designing organisms with novel traits raises biosafety concerns regarding accidental release or ecological disruption. Dual-use research of concern (DURC) poses security risks if designs are misapplied for harmful purposes. Ethical questions arise around creating synthetic life and the potential consequences for society and natural ecosystems. Regulatory frameworks and responsible innovation practices must evolve alongside technological capabilities, ensuring that BioCAD applications are safe, ethical, and socially acceptable. Public engagement and education are essential for informed decision-making in this rapidly advancing field.

2.3.2 DNA Sequencing and Synthetic Biology Databases

DNA sequencing and synthetic biology databases represent a synergistic alliance that underpins the modern life sciences. Sequencing provides the raw material—the genetic code—while databases organize, analyze, and make this information actionable for engineering new biological functions.

Together, they enable the rational design of organisms with customized properties, the discovery of novel biomolecules, and the acceleration of innovation in biotechnology and medicine. As technology continues to advance, the boundaries between sequencing, computation, and synthesis will blur even further, paving the way for a future in which biological design is as precise, predictable, and scalable as modern engineering disciplines. The continued development of robust, secure, and interoperable databases will be essential to harness the full potential of DNA sequencing in shaping the next generation of synthetic biology research and applications.

Evolution of DNA Sequencing Technologies

The development of DNA sequencing has progressed through multiple generations of technologies, each offering improvements in accuracy, speed, and cost-effectiveness. The first generation of sequencing, known as Sanger sequencing, was introduced in the 1970s and relied on chain termination methods. It became the foundation for the Human Genome Project, which was completed in 2003 and represented one of the greatest scientific achievements in modern history. However, Sanger sequencing was labor-intensive and costly, prompting the development of next-generation sequencing (NGS) technologies.

NGS platforms, such as Illumina, Ion Torrent, and SOLiD, allowed massive parallel sequencing, enabling billions of DNA fragments to be read simultaneously. This revolutionized genomics by drastically reducing the cost per base and increasing throughput. The emergence of third-generation sequencing, such as PacBio and Oxford Nanopore technologies, has further advanced the field by enabling long-read sequencing that captures structural variations and epigenetic modifications more effectively. The ability to sequence entire genomes quickly and affordably has made DNA sequencing indispensable to synthetic biology, as it provides the blueprint required for designing and constructing new genetic systems.

Principles and Process of DNA Sequencing

DNA sequencing typically involves several critical steps, beginning with the extraction of DNA from biological samples. Once purified, the DNA is fragmented into smaller pieces and prepared into a sequencing library through adapter ligation and amplification. Depending on the sequencing platform, the DNA fragments are then read through chemical or optical methods that detect the order of nucleotides. The data obtained are computationally processed to assemble reads into contiguous sequences and align them with reference genomes. Quality control measures are employed to ensure data accuracy and to eliminate sequencing errors. Bioinformatics tools are essential at every stage of this process, converting raw sequence data into meaningful biological information that can be integrated into databases for further use in synthetic biology research.

Synthetic Biology and Data-Driven Design

Synthetic biology relies on a data-driven framework for the rational design and assembly of biological systems. The availability of detailed genetic sequences enables researchers to identify functional elements such as promoters, ribosome binding sites, and coding sequences. These components can be combined in new configurations to achieve desired behaviors in engineered organisms. The design-build-test-learn cycle in synthetic biology is heavily dependent on accurate sequence data and computational models that predict system behavior. Databases that store annotated genetic information and standardized biological parts play an essential role in enabling reproducibility and interoperability across research laboratories. Through these data resources, scientists can rapidly identify compatible genetic parts, predict outcomes, and minimize experimental errors.

Major Synthetic Biology Databases

Several specialized databases have been established to support the growing field of synthetic biology. The Registry of Standard Biological Parts is one of the most prominent repositories, providing a catalog of well-characterized DNA sequences that can be assembled into synthetic constructs. Each entry includes information on function, sequence, and experimental validation, allowing researchers to reuse and modify existing components efficiently.

The iGEM Parts Registry, developed for the International Genetically Engineered Machine competition, has contributed thousands of genetic elements that facilitate educational and research applications. The Synthetic Biology Open Language (SBOL) repository standardizes the exchange of design data, ensuring that genetic constructs can be shared across platforms and institutions. Other databases such as SynBioHub, Benchling, and Addgene serve as collaborative spaces for sharing plasmids, vectors, and engineered genes. The integration of these databases with sequencing data enhances their utility, allowing for direct connections between genotype and phenotype information.

Integration of Sequencing Data into Synthetic Biology Workflows

The connection between DNA sequencing and synthetic biology databases enables continuous feedback between experimental results and design refinement. Sequencing data can be used to verify that engineered constructs match their intended designs and to identify mutations or errors introduced during synthesis. Moreover, sequencing facilitates the monitoring of dynamic changes in engineered organisms, such as variations in gene expression or genome stability. When integrated into databases, these data contribute to the cumulative knowledge base that drives innovation in synthetic biology. The interoperability between sequencing technologies, bioinformatics pipelines, and databases ensures that information flows seamlessly from data generation to application, supporting the automation and scalability of synthetic biology workflows.

Applications in Medicine and Biotechnology

DNA sequencing and synthetic biology databases are playing a transformative role in medical and biotechnological applications. In medicine, sequencing enables the identification of disease-associated genes, guiding the design of targeted therapies and personalized treatments. Synthetic biology applies this information to develop engineered microorganisms that produce therapeutic compounds, biosensors for disease detection, and genetic circuits for controlled drug delivery.

In biotechnology, databases facilitate the design of microorganisms optimized for biofuel production, environmental remediation, and agricultural improvement. For instance, synthetic biologists use sequence data to engineer yeast or bacteria that produce complex molecules such as insulin, antibiotics, or bio-based plastics. The integration of large-scale genomic and functional data supports the development of more efficient and sustainable biomanufacturing systems.

Ethical and Security Considerations

The increasing accessibility of DNA sequencing and synthetic biology databases raises important ethical, legal, and security issues. Open access to genetic data fosters collaboration and scientific progress, but it also presents potential risks related to privacy, misuse, and biosecurity.

The possibility of synthesizing pathogenic sequences or creating genetically modified organisms with unpredictable effects necessitates stringent oversight. Ethical frameworks emphasize responsible data sharing, informed consent, and the protection of sensitive genetic information. Synthetic biology communities are actively developing policies and technical safeguards to prevent the misuse of sequencing and design data while ensuring that legitimate research can continue to flourish.

Computational Tools and Database Architecture

The architecture of synthetic biology databases is designed to handle vast quantities of complex data efficiently. These systems rely on relational databases, cloud-based storage, and distributed computing infrastructures to ensure scalability and accessibility. Metadata standards such as SBOL, GenBank format, and FASTA enable consistent representation and interoperability of sequence information. Computational tools integrated within databases allow users to perform sequence alignments, motif searches, and pathway modeling directly within the platform. Machine learning and artificial intelligence are increasingly being incorporated to predict gene function, optimize metabolic pathways, and design genetic circuits. The convergence of data analytics, automation, and cloud computing enhances the capacity of databases to serve as dynamic resources for scientific discovery and engineering.

2.4 METABOLIC ENGINEERING FOR SYNTHETIC BIOLOGY

2.4.1 Pathway Construction and Optimization

Pathway construction and optimization exemplify the shift from passive observation to active design in biology, redefining the practice of metabolic engineering and synthetic biotechnology. Through careful assembly, balancing, and iterative improvement, scientists turn cells into programmable biofactories capable of producing a growing array of valuable compounds. The multidisciplinary integration of genetics, chemistry, computational modeling, and process engineering provides ever-greater capabilities for manipulating life at the molecular level. While challenges persist related to host complexity, metabolic burden, and industrial scale-up, the future of pathway optimization is bright, promising more sustainable, efficient, and creative solutions across biotechnology and beyond.

Principles of Pathway Construction

Central to pathway construction is the identification and assembly of sequential enzyme-catalyzed reactions that convert precursor molecules into desired products. Synthetic biologists start by mapping naturally occurring biosynthetic routes—sometimes called blueprints—within donor organisms and selecting specific genes encoding the enzymes necessary for these transformations. The chosen genes are then clustered into a biosynthetic gene cluster or assembled into a synthetic operon, often with tailored promoter and regulatory elements to ensure controlled expression. Advances in DNA synthesis and high-precision cloning enable seamless assembly of multi-gene pathways. The modular nature of pathway construction allows for swapping individual enzymes or regulatory parts, providing a flexible framework for exploring alternative route designs and maximizing yields.

Host Selection and Genetic Integration

Selection of an appropriate host organism is crucial for successful pathway implementation. Common microbial hosts such as Escherichia coli, Saccharomyces cerevisiae, and various actinobacteria provide well-characterized genetic backgrounds, rapid growth, and ease of manipulation. Synthetic pathways may be integrated into native genomes or carried on plasmids, depending on stability and expression needs. Genome-wide assessment of integration sites—known as "safe harbors"—is now common practice to achieve optimal expression and genetic stability. These considerations extend into choosing appropriate ribosome binding sites, terminators, and scaffolding sequences that enable balanced translation across all pathway enzymes, minimizing the risk of toxic intermediates or byproduct accumulation.

Refactoring Native Pathways

Pathway refactoring involves reengineering native biosynthetic routes by replacing e ndogenous regulatory elements with artificial control mechanisms, such as synthetic promoters and riboswitches. This process decouples gene expression f rom native cellular regulation and enables precise tuning of metabolic flux.

Refactored pathways are redesigned to avoid transcriptional and translational bottlenecks, increase enzyme compatibility, and facilitate the installation of new metabolic capabilities in surrogate hosts. The use of homologous recombination or CRISPR-mediated genome editing facilitates rapid replacement of regulatory sequences, while iterative rounds of design-build-test-learn cycles enable stepwise improvement of pathway performance.

Metabolic Balancing and Flux Optimization

Efficient pathway function relies on balanced expression and activity of all constituent enzymes. If upstream reactions generate intermediates faster than downstream enzymes can consume them, the intermediates may accumulate to toxic levels, inhibiting cell growth and pathway throughput. Metabolic balancing is achieved by careful control of gene copy number, promoter strength, translational efficiency, and mRNA stability. Mathematical models and flux balance analysis help quantify metabolite flow and predict bottlenecks, guiding rational adjustment of genetic constructs. Dynamic regulation—using feedback sensors, inducible promoters, and synthetic genetic circuits—further refines pathway operation by responding to real-time changes in cellular status and environmental conditions.

Protein Engineering in Pathway Optimization

Improving the catalytic efficiency, specificity, and stability of pathway enzymes is a central focus of optimization. Protein engineering techniques such as site-directed mutagenesis and directed evolution enable scientists to create enzyme variants with enhanced activity or altered substrate preference. By selecting mutants with improved kinetic properties, roles in rate-limiting steps are amplified, allowing for greater control over overall pathway performance. In some cases, enzyme cyclization and fusion to scaffolding proteins are employed to increase spatial proximity among pathway components, mimicking nature's strategies for channeling intermediates and reducing side reactions.

Systems Biology and Omics Integration

Integration of omics data—genomics, transcriptomics, proteomics, and metabolomics—facilitates comprehensive understanding of host cell responses and pathway interactions. Systems biology provides global context for metabolic engineering, enabling prediction of how pathway modifications impact cellular physiology and homeostasis. Machine learning and artificial intelligence harness large datasets to direct metabolic strain design, predict significant gene targets for pathway alteration, and maximize metabolite output. These computational approaches accelerate development by identifying optimal phenotypes and tailoring pathway construction strategies to individual host capabilities.

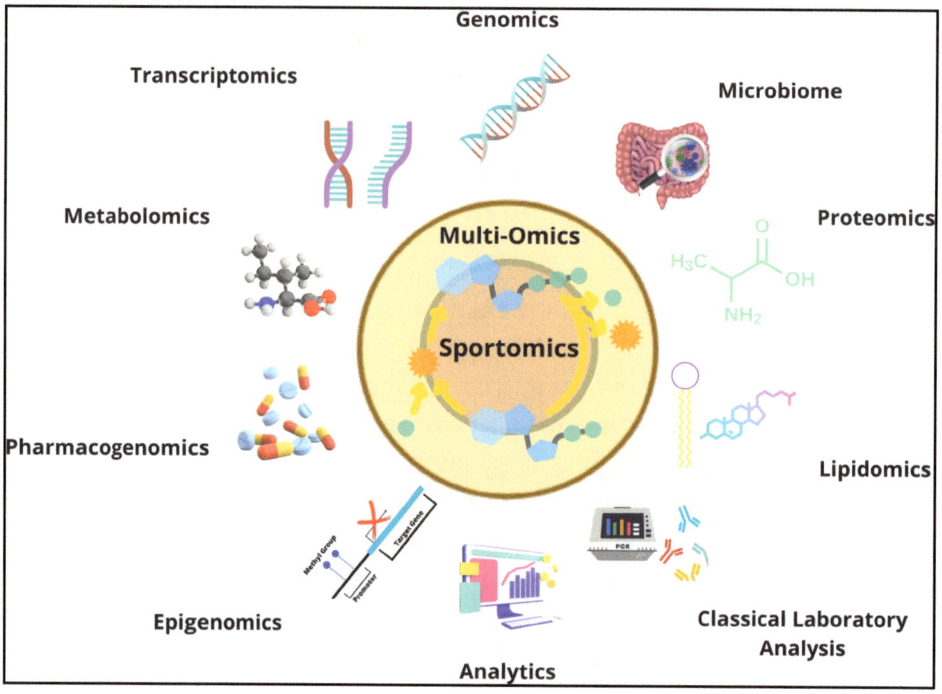

Omics, Multiomics, and Sportomics Approaches

Automated and Combinatorial Approaches

Recent advances in automation and combinatorial biology have revolutionized pathway assembly and optimization. Robotic platforms now assemble hundreds to thousands of pathway variants in parallel, driving high-throughput prototyping of genetic designs. Combinatorial pathway optimization leverages mix-and-match approaches, screening multiple enzyme homologs, regulatory parts, and organizational strategies to identify the best-performing configurations. Retrobiosynthetic techniques allow the exploration of alternative routes borrowed from diverse organisms, broadening the range of achievable products and overcoming limitations imposed by native enzyme specificity.

Dynamic Control and Artificial Enzyme Complexes

Spatial organization of pathway enzymes into artificial complexes—also known as synthetic multienzyme channels—mimics natural approaches for enhancing metabolic efficiency. By co-localizing enzymes, cells reduce the diffusion distance between active sites, promoting faster turnover of intermediates and suppressing off-target side reactions. Dynamic pathway control employs metabolite-responsive sensors, synthetic genetic switches, and feedback loops to adjust enzyme expression in response to cellular state. This level of sophistication increases robustness, prevents toxicity, and enables production of complex molecules with high yield and purity.

Challenges and Bottlenecks in Pathway Optimization

Despite technological advances, pathway optimization faces significant challenges. Metabolic burden caused by excessive protein expression can strain cellular resources, reducing growth rates and genetic stability. Achieving balanced flux remains complex, as misregulated pathways may generate toxic byproducts, leading to cell death or impaired performance. Limited knowledge of complex cellular networks, cross-talk between metabolic routes, and unintended interactions with host physiology complicate optimization efforts. Large-scale fermentation introduces additional hurdles related to stress adaptation, heterogeneity, and mass transfer limitations that do not appear during laboratory-scale validation. Better coordination between molecular biologists and process engineers is increasingly recognized as critical for translating optimized pathways from bench to industry.

Computational Design and Predictive Modeling

Computational models—ranging from simple stoichiometric analyses to entire genome-scale metabolic reconstructions—guide pathway design and optimization by predicting performance, flux distributions, and system responses to genetic changes. Artificial intelligence and machine learning streamline enzyme selection, parameter optimization, and automated design of genetic parts. These tools generate in silico libraries of pathway variants, forecast bottlenecks, and prioritize experimental efforts. Digital twins of cells may one day allow for simulation of metabolic processes with unprecedented accuracy, facilitating rational design of cell factories.

Innovations for Industrial Translation

Efficient transfer of optimized metabolic pathways to large-scale industrial settings requires careful attention to process techno-economics, cellular fitness, and regulatory compliance. Strain performance under fermentation conditions is predicted by models integrating genetic stability, stress resistance, and product tolerance. Genome editing protocols now focus on integrating pathway genes into genomic safe harbors that confer both high stability and flexibility. High-throughput screening and modeling are employed early in strain development to ensure that novel biotechnological potential is maintained during scale-up. Advanced monitoring systems, real-time process feedback, and adaptive control strategies further enable successful transfer from bench research to industrial application.

2.4.2 Industrial Microorganisms for Bioproduction

Industrial microorganisms are the unsung heroes of the modern bioeconomy. From the food we eat and the medicines that heal us to the materials we use and the fuels that power our world, they provide efficient, sustainable, and often ingenious solutions to complex manufacturing challenges. Through relentless scientific advancement in microbiology, genetic engineering, and bioprocess technology, we continue to unlock the vast potential encoded within these tiny life forms. While challenges related to scale, cost, and regulation persist, the ongoing revolution in tools like CRISPR, AI, and synthetic biology is rapidly addressing them. The future points towards increasingly sophisticated microbial cell factories that are not only more productive and versatile but also seamlessly integrated into circular, low-carbon industrial systems. As we strive towards a more sustainable and resilient future, the role of industrial microorganisms in bioproduction will only become more critical and transformative, proving that some of the most powerful engines of industry are indeed alive.

Microbial Workforce: Types of Industrial Microorganisms

A diverse consortium of microorganisms is employed in industrial bioproduction, each chosen for specific traits suited to particular products and processes. Bacteria are perhaps the most widely used group. Escherichia coli (E. coli), a common gut bacterium, is a workhorse for producing recombinant proteins, enzymes, and metabolites due to its well-understood genetics, rapid growth, and ease of manipulation. Bacillus species, like B. subtilis, are favored for secreting large quantities of enzymes (e.g., proteases, amylases) into the growth medium and are generally regarded as safe.

Corynebacterium glutamicum is legendary for its ability to produce massive amounts of amino acids, particularly L-glutamate and L-lysine, which are essential for animal feed and human food. Fungi also play pivotal roles. Saccharomyces cerevisiae (baker's yeast) is not only crucial for baking and brewing but is also a premier host for producing bioethanol, wine, and a growing list of biopharmaceuticals and chemicals. Filamentous fungi, such as Aspergillus niger, are prolific producers of organic acids (like citric acid, a major food additive) and enzymes (e.g., pectinases, glucoamylases). Penicillium species are the original sources of penicillin and other beta-lactam antibiotics.

Algae and cyanobacteria represent another important category, particularly for applications involving light-driven processes. Microalgae like Chlorella and Spirulina are grown for high-value compounds like omega-3 fatty acids, pigments (astaxanthin), and as potential food sources. Cyanobacteria are being engineered to produce biofuels, bioplastics, and other chemicals directly from CO2 and sunlight, offering promising routes for carbon capture and utilization. Actinomycetes, a group of bacteria resembling fungi, are prolific sources of antibiotics, antifungals, and other secondary metabolites. Streptomyces species, for instance, produce over two-thirds of the clinically used antibiotics. The selection of the optimal microorganism depends entirely on the target product, the required metabolic pathway, and the economic and practical constraints of large-scale fermentation.

Engineering Excellence: Key Characteristics of Industrial Strains

Not every microorganism is suitable for industrial bioproduction. Strains must possess a specific set of characteristics to be effective and economically viable at scale. High growth rate and yield are paramount; the microorganism must grow rapidly to high cell densities and efficiently convert the carbon and energy source into the desired product with minimal byproducts. Genetic stability is critical for maintaining consistent production over many generations in large fermenters, preventing the loss of engineered traits or productivity. Non-pathogenicity is a non-negotiable safety requirement, ensuring the microorganism poses no risk to human health or the environment, especially for products like food additives or pharmaceuticals.

Ability to utilize cheap substrates is essential for economic feasibility. Industrial strains are often selected or engineered to thrive on inexpensive, readily available feedstocks like molasses, corn steep liquor, lignocellulosic hydrolysates (from agricultural residues), or even methane and syngas, rather than costly refined sugars. Tolerance to process conditions is another vital attribute. Industrial fermentations involve specific conditions–often high temperatures, extremes of pH, high osmotic pressure, or the presence of inhibitory compounds–that can stress or kill microorganisms. Robust strains must be able to withstand these conditions to maintain productivity. Furthermore, ease of product recovery is desirable. Some microorganisms secrete the product into the surrounding medium (extracellular), simplifying downstream purification, while others require cell lysis (intracellular). The ideal industrial microorganism is a carefully balanced package of these traits, optimized through years of selection, classical mutagenesis, and increasingly, sophisticated genetic engineering.

Precision Tools: Genetic Engineering and Strain Improvement

The capabilities of industrial microorganisms have been dramatically expanded through genetic engineering and advanced strain improvement techniques. Metabolic engineering involves the deliberate re-design of an organism's metabolic pathways to enhance the production of a desired compound or to enable the synthesis of novel molecules. This is achieved by overexpressing key enzymes, deleting genes that divert resources to unwanted byproducts, or introducing entirely new pathways from other organisms. Synthetic biology takes this a step further, applying engineering principles to biology. It involves designing and constructing new biological parts, devices, and systems, or re-designing existing natural systems for useful purposes. This includes creating genetic circuits for optimal expression, building synthetic genomes, and programming cells with novel functions. The development of CRISPR-Cas9 and other precise genome editing tools has revolutionized strain engineering, allowing for targeted modifications with unprecedented speed, efficiency, and accuracy. These tools enable multiplexed editing–making multiple changes in a single step–and are significantly accelerating the design-build-test-learn cycle in strain development. Directed evolution is another powerful complementary technique, mimicking natural evolution in the lab. It involves creating genetic diversity (e.g., through random mutagenesis) and applying selective pressure to enrich for mutants with improved properties, such as higher product yield, better substrate utilization, or increased tolerance. These techniques are continuously evolving, allowing scientists to push the boundaries of what industrial microorganisms can produce and how efficiently they do it.

Flask to Factory: Bioproduction Processes

Transforming a microorganism's potential into a commercial product involves sophisticated bioproduction processes, primarily centered around fermentation. Submerged fermentation (SmF) is the most common method, where microorganisms are grown in large, sterile vessels (bioreactors) containing a liquid nutrient medium.

The bioreactor provides a controlled environment–regulating temperature, pH, dissolved oxygen, and nutrient levels–to optimize growth and product formation. Agitation ensures mixing and oxygen transfer. This method is ideal for bacteria, yeast, and many fungi, and is used for products like antibiotics, amino acids, and enzymes. Solid-state fermentation (SSF) involves growing microorganisms on moist solid substrates, like agricultural wastes or inert materials. This mimics natural habitats for some fungi and is particularly advantageous for producing certain enzymes, spores, or secondary metabolites where the solid matrix provides specific benefits, such as higher product concentrations or lower water usage. However, scale-up and process control can be more challenging than SmF. Following fermentation, downstream processing (DSP) is crucial to recover and purify the product from the complex mixture in the bioreactor.

DSP typically involves multiple steps: separating cells from the broth (e.g., via filtration or centrifugation), concentrating the product (e.g., evaporation, membrane filtration), and finally purifying it to the required specification (e.g., chromatography, crystallization). DSP can be a significant portion of the total production cost, especially for high-value products like pharmaceuticals. The design and operation of bioreactors are critical to the success of any bioproduction process. They range from small laboratory-scale vessels to massive industrial fermenters holding tens of thousands of liters. Key types include stirred-tank reactors (most common), airlift reactors, and packed-bed reactors, each with advantages depending on the microorganism and product.

Bounty of Biology: Major Products from Industrial Microorganisms

The range of products manufactured using industrial microorganisms is vast and continually expanding, impacting nearly every aspect of modern life. Primary metabolites are essential compounds synthesized during the growth phase of the microorganism, directly involved in its central metabolism. Examples include ethanol (produced by yeast via fermentation, used as a biofuel and solvent), organic acids like citric acid (by Aspergillus niger, a major food acidulant), lactic acid (by Lactobacillus, for bioplastics and food), and amino acids like L-glutamate (by Corynebacterium, a flavor enhancer) and L-lysine (for animal feed).

Secondary metabolites are compounds not essential for growth but often conferring ecological advantages. They are typically produced during the stationary phase and include a treasure trove of antibiotics (penicillin from Penicillium, streptomycin from Streptomyces), pigments (e.g., beta-carotene from algae, monascus pigments from fungi), immunosuppressants, and anticancer agents. Enzymes and biocatalysts are another major category. Industrial enzymes are produced by bacteria (e.g., proteases, amylases), fungi (e.g., cellulases, pectinases), and yeasts, and are used extensively in food processing (cheese making, baking), textiles (stone-washing denim), detergents (breakdown of stains), pulp and paper, and biofuel production (cellulases for ethanol).

Biopolymers and biomaterials represent a rapidly growing sector. This includes bioplastics like polyhydroxyalkanoates (PHAs, produced by bacteria like Cupriavidus necator) and polylactic acid (PLA, derived from fermented lactic acid), offering biodegradable alternatives to conventional plastics. Biofuels are also a key product, with bioethanol (from yeast fermentation of sugars) and biodiesel (from algal oils or microbial lipids) being prominent examples, alongside emerging advanced biofuels from engineered microbes. The list continues to grow, including vitamins, vaccines, diagnostic proteins, and even electronic materials.

Navigating Hurdles: Challenges in Industrial Bioproduction

Despite its immense promise, industrial bioproduction faces significant challenges that must be overcome for wider adoption and economic competitiveness. Contamination risks are a constant threat in large-scale fermentations, where unwanted microorganisms can invade and outcompete the production strain, consuming nutrients and producing contaminants that ruin the batch. Strict sterility protocols and robust strain design are essential but add complexity and cost. Scale-up difficulties are notorious. What works well in a small laboratory flask often performs poorly in a 100,000-liter industrial reactor. Factors like mixing efficiency, oxygen transfer, shear stress, and heterogeneity in nutrient and temperature distribution become critical and can drastically affect cell physiology and product formation. Predicting and managing these changes requires sophisticated engineering and scale-down models. Regulatory hurdles present another major barrier, particularly for products destined for human consumption or medical use. Genetically modified organisms (GMOs) face stringent and often costly regulatory pathways for approval, varying significantly by country. Even non-GMO products require rigorous safety and quality assessments. Economic viability is perhaps the most persistent challenge. While bioproduction offers environmental benefits, it must also be cost-competitive with established chemical processes. High costs associated with feedstocks (especially if relying on food crops), energy consumption, downstream processing, and capital investment for fermentation facilities can make bioprocesses economically challenging. Fermentation processes often have lower volumetric productivities and higher water usage compared to chemical plants. Continuous innovation in strain engineering, fermentation technology, and DSP is required to drive down costs and improve efficiency.

Bio-Revolution Continues

Industrial bioproduction is exceptionally bright, driven by rapid advancements in science and technology. Emerging technologies are poised to reshape the field. Cell-free systems involve using the molecular machinery of cells (enzymes, ribosomes) in vitro to perform specific reactions, bypassing the need to maintain living cells. This offers advantages for toxic products or multi-step pathways that are difficult to achieve in vivo. The integration of Artificial Intelligence (AI) and machine learning is revolutionizing strain design and process optimization. AI algorithms can analyze vast datasets to predict optimal genetic modifications, design novel pathways, and model fermentation processes, accelerating the design-build-test-learn cycle exponentially. Synthetic biology continues to push boundaries, enabling the creation of "cellular foundries" programmed to produce an ever-wider array of complex molecules, including those impossible to synthesize chemically.

The drive towards sustainable production and a circular bioeconomy is a major catalyst. Industrial biotechnology is key to utilizing renewable resources, waste streams (e.g., CO2, agricultural residues), and renewable energy to produce chemicals, materials, and fuels, reducing greenhouse gas emissions and creating closed-loop systems. This includes engineering microbes to convert plastic waste into valuable products or to capture and utilize CO2 directly. Personalized medicine and biopharmaceuticals represent another frontier. Engineered microorganisms are being used to produce highly specific therapeutic proteins, vaccines, and diagnostic tools tailored to individual patients. The production of complex biologics like monoclonal antibodies and viral vectors via microbial systems (though often using mammalian cells, microbial systems are advancing for some products) is becoming more efficient. As these technologies mature and converge, industrial microorganisms will play an even more central role in building a more sustainable, healthy, and prosperous future.

CHAPTER 3

Synthetic Biology Applications

3.1 BIOMANUFACTURING AND INDUSTRIAL BIOTECHNOLOGY

3.1.1 Production of Pharmaceuticals and Biologics

The production of pharmaceuticals and biologics lies at the heart of modern medicine, reflecting a remarkable convergence of biological science, chemical engineering, and computational innovation. While traditional pharmaceuticals remain crucial for many therapeutic areas, biologics have ushered in a new era of targeted, efficient, and personalized treatment options. From recombinant proteins and monoclonal antibodies to RNA-based vaccines and gene therapies, these products have transformed healthcare globally. Despite challenges related to cost, complexity, and regulation, continuous innovation in bioprocessing, automation, and sustainability ensures that the production of pharmaceuticals and biologics will remain one of the most dynamic and impactful fields in science and technology. The ongoing integration of advanced manufacturing, synthetic biology, and digital tools promises a future where life-saving therapies are developed faster, produced more sustainably, and delivered to patients worldwide with unprecedented precision and reliability.

Historical Background

The history of pharmaceutical production dates back centuries, but the integration of biology into drug production began in the late 20th century. Traditional pharmaceuticals relied on chemical synthesis, but many therapeutic molecules proved too complex to produce chemically. The discovery of recombinant DNA technology in the 1970s marked a major turning point, allowing genes encoding therapeutic proteins to be inserted into host cells for biological production. The first recombinant drug, human insulin (Humulin), produced by genetically modified Escherichia coli, was approved in 1982. This success initiated an era of biotechnology-driven drug development, leading to the rise of monoclonal antibodies, recombinant enzymes, and, more recently, advanced therapies such as RNA-based vaccines and cell therapies.

Pharmaceutical Production

Pharmaceutical production typically involves the chemical synthesis or extraction of small, well-defined molecules that interact with specific biological targets in the body. These drugs are designed to modulate enzymes, receptors, or signaling pathways to achieve therapeutic effects. Production processes emphasize purity, reproducibility, and scalability. Chemical synthesis remains dominant for small-molecule drugs, although new technologies such as automated synthesis, flow chemistry, and computational drug design have enhanced efficiency and precision. Quality control and regulatory compliance are crucial, as even small variations in composition can affect safety and efficacy.

Biologics: Definition and Characteristics

Biologics differ fundamentally from traditional pharmaceuticals. They are typically large, complex molecules such as proteins, nucleic acids, or cells, produced through biological processes rather than chemical synthesis. Examples include monoclonal antibodies, vaccines, hormones, cytokines, and gene therapies. Unlike small molecules, biologics are sensitive to environmental conditions such as temperature and pH, requiring specialized manufacturing and storage. Their production involves living organisms that express recombinant proteins under controlled conditions. The complexity of biologics allows them to target diseases with high specificity, but it also introduces challenges in standardization, scalability, and cost.

Biotechnological Production Systems

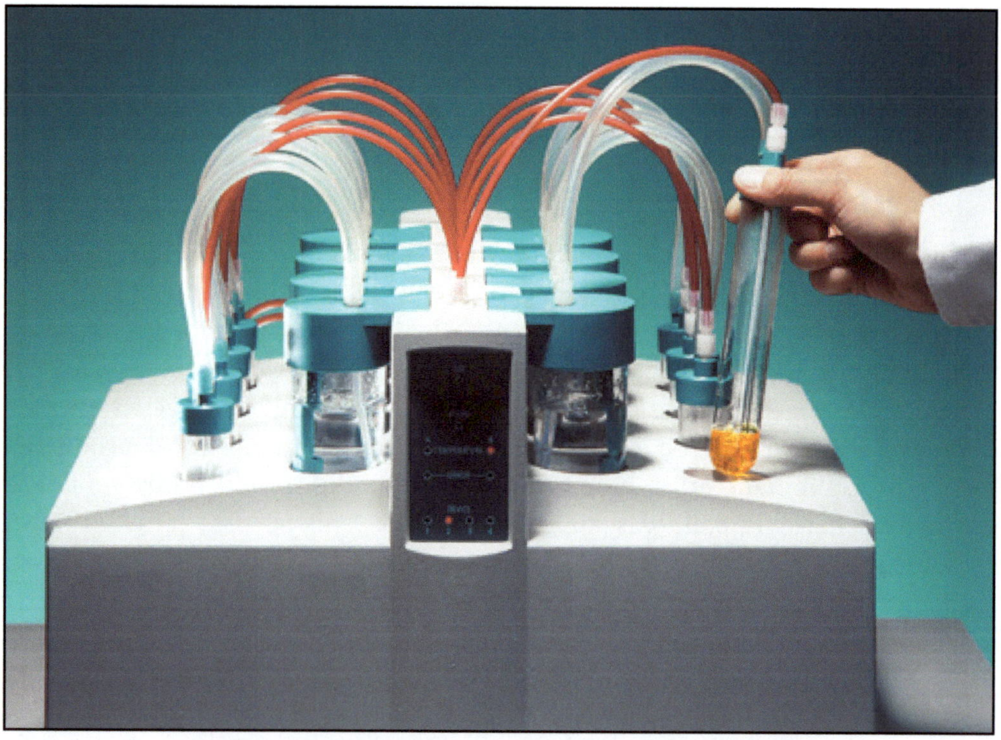

Engineered Systems for Biotechnological Production

The production of biologics relies on biotechnological systems that utilize living cells as factories for complex molecules. The choice of host system depends on the desired product, yield, and post-translational modifications. Bacterial systems like E. coli are widely used for producing simple proteins and enzymes due to their rapid growth and cost-effectiveness.

Yeast systems, such as Pichia pastoris, offer higher yields and eukaryotic protein folding capabilities. Mammalian cell lines, including CHO (Chinese Hamster Ovary) cells, are the gold standard for complex proteins and antibodies requiring glycosylation. In addition, insect cell systems and plant-based expression platforms are emerging alternatives, providing scalable and flexible production options.

Steps in Biopharmaceutical Production

The process of producing biologics involves several critical stages, each requiring precise control and optimization. It begins with gene cloning and vector construction, where the gene encoding the desired protein is inserted into a suitable expression vector. The next stage is host cell transformation and selection, ensuring stable expression of the recombinant gene.

Cell culture or fermentation follows, where cells are grown in bioreactors under optimized conditions. The expressed protein is then harvested and purified using chromatography, filtration, and concentration methods. Finally, the purified biologic undergoes formulation, stability testing, and quality control to ensure safety and efficacy.

Quality Control and Regulatory Compliance

The production of pharmaceuticals and biologics is tightly regulated to ensure product safety, efficacy, and consistency. Regulatory agencies such as the U.S. Food and Drug Administration (FDA) and the European Medicines Agency (EMA) oversee manufacturing standards under Good Manufacturing Practices (GMP). Quality control measures include testing for purity, potency, sterility, and stability at every stage of production. For biologics, additional tests are required to confirm proper folding, glycosylation, and absence of contaminants such as viruses or endotoxins. Batch-to-batch consistency is essential, and advanced analytical techniques like mass spectrometry, high-performance liquid chromatography (HPLC), and bioassays are used for validation.

Technologies and Techniques

The production of pharmaceuticals and biologics integrates several advanced technologies and techniques:

- **Recombinant DNA Technology:** Enables insertion of therapeutic genes into host cells for protein expression.
- **Cell Culture Systems:** Use of mammalian, bacterial, or yeast cells to produce biologics in controlled environments.

- **Bioreactors:** Provide optimal growth conditions for large-scale production of cells and proteins.
- **Chromatography:** Purification of proteins and biomolecules with high specificity.
- **Downstream Processing:** Includes filtration, concentration, and formulation for final product preparation.
- **Protein Engineering:** Modifies structure for improved stability, activity, or reduced immunogenicity.
- **Continuous Manufacturing:** Increases efficiency and reduces production time.
- **Analytical Testing:** Ensures product quality, potency, and consistency through advanced assays.
- **Automation and Process Control:** Enhances reproducibility and minimizes human error.

Cryopreservation and cold chain logistics: Maintain stability during storage and distribution.

Applications in Medicine and Healthcare

Pharmaceuticals and biologics have transformed the treatment landscape across multiple medical fields. In oncology, monoclonal antibodies and immune checkpoint inhibitors have enabled targeted cancer therapy. In endocrinology, recombinant insulin and growth hormones provide effective management of diabetes and growth disorders.

Autoimmune diseases are treated with biologics such as TNF inhibitors for rheumatoid arthritis and multiple sclerosis. Vaccines, including mRNA-based COVID-19 vaccines, represent one of the most successful biologic applications, offering rapid response to emerging pathogens. Gene therapies are now correcting genetic defects at their source, while cell therapies like CAR-T cells are revolutionizing personalized medicine.

Challenges in Production

The production of pharmaceuticals and biologics is not without challenges. The complexity of biologics demands high-cost facilities, specialized expertise, and stringent quality controls. Maintaining consistency in living cell-based production is difficult, as minor changes in culture conditions can affect yield and structure. Scale-up from laboratory to industrial level requires significant optimization to maintain product integrity. Regulatory approval processes are lengthy and expensive, reflecting the need for extensive safety data. Moreover, global supply chain disruptions and cold chain requirements can complicate distribution, particularly for vaccines and temperature-sensitive products.

Sustainability and Innovation in Biomanufacturing

Sustainability and innovation are emerging priorities in pharmaceutical and biologic production. Advances in continuous manufacturing reduce waste, increase efficiency, and lower environmental impact. Synthetic biology enables the creation of optimized cell factories capable of higher yields and reduced resource use. Green chemistry principles are applied to reduce hazardous by-products in small-molecule synthesis. The rise of 3D bioprinting, microfluidic bioreactors, and AI-driven process optimization promises further improvements in production efficiency. In addition, the move toward decentralized biomanufacturing and single-use bioreactors supports flexible and scalable production for global health needs.

3.1.2 Biofuels and Renewable Energy

Biofuels and renewable energy together embody the vision of a cleaner, more sustainable future. As global energy demands continue to rise, the need for diverse and low-carbon energy solutions becomes increasingly urgent. Biofuels provide a practical pathway for decarbonizing transportation and industrial sectors, while renewable energy sources offer reliable and inexhaustible alternatives for electricity and heating. The synergy between these technologies represents a holistic approach to sustainability, combining innovation, policy support, and environmental stewardship. Moving forward, the challenge lies not only in technological advancement but also in fostering global cooperation, equitable access, and responsible resource management. Through these collective efforts, biofuels and renewable energy will continue to shape the transition toward a resilient, low-carbon, and inclusive global energy future.

Global Energy Landscape

The modern energy landscape is dominated by fossil fuels—coal, oil, and natural gas—which together supply over 80 percent of the world's primary energy needs. However, their extensive use has resulted in environmental degradation, air pollution, and rising levels of carbon dioxide in the atmosphere. The urgent need to transition to cleaner alternatives has driven global investment in renewable energy technologies. Biofuels have emerged as an important component of this transition because they can be directly integrated into existing transportation and industrial systems. Unlike solar or wind energy, which generate electricity, biofuels can power internal combustion engines, aircraft, and heavy vehicles, providing a practical bridge between traditional and sustainable energy systems. Countries across the world are implementing policies and incentives to promote renewable energy adoption, creating a dynamic environment for research, development, and commercialization.

Biofuels: Definition and Classification

Biofuels are fuels produced directly or indirectly from organic matter, collectively referred to as biomass. This includes plant materials, agricultural residues, algae, and even animal waste. Biofuels are classified into several generations based on their feedstock and production technology. First-generation biofuels are derived from food crops such as corn, sugarcane, and soybeans, and include ethanol and biodiesel. While they are easy to produce, they compete with food production and can contribute to deforestation. Second-generation biofuels are produced from non-food biomass like agricultural residues, wood chips, and grasses, offering a more sustainable alternative. Third-generation biofuels are derived from microalgae, which have high lipid content and can be cultivated in non-arable areas. Emerging fourth-generation biofuels incorporate genetic engineering and carbon capture technologies, aiming to create carbon-neutral or even carbon-negative fuels. This progression reflects the growing sophistication and sustainability of biofuel production.

Production Processes and Technologies

The production of biofuels involves a series of biochemical and thermochemical processes that convert biomass into usable fuel. Ethanol, one of the most common biofuels, is typically produced through the fermentation of sugars using yeast or bacteria. Biodiesel is produced via transesterification, a chemical reaction that converts oils and fats into fatty acid methyl esters suitable for diesel engines. Advanced technologies such as gasification, pyrolysis, and hydrothermal liquefaction can convert lignocellulosic biomass into synthetic fuels and bio-oils. Microalgae cultivation systems, which can be integrated with wastewater treatment, provide another efficient route for biofuel production. These technologies continue to evolve, driven by improvements in enzyme engineering, microbial metabolism, and process optimization, all aimed at enhancing yield and reducing production costs.

Renewable Energy Sources and their Synergy with Biofuels

Renewable energy encompasses a wide range of naturally replenished sources that include solar radiation, wind, water flow, geothermal heat, and biomass. Solar and wind energy have witnessed unprecedented growth due to technological advancements and declining costs. Hydropower remains the most widely used renewable energy source, providing reliable electricity generation. Geothermal energy utilizes heat from the Earth's interior to produce power and heating. Biofuels occupy a unique position among renewable energies because they provide a liquid or gaseous form of energy suitable for transportation and industrial heating—sectors that are difficult to electrify. Integrating biofuels with other renewable energy systems can enhance energy efficiency and reliability. For example, excess renewable electricity can be used to power biofuel production facilities, while bioenergy plants can provide stable energy supply when solar or wind output fluctuates. This synergy contributes to a more resilient and balanced energy infrastructure.

Environmental and Economic Impacts

The environmental benefits of biofuels and renewable energy are substantial, though not without complexity. Biofuels significantly reduce greenhouse gas emissions compared to fossil fuels, especially when produced from waste materials or non-food crops. They also help in reducing particulate emissions and sulfur oxides, improving air quality. However, large-scale biofuel production can have negative environmental consequences if not managed properly, such as deforestation, loss of biodiversity, and excessive water use. Economically, biofuel production stimulates rural development by creating jobs in agriculture and processing industries. It can also reduce dependence on imported oil, improving energy security. The renewable energy sector as a whole has become a major driver of economic growth, with millions of people employed in solar, wind, and bioenergy industries worldwide. Nevertheless, balancing environmental sustainability with economic viability remains a critical challenge for policymakers and researchers.

Applications of Biofuels in Modern Energy Systems

Biofuels are primarily used in the transportation sector, where they serve as substitutes or additives to gasoline and diesel. Ethanol is commonly blended with gasoline to create formulations like E10 or E85, depending on the percentage of ethanol content. Biodiesel can be used in standard diesel engines with little or no modification, reducing engine emissions and improving lubrication. Aviation biofuels, also known as sustainable aviation fuels (SAF), have gained significant attention due to the aviation sector's heavy reliance on fossil fuels. Marine biofuels are also being explored as alternatives for shipping industries aiming to comply with international emission standards. Beyond transportation, biofuels can be used in power generation and heating, particularly in remote or rural areas where grid connectivity is limited. These diverse applications highlight the versatility and adaptability of biofuels within the global energy system.

Technological Innovations Driving the Future of Bioenergy

The future of biofuels is being shaped by advances in biotechnology, synthetic biology, and systems engineering. Genetic modification of microorganisms and plants allows for improved feedstock yield and enhanced conversion efficiency.

Synthetic biology enables the design of microbial strains capable of directly producing hydrocarbons or alcohols from carbon dioxide or waste gases.

Integration of artificial intelligence and machine learning in biofuel research accelerates the discovery of optimal metabolic pathways and process parameters. Additionally, innovations in biorefinery design promote circular economy principles, where waste streams from one process are utilized as inputs for another, minimizing environmental impact. The convergence of these technologies is paving the way for next-generation biofuels that are more efficient, cost-effective, and environmentally sustainable.

Table: Comparison of Major Biofuel Types and Characteristics

Biofuel Type	Feedstock Source	Production Process	Energy Output (MJ/L)	Primary Application
Ethanol	Sugarcane, Corn	Fermentation	21	Transportation (gasoline blend)
Biodiesel	Vegetable Oils, Algae	Transesterification	33	Transportation (diesel engines)
Biogas	Organic Waste, Manure	Anaerobic Digestion	23	Power Generation and Heating
Bioethanol (2nd Gen)	Lignocellulosic Biomass	Enzymatic Hydrolysis	25	Transportation
Biohydrogen	Algae, Bacteria	Photolysis/Fermentation	120	Industrial and Power Generation

Challenges in Biofuel Development and Deployment

Despite their potential, biofuels face several technical, economic, and environmental challenges. Feedstock availability remains a major issue, as large-scale cultivation of energy crops can compete with food production and land use. Production costs are still relatively high for advanced biofuels, limiting their commercial competitiveness.

Moreover, the energy balance and carbon footprint of biofuel production depend heavily on agricultural practices, processing methods, and transportation logistics. Infrastructure compatibility also poses difficulties, particularly in the case of high ethanol blends or alternative fuels that require engine modifications. Regulatory and policy uncertainties further hinder investment in biofuel industries. Addressing these challenges requires coordinated efforts among governments, industries, and research institutions to promote sustainable feedstock management, technological innovation, and supportive policy frameworks.

Role of Policy and International Collaboration

Policy frameworks play a crucial role in shaping the development of biofuels and renewable energy. Many countries have adopted renewable fuel standards, blending mandates, and tax incentives to encourage biofuel production and consumption. The European Union's Renewable Energy Directive and the United States Renewable Fuel Standard are prominent examples of such policies. International collaborations, such as the International Renewable Energy Agency (IRENA) and the Global Bioenergy Partnership (GBEP), facilitate knowledge exchange, technology transfer, and investment in sustainable energy infrastructure. Developing countries, particularly those in tropical regions, have significant potential for biofuel production due to favorable climatic conditions and abundant biomass resources. Global cooperation ensures that biofuel development aligns with sustainability goals and contributes to equitable energy access.

3.1.3 Specialty Chemicals and Materials

The rise of specialty chemicals and advanced materials through synthetic biology and biomanufacturing represents a transformative leap for global industry and sustainability. Moving past traditional chemical synthesis, engineered biological systems now provide tailored, eco-friendly solutions for diverse market needs, ranging from pharmaceuticals and food additives to bioplastics, textiles, and electronics. With advances in metabolic engineering, protein design, and computational optimization, specialty chemicals and materials are increasingly accessible, customizable, and environmentally benign. These developments not only deliver functional advantages but also align with circular economy principles and climate goals. As research advances and market acceptance grows, synthetic biology will continue to innovate specialty chemicals and materials, reshaping production paradigms and future-proofing industries for the challenges ahead.

Specialty Chemicals: Definitions and Significance

Specialty chemicals are high-value compounds designed for specific functions, often present as additives, catalysts, surfactants, pigments, flavors, fragrances, or active pharmaceutical ingredients. Unlike bulk chemicals, their production hinges on tailored compositions that support targeted applications across multiple industries.

In pharmaceuticals, specialty chemicals provide complex intermediates and enablers of drug synthesis, while in agriculture, they support crop protection, nutrient management, and food preservation. The automotive sector leverages specialty chemicals for lubricants and paints, electronics rely on advanced dielectrics and semiconductors, and consumer goods are enhanced through unique polymers and stabilizers. The underpinning requirement for specialty chemicals is their ability to serve critical, high-performance functions, which increasingly rely on precision engineering, sustainability, and biocompatibility.

Synthetic Biology and Biomanufacturing: Redefining Production

Traditional chemical synthesis relies on petroleum feedstocks and energy-intensive processes, often limited by inefficiencies and environmental impact. Synthetic biology introduces engineered biological systems—microbes, plants, or mammalian cells—as living factories that produce specialty chemicals from renewable feedstocks. By manipulating metabolic pathways and optimizing biological machinery, scientists can now assemble multi-step biosynthetic routes to generate molecules once accessible only through complex chemical synthesis. Sugars, agricultural residues, or even captured CO2 can be converted into high-value chemicals using well-orchestrated biocatalysis, bypassing many limitations of conventional chemistry. This paradigm shift emphasizes both scalability and sustainability, heralding a new era where biology drives industrial change.

Engineering Biological Systems for Material Innovation

Biomanufacturing goes beyond molecules to embrace advanced materials, utilizing microbial domains as foundries for polymers, nanomaterials, fibers, adhesives, and responsive biomaterials. Genetic engineering allows for precise control of monomer sequences, branching, and crosslinking, endowing biopolymers with tailored mechanical, chemical, and functional properties. Natural polymers such as polylactic acid (PLA), polyhydroxyalkanoates (PHAs), and bio-based nylons emerge from engineered metabolic pathways in bacteria and yeast, replacing fossil-derived plastics with biodegradable alternatives. Nanobiotechnology leverages engineered viruses, protein cages, and biological vesicles to fabricate drug delivery vehicles, sensors, and magnetic or electronic nanomaterials. Biological systems present unique opportunities for customizing material architecture, integrating sensing and actuation, and scaling up production with reduced toxicity and improved biocompatibility.

Applications in Multiple Industry Sectors

The impact of specialty chemicals and materials manufactured via synthetic biology spans numerous sectors. In healthcare, engineered biological systems synthesize antimicrobial peptides, enzyme-based therapeutics, smart drug delivery scaffolds, and regenerative materials. Industrial biotechnology leverages living cells to produce pigments, flavor enhancers, surfactants, and biolubricants for food, cosmetics, and environmental treatment. The packaging and textile industries benefit from biodegradable or recyclable polymers, moving toward circular economies and reduced plastic pollution. Electronics increasingly incorporate bio-inspired conductive polymers and organic semiconductors, enhancing device efficiency and flexibility. Automotive and aerospace industries pursue lightweight, corrosion-resistant, and high-strength biocomposites, contributing to energy efficiency and sustainability.

Sustainability and Circular Economy Potential

Environmental imperatives are driving the shift toward bio-based specialty chemicals and materials. Biomanufacturing harnesses renewable feedstocks, supports closed-loop production, and reduces greenhouse gas emissions compared to conventional chemistry. Waste products from other industries—such as lignocellulosic biomass, agricultural side streams, and food residues—are valorized as raw materials for specialty chemical synthesis. The capacity for engineered microbes to process unconventional carbon sources expands the pool of available materials while decreasing the reliance on finite petroleum. Biodegradable polymers and circular material flows integrate seamlessly with global sustainability objectives, reducing landfill accumulation and enabling new modes of resource recovery and product lifecycle management.

Design, Construction, and Optimization Challenges

Engineering biological systems for specialty chemical and material production requires mastery of genetic and metabolic design. The complex biosynthetic pathways necessary for producing these compounds must be assembled and optimized within host cells, often requiring the integration of many genes, fine-tuned regulation, and balanced metabolic flux. High-throughput DNA synthesis, robust cloning techniques, and advanced genome-editing tools such as CRISPR/Cas systems facilitate assembly of multi-step pathways, but achieving industrial-scale yield and purity remains a formidable challenge. Predictive computational models, bioinformatics, and machine learning are increasingly indispensable for optimizing strain design, improving pathway efficiency, and navigating the immense design space.

Technological Innovations in Material Fabrication

Synthetic biology has enabled the creation of novel classes of advanced materials unattainable by traditional methods. Engineered protein scaffolds can self-assemble into films, fibers, or gels imbued with programmable strength, elasticity, and biological activity. Living materials, where cells are incorporated into composite structures, offer self-healing and adaptive properties for applications ranging from biomedicine to construction. Biogenic nanomaterials, fabricated by microbial or plant hosts, find use as catalysts, sensor components, and agents of targeted drug delivery. These innovations hinge on the ability to control molecular structure at every level, from atomic detail to bulk architecture, utilizing biological templates for precision assembly. As the toolbox of synthetic biology expands, so too does the ability to integrate organic and inorganic features, bridging the gap between living systems and engineered materials.

Economic Considerations and Market Growth

The specialty chemical and advanced materials industries are witnessing strong growth in bio-based product development, fueled by market demands for sustainability, unique performance features, and regulatory pressures. Europe has emerged as a global leader in supporting bio-based material adoption, spurred by governmental policies and investments in green technology.

However, cost-competitiveness remains a key concern, with biomanufactured specialty chemicals often carrying higher production costs than their petroleum-derived analogs. Advances in process optimization, strain stabilization, and downstream purification are beginning to close the gap, while the introduction of innovative biocatalysts and efficient fermentation processes are reducing barriers to commercialization. As industries respond to escalating sustainability targets and consumer preference for green products, the demand for bio-based specialty chemicals and materials is expected to accelerate.

3.2 ENVIRONMENTAL APPLICATIONS

3.2.1 Bioremediation and Pollution Control

Bioremediation stands as a testament to nature's remarkable ability to restore balance to damaged ecosystems when properly harnessed and supported by human innovation. The continued evolution of this field from empirical observation to sophisticated biotechnology demonstrates the power of interdisciplinary collaboration in addressing complex environmental challenges. As pollution continues to threaten environmental and human health globally, bioremediation offers hope for sustainable solutions that work with natural processes rather than against them. The integration of cutting-edge technologies with traditional biological principles creates unprecedented opportunities for treating previously intractable contamination while minimizing environmental impact and resource consumption. Success in bioremediation requires continued investment in research, development of supportive regulatory frameworks, and public education to build understanding and acceptance of biological treatment technologies. The path forward demands recognition that bioremediation represents not just a technical solution but a fundamental shift in how humanity approaches environmental restoration, embracing biological diversity and natural processes as allies in the fight against pollution.

Fundamental Principles of Bioremediation

The scientific foundation of bioremediation rests on the principle that living organisms possess enzymatic systems capable of catalyzing the transformation of pollutants through various metabolic pathways. These biological processes involve either the complete mineralization of contaminants into harmless end products such as carbon dioxide and water, or their conversion into less toxic intermediates that can be further processed by other organisms.

The success of bioremediation depends on several critical factors including the bioavailability of contaminants, presence of appropriate degrading organisms, suitable environmental conditions, and adequate nutrients to support microbial growth and metabolism. Temperature, pH, oxygen availability, and moisture content significantly influence the rate and extent of biodegradation, requiring careful optimization for effective treatment. The thermodynamic favorability of degradation reactions determines whether organisms can derive energy from breaking down specific pollutants, making some compounds readily biodegradable while others resist biological treatment. Understanding these fundamental principles enables environmental scientists to design targeted bioremediation strategies that maximize pollutant removal while minimizing treatment time and cost. The interplay between biological, chemical, and physical processes in contaminated environments creates complex systems that require sophisticated monitoring and management to achieve successful remediation outcomes.

Types and Mechanisms of Bioremediation

Bioremediation encompasses diverse strategies that can be broadly categorized based on the treatment location and the degree of human intervention required. In-situ bioremediation treats contamination directly at the polluted site without excavation or removal, making it particularly suitable for extensive contamination or sensitive locations where disturbance must be minimized. This approach includes techniques such as bioventing, which enhances aerobic degradation by supplying oxygen to subsurface soils, and biosparging, which injects air into groundwater to stimulate indigenous microorganisms.

Ex-situ bioremediation involves removing contaminated material for treatment elsewhere, offering greater control over environmental conditions but requiring additional costs for excavation and transportation. Biostimulation enhances the activity of indigenous microorganisms by adding nutrients, electron acceptors, or other amendments that limit degradation rates. Bioaugmentation introduces specialized microorganisms with superior degradation capabilities to supplement existing populations, particularly useful when native communities lack the necessary metabolic pathways. Natural attenuation relies on inherent biological processes to reduce contaminant concentrations over time without active intervention, though it requires extensive monitoring to ensure adequate progress.

Phytoremediation employs plants and their associated microorganisms to extract, accumulate, or transform pollutants, combining aesthetic benefits with effective treatment for certain contaminants.

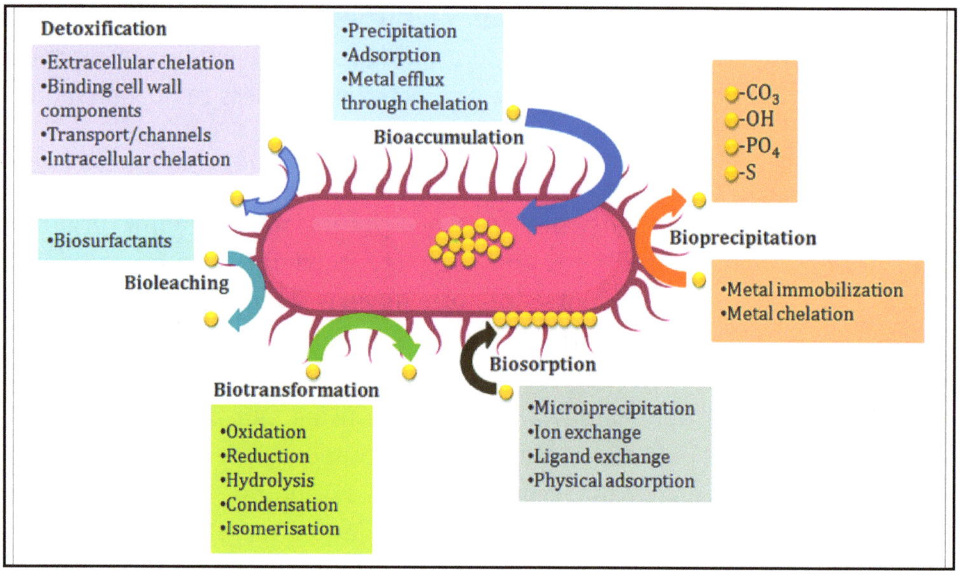

Biological Mechanisms in Environmental Cleanup

Microbial Agents in Bioremediation

The success of bioremediation fundamentally depends on the metabolic versatility of microorganisms that serve as the primary agents of pollutant degradation. Bacteria dominate many bioremediation applications due to their rapid growth rates, genetic adaptability, and diverse metabolic capabilities that enable them to utilize various pollutants as carbon and energy sources. Pseudomonas species exemplify this versatility, degrading hydrocarbons, pesticides, and solvents through multiple enzymatic pathways.

Fungi, particularly white-rot fungi, produce powerful extracellular enzymes such as laccases and peroxidases that break down complex aromatic compounds including lignin-like pollutants and recalcitrant dyes. Algae contribute to bioremediation through biosorption of heavy metals and transformation of nutrients in eutrophic waters, while also producing oxygen that supports aerobic degradation processes.

Consortia of multiple species often prove more effective than single organisms, as different members contribute complementary metabolic capabilities and create syntrophic relationships that enhance overall degradation efficiency. Genetic engineering has produced modified organisms with enhanced degradation capabilities, expanded substrate ranges, and improved survival in harsh environments, though their environmental release remains controversial. The isolation and characterization of novel degrading organisms from extreme environments continues to expand the toolkit available for addressing emerging contaminants and challenging remediation scenarios.

Applications in Different Environmental Contexts

Bioremediation technologies find widespread application across diverse environmental settings, each presenting unique challenges and opportunities for biological treatment. Petroleum hydrocarbon contamination from oil spills, leaking underground storage tanks, and refinery operations represents one of the most common targets for bioremediation, with numerous successful demonstrations from the Exxon Valdez spill to routine treatment at thousands of contaminated sites worldwide. Agricultural chemicals including pesticides, herbicides, and fertilizers contaminate vast areas of soil and water, where bioremediation offers a cost-effective alternative to excavation and chemical treatment.

Heavy metal contamination from mining, smelting, and industrial processes poses particular challenges since metals cannot be degraded, requiring biological strategies that focus on immobilization, transformation to less toxic forms, or bioaccumulation for subsequent removal. Chlorinated solvents used in dry cleaning and degreasing operations create dense plumes in groundwater that respond well to anaerobic bioremediation through reductive dechlorination. Municipal and industrial wastewater treatment increasingly incorporates specialized biological processes to remove emerging contaminants such as pharmaceuticals, personal care products, and endocrine-disrupting compounds. Marine environments contaminated with plastics benefit from newly discovered organisms capable of degrading synthetic polymers, though complete mineralization remains challenging for many plastic types.

Technological Advances and Innovations

Recent technological innovations have dramatically expanded the capabilities and applications of bioremediation, transforming it from a passive treatment option to a sophisticated, engineered solution. Molecular biology tools enable rapid identification and characterization of degrading organisms through metagenomics, while transcriptomics and proteomics reveal the specific pathways and enzymes involved in pollutant degradation.

Nanotechnology enhances bioremediation through the development of nano-carriers for enzyme immobilization, nanoparticles that improve contaminant bioavailability, and nano-sensors for real-time monitoring of degradation progress. Synthetic biology approaches design novel metabolic pathways and create designer organisms optimized for specific pollutants, though regulatory and safety concerns limit field applications. Bioelectrochemical systems combine biological and electrochemical processes to accelerate degradation rates and enable treatment of previously recalcitrant compounds. Advanced monitoring technologies including biosensors, remote sensing, and isotope analysis provide unprecedented insights into bioremediation processes and enable adaptive management strategies. Computational modeling and artificial intelligence predict treatment outcomes, optimize operational parameters, and identify promising organisms for specific applications. The integration of bioremediation with other treatment technologies creates hybrid systems that overcome individual limitations and achieve superior treatment efficiency.

Challenges and Limitations

Despite its numerous advantages, bioremediation faces significant challenges that limit its application in certain scenarios and require ongoing research to address. The slow rate of biodegradation for some compounds means treatment can take months or years, making bioremediation unsuitable for urgent remediation needs. Incomplete degradation may produce metabolites more toxic than parent compounds, requiring careful monitoring and management to prevent secondary contamination. Environmental conditions at contaminated sites often prove hostile to degrading organisms, with extreme pH, temperature, or salinity limiting biological activity.

Bioavailability restrictions prevent organisms from accessing pollutants bound to soil particles or trapped in non-aqueous phase liquids, reducing treatment effectiveness. Competition from indigenous organisms, predation, and viral infections can eliminate introduced organisms before they establish effective populations. Regulatory frameworks often lag behind technological developments, creating uncertainty about the approval and implementation of novel bioremediation approaches. Public perception and acceptance vary widely, with concerns about genetically modified organisms and potential unintended consequences influencing policy decisions and project implementation.

3.2.2 Carbon Capture and Sequestration

Carbon Capture and Sequestration represents a cornerstone technology in the global effort to combat climate change. By capturing CO_2 at its source or directly from the air and safely storing or reusing it, CCS bridges the gap between current fossil fuel dependence and a sustainable, low-carbon future.

Its applications span power generation, manufacturing, and carbon utilization industries, providing flexible pathways to decarbonization. Although challenges remain in terms of cost, scalability, and perception, the steady advancement of capture materials, storage monitoring, and policy support offers a pathway toward wide-scale adoption. In the long term, the integration of CCS with renewable energy systems and circular carbon economies will be pivotal in achieving net-zero emissions and stabilizing the Earth's climate. Through global cooperation and technological innovation, carbon capture and sequestration can transform from a niche strategy into a mainstream solution for sustainable planetary stewardship.

Need for Carbon Capture

The urgency of implementing carbon capture technologies arises from the accelerating pace of climate change. Fossil fuel-based industries, power plants, and transportation sectors are major contributors to atmospheric CO_2. Despite the global transition toward renewable energy, fossil fuels remain the dominant energy source in many countries.

Thus, CCS offers a critical transitional solution by allowing continued use of fossil fuels while minimizing environmental impact. Without such technologies, it is nearly impossible to meet the goals outlined in the Paris Agreement, which seeks to limit global temperature rise to below 1.5°C above pre-industrial levels. In addition to mitigating emissions, CCS plays a vital role in decarbonizing hard-to-abate sectors such as cement, steel, and chemical manufacturing, where alternative low-carbon solutions are not yet fully viable.

Principles of Carbon Capture and Sequestration

Carbon Capture and Sequestration operates on three fundamental stages: capture, transport, and storage or utilization. The capture phase involves separating CO_2 from other gases produced in industrial or power generation processes. The captured CO_2 is then compressed and transported, usually through pipelines, to designated storage sites. The final stage involves the secure injection of CO_2 into geological formations, where it is stored for long periods, or its conversion into useful products through carbon utilization technologies. Each stage demands precise engineering, strong safety protocols, and regulatory oversight to ensure efficiency and prevent environmental risks.

Carbon Capture Techniques

Several methods are used to capture CO_2, each suited to specific industrial processes and concentrations of emissions. The most common approach is post-combustion capture, where CO_2 is removed from flue gases after fossil fuel combustion using solvents like amines. This method can be retrofitted to existing power plants, making it highly versatile. Pre-combustion capture involves converting fuel into a mixture of hydrogen and CO_2 before combustion, allowing CO_2 to be separated under high pressure. Oxy-fuel combustion burns fuel in pure oxygen, resulting in a concentrated CO_2 stream that is easier to capture. Emerging technologies such as direct air capture (DAC) extract CO_2 directly from the atmosphere, offering potential for negative emissions, though they currently face challenges in energy use and cost.

CO_2 Transportation Methods

Once captured, CO_2 must be safely transported to storage or utilization sites. The most efficient and widely used method is pipeline transportation, similar to natural gas pipelines but designed for supercritical CO_2, which behaves as both a gas and a liquid under high pressure. Pipelines can efficiently move large volumes over long distances, provided they comply with strict safety standards. For regions where pipelines are impractical, shipping offers an alternative, particularly for offshore storage. Transporting CO_2 requires constant monitoring of pressure, temperature, and flow rates to prevent leaks and ensure operational integrity. The infrastructure for CO_2 transport is expanding globally, with networks being developed in North America, Europe, and Asia to support industrial carbon management.

Storage and Sequestration

The storage phase of CCS is crucial for ensuring that the captured carbon remains permanently isolated from the atmosphere. Geological sequestration is the most established approach, involving the injection of CO_2 into deep underground rock formations. These formations include depleted oil and gas reservoirs, deep saline aquifers, and unmineable coal seams. At depths greater than 800 meters, CO_2 remains in a dense, supercritical state, allowing it to occupy less space and remain trapped by impermeable rock layers. Over time, CO_2 becomes further immobilized through physical and chemical processes such as residual trapping, dissolution trapping, and mineralization, ensuring long-term stability. Ongoing research focuses on enhancing the monitoring and verification of storage sites using seismic imaging, pressure sensors, and satellite-based surveillance to guarantee environmental safety.

Technologies in CCS

The success of CCS depends on a range of advanced technologies that ensure efficiency, safety, and scalability. These include:

- Amine-based absorption systems for capturing CO_2 from industrial gases.
- Membrane separation technologies that selectively filter CO_2 molecules.
- Cryogenic distillation, which separates CO_2 through cooling and condensation.
- Solid sorbents, including zeolites and metal-organic frameworks (MOFs), for selective capture.
- Direct Air Capture (DAC) units that extract CO_2 directly from the atmosphere.
- Enhanced Oil Recovery (EOR) using injected CO_2 to improve oil extraction while storing carbon.

- Mineral carbonation, converting CO_2 into stable mineral compounds.
- Geological monitoring tools such as seismic surveys and borehole sensors.
- Pipeline integrity management systems for safe transport.
- AI and data analytics to optimize capture efficiency and monitor storage stability.

These technologies collectively enhance the feasibility and cost-effectiveness of CCS operations, making it possible to scale them up across industries.

Carbon Utilization and Circular Economy

Beyond storage, carbon utilization represents an emerging frontier in CCS technology. Instead of permanently storing CO_2, captured carbon can be repurposed into valuable products, thereby creating a circular carbon economy. CO_2 can be used as a feedstock in the production of synthetic fuels, building materials, and chemicals such as methanol and urea. In agriculture, CO_2 is used to enhance greenhouse crop yields, while in the food and beverage industry, it serves as a carbonation agent. Research in carbon-to-value technologies explores the conversion of CO_2 into polymers, carbonates, and even sustainable aviation fuels through catalytic or electrochemical reactions. By turning waste carbon into economic resources, utilization strategies can offset the costs of capture and support broader climate goals.

Environmental and Economic Implications

While CCS offers significant environmental benefits, its large-scale implementation involves economic, technical, and social considerations. The environmental advantage is clear—reducing CO_2 emissions from heavy industries and power plants directly contributes to climate mitigation. However, the high cost of capture and storage, estimated at $50 to $120 per ton of CO_2, remains a barrier. Infrastructure investments, energy requirements, and public acceptance are additional challenges. There are also environmental risks associated with storage, such as potential leakage or induced seismic activity, though these risks are minimized through careful site selection and monitoring. Economically, CCS can create new job opportunities in engineering, geology, and environmental management, while also sustaining existing industries during the energy transition.

Global Policies and Implementation

Governments and international organizations recognize the strategic importance of CCS in achieving climate targets. Policies promoting CCS include carbon pricing, tax credits, and emission trading systems that make carbon reduction financially viable. The U.S. 45Q tax credit, for instance, provides incentives for companies that capture and store CO_2. The European Union's Innovation Fund and the Carbon Border Adjustment Mechanism (CBAM) further encourage adoption. Countries like Norway, Canada, Japan, and Australia are leading in CCS research and demonstration projects. Flagship initiatives such as Sleipner in the North Sea, Boundary Dam in Canada, and Gorgon CCS in Australia showcase real-world success in large-scale carbon storage. Collaborative international frameworks are essential to expand infrastructure, share best practices, and achieve global deployment.

3.3 AGRICULTURE AND FOOD SECURITY

3.3.1 Synthetic Biology for Crop Improvement

Synthetic biology represents a paradigm shift in the science of crop improvement. By combining genetic precision with engineering design, it enables the creation of plants that are more productive, resilient, and sustainable. From enhancing photosynthesis and nutrient content to enabling nitrogen fixation and pest resistance, the applications of synthetic biology span every dimension of agriculture. However, its success will depend not only on technological breakthroughs but also on responsible governance, public trust, and equitable access.

As research advances, synthetic biology will play an increasingly vital role in feeding the world's growing population while safeguarding the planet's ecological balance. Through continuous innovation and ethical stewardship, synthetic biology holds the promise of ushering in a new era of intelligent, sustainable, and resilient agriculture.

Foundations of Synthetic Biology in Agriculture

The foundation of synthetic biology in agriculture lies in the ability to manipulate genetic material precisely and predictably. Unlike conventional genetic modification, which typically involves inserting a single gene from one organism into another, synthetic biology enables the design of entirely new genetic circuits and regulatory networks. These circuits can be tailored to control when, where, and how genes are expressed in plants, offering unprecedented control over biological processes. Through modular design principles, genetic parts such as promoters, coding sequences, and regulatory elements can be standardized and reused, allowing the assembly of complex pathways that govern plant metabolism, growth, and stress responses. This approach mirrors the engineering of electronic systems, where standardized components are combined to perform specific functions, but in this case, the systems are living cells and tissues.

Genetic Circuit Design and Metabolic Engineering

One of the most powerful aspects of synthetic biology in crop improvement is the construction of synthetic genetic circuits that regulate plant metabolism.

Metabolic engineering involves redirecting or enhancing natural pathways to increase the production of valuable compounds, such as vitamins, amino acids, or bioactive molecules.

For example, synthetic circuits can be designed to increase photosynthetic efficiency by optimizing carbon fixation pathways or to produce secondary metabolites that enhance nutritional quality or pest resistance. Using computational models, researchers can predict the behavior of these circuits before introducing them into plants, reducing trial-and-error experimentation. Moreover, multi-gene constructs can be assembled and controlled in coordinated ways, allowing for the modification of entire metabolic networks rather than single traits. This holistic approach enables the development of crops with improved yields, enhanced resilience, and reduced dependence on chemical fertilizers and pesticides.

Enhancing Stress Tolerance in Crops

Climate change has intensified abiotic stresses such as drought, salinity, and extreme temperatures, posing serious threats to crop productivity. Synthetic biology provides innovative strategies to engineer stress-tolerant crops by rewiring signaling pathways and introducing synthetic stress-response modules. Scientists have developed synthetic promoters that respond dynamically to environmental stimuli, triggering protective responses when plants experience water scarcity or high salt levels.

For instance, synthetic regulatory elements can activate genes associated with osmoprotectants or antioxidant enzymes, enhancing the plant's capacity to maintain cellular homeostasis under stress conditions. Additionally, synthetic biology facilitates the introduction of synthetic transcription factors that coordinate complex gene networks involved in stress adaptation. These engineered systems not only improve plant resilience but also allow crops to thrive in marginal environments, expanding the agricultural frontier into areas previously unsuitable for cultivation.

Improving Nutritional Quality and Food Security

Nutritional enhancement is a major focus of synthetic biology in agriculture, particularly in addressing malnutrition and micronutrient deficiencies prevalent in developing countries. Traditional breeding approaches have limited capacity to enhance specific nutrients without affecting other desirable traits. Through synthetic biology, targeted pathways can be modified to increase the accumulation of vitamins, minerals, and essential fatty acids in edible plant parts. A notable example is the development of "Golden Rice," which was engineered to produce beta-carotene, a precursor of vitamin A, in rice endosperm. Similar approaches are now being applied to other staple crops such as maize, cassava, and wheat.

Beyond micronutrients, synthetic biology enables the design of plants that produce novel bioactive compounds with health benefits, such as antioxidants and anti-inflammatory agents. By enhancing the nutritional content of crops, synthetic biology contributes not only to food quantity but also to food quality, promoting global health and well-being.

Pest and Disease Resistance through Synthetic Pathways

Pests and pathogens continue to cause significant crop losses worldwide, prompting heavy reliance on chemical pesticides. Synthetic biology offers environmentally friendly solutions by enabling plants to produce their own protective compounds or activate defense mechanisms upon detecting pathogens. By integrating biosensors into plant cells, researchers can create plants that sense specific molecular signals associated with infection and respond by synthesizing antimicrobial peptides or signaling molecules that inhibit pathogen growth.

For example, engineered pathways can mimic immune responses by producing small interfering RNAs that silence pathogen genes or by generating synthetic phytoalexins—natural compounds involved in plant defense. This programmable defense system provides a sustainable alternative to chemical control, reducing environmental impact while enhancing crop productivity.

Synthetic Symbiosis and Nitrogen Fixation

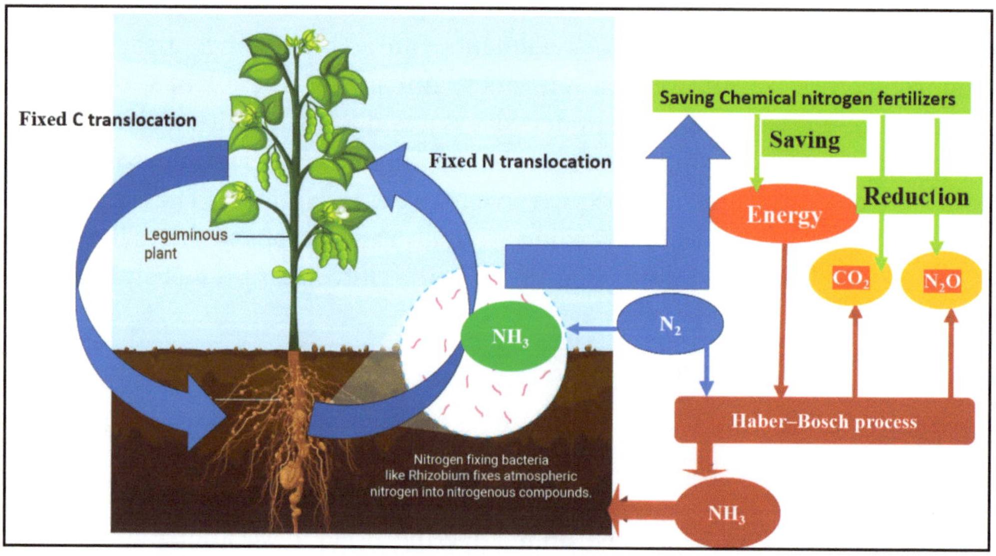

Synthetic Biological Symbiosis and Nitrogen Fixing Mechanisms

Nitrogen is an essential nutrient for plant growth, yet synthetic fertilizers used to provide it are energy-intensive and environmentally damaging. Synthetic biology is paving the way for crops that can form synthetic symbioses with nitrogen-fixing bacteria or even fix atmospheric nitrogen independently. Scientists are transferring key genes from nitrogen-fixing microorganisms into non-leguminous crops like rice, wheat, and maize. This involves the design of synthetic operons that encode nitrogenase enzymes and associated regulatory elements capable of functioning in plant cells. In parallel, microbial engineering is being used to create beneficial bacteria with enhanced nitrogen-fixing capabilities and the ability to colonize plant roots efficiently. These advancements have the potential to drastically reduce fertilizer use, lower greenhouse gas emissions, and improve soil health, creating a more sustainable agricultural ecosystem.

Table: Applications of Synthetic Biology in Crop Improvement

Application Area	Synthetic Strategy	Example or Outcome
Stress Tolerance	Synthetic promoters and gene circuits	Drought-resistant maize and rice
Nutritional Enhancement	Metabolic pathway modification	Beta-carotene-enriched Golden Rice
Pest Resistance	Synthetic biosensors and defense peptides	Pathogen-sensing tomato plants
Nitrogen Fixation	Gene transfer and synthetic symbiosis	Engineered rice with nitrogenase genes
Photosynthesis Efficiency	Carbon fixation optimization	Improved light-use efficiency in wheat

Synthetic Photosynthesis and Yield Enhancement

Photosynthesis is the foundation of plant productivity, yet natural photosynthetic efficiency remains relatively low. Synthetic biology provides tools to redesign the photosynthetic process by introducing optimized enzymes, artificial carbon fixation pathways, or light-harvesting components. Scientists have constructed synthetic versions of the Calvin cycle and other carbon assimilation routes that improve carbon dioxide utilization and energy conversion. Additionally, the integration of synthetic photoreceptors enables plants to adjust their light absorption dynamically, maximizing productivity under changing environmental conditions. Such innovations can significantly increase crop yields without expanding agricultural land, helping to meet the food demands of a growing population. The combination of synthetic photosynthesis with other engineered traits creates a new generation of high-performance crops that can thrive in diverse environments.

Genome Editing and Synthetic Tools

The success of synthetic biology in agriculture is largely supported by advanced genome editing technologies such as CRISPR-Cas systems, TALENs, and zinc-finger nucleases. These tools allow precise modification of plant genomes, enabling the insertion, deletion, or regulation of specific genes. Synthetic biologists often combine CRISPR-based systems with programmable transcription factors to create dynamic gene circuits that control plant development and metabolism. Moreover, de novo genome synthesis and chromosome engineering are emerging as next-generation approaches, allowing the design of entirely new genomes with optimized architecture. The integration of computational modeling and machine learning enhances predictive capabilities, ensuring that engineered modifications yield stable and desired outcomes. These technological advancements collectively make synthetic biology a powerful and reliable approach for crop improvement.

Challenges and Ethical Considerations

Despite its immense promise, synthetic biology for crop improvement raises important ethical, ecological, and socio-economic concerns. The deliberate modification of plant genomes may lead to unintended effects on ecosystems, biodiversity, or food safety. Gene flow from engineered crops to wild relatives could alter natural populations, requiring rigorous biosafety assessments. There are also ethical debates surrounding intellectual property rights, access to biotechnology, and the potential dominance of large corporations in agricultural innovation. Public perception and regulatory acceptance remain critical factors influencing the adoption of synthetic crops. Ensuring transparency, stakeholder engagement, and equitable access to synthetic biology technologies is essential for their responsible deployment in agriculture. Robust ethical frameworks and international cooperation are necessary to balance innovation with societal and environmental well-being.

3.3.2 Disease-Resistant Crops and Livestock

The development of disease-resistant crops and livestock through classical breeding, genetic engineering, and genome editing represents a cornerstone of modern agricultural innovation. These technologies significantly reduce crop losses and animal morbidity caused by pathogens, decrease reliance on chemical treatments, and promote sustainable farming practices. Achievements such as CRISPR-derived resistant tomatoes, genetically modified potatoes, virus-resistant cucumbers, and virus-resistant livestock underscore the power and versatility of modern molecular tools. While challenges related to pathogen evolution, safety, regulatory acceptance, and equitable access remain, rapid technological progress and integrative approaches promise to overcome these barriers. Ultimately, robust disease resistance in agriculture is essential for ensuring food security, protecting natural resources, and building sustainable farming systems for future generations.

Disease-Resistant Crops: Traditional and Modern Approaches

Historically, disease resistance in crops was primarily achieved through classical breeding, involving the selection of naturally resistant varieties and their crosses with high-yielding cultivars to combine traits. Resistance genes, or R genes, which confer specific recognition and defense against pathogens, have been identified, cloned, and introgressed into commercial varieties. Notable examples include the introduction of the Rxo1 gene from maize into rice, which protects against bacterial streak disease, and the use of pepper-derived Bs2 gene in tomato conferring resistance to bacterial spot disease. These successes have showcased the feasibility of improving resistance across species boundaries, enhancing crop protection in the field.

Modern genome engineering technologies have accelerated this process. For example, CRISPR-Cas9 has been used to knock out susceptibility genes or precisely insert protective genes without introducing foreign DNA, thereby generating non-transgenic mutants indistinguishable from natural variants. In tomatoes, targeted disruption of the SlMLO1 gene by CRISPR resulted in resistance to powdery mildew without compromising plant vigor. Similar approaches have produced resistance in cucumbers, rice, and Arabidopsis against viruses and fungal pathogens by mutating translation initiation factors or susceptibility loci. This precision breeding facilitates rapid development of resistant cultivars tuned for specific diseases and environmental conditions, reducing timeframes from years to months or weeks.

Genetic Engineering and Biotechnological Innovations for Disease Resistance

Genetic engineering allows the introduction of novel resistance traits by transferring genes from unrelated species or designing synthetic constructs. Transgenic crops expressing antifungal proteins, toxins targeting insect pests, or RNA interference molecules that silence essential pathogen genes have been commercially developed and widely adopted. For instance, late blight–resistant potatoes expressing the RB resistance gene demonstrate enhanced protection against Phytophthora infestans, reducing fungicide applications and boosting yield stability. Citrus crops modified with CRISPR to inhibit susceptibility genes show reduced incidence of bacterial canker, an economically devastating disease.

Besides gene addition, genome editing strategies now exploit multiplex editing to stack multiple resistance genes simultaneously, creating broad-spectrum and durable protection. Wheat cultivars have been engineered to carry genes conferring resistance to powdery mildew, rust diseases, and other fungal threats. Such combinatorial gene editing efforts not only protect crops from individual pathogens but also mitigate the risk of resistance breakdown due to pathogen evolution. Epigenome editing and transcriptional activation techniques further refine resistance by modulating host defense pathways, customizing responses to diverse biotic stresses.

Disease Resistance in Livestock

Livestock disease resistance is equally vital to global food security. Infectious diseases in animals cause significant economic losses, threaten animal welfare, and raise food safety concerns. Traditional breeding for disease-resistant traits has met with limited success due to complex genetic architectures and long generation times. Genetic engineering now provides alternatives by directly modifying host immune components or eliminating receptors exploited by pathogens.

For example, genetically engineered pigs lacking the receptor for porcine reproductive and respiratory syndrome virus (PRRSV) have demonstrated resistance to this devastating disease. Gene-editing techniques also enable the production of cattle resistant to bovine tuberculosis or mastitis by enhancing immune system genes or removing susceptibility loci. These approaches reduce dependency on antibiotics and vaccines, which carry risks of resistance development and environmental contamination. Moreover, advances in synthetic biology allow the design of immune cells with enhanced targeting ability against specific pathogens, development of messenger RNA vaccines, and engineering of microbiomes to support animal health. These interventions promote healthier livestock, improved yield, and reduced environmental footprints associated with animal husbandry.

Integrated Pest and Disease Management and Crop-Livestock Systems

Developing disease-resistant crops and livestock is part of a larger integrated pest and disease management strategy aimed at sustainable agriculture. Combining genetic resistance with agronomic practices, biological control agents, and reduced chemical inputs minimizes environmental impact while enhancing productivity. Resistant varieties reduce the frequency and severity of disease outbreaks, lowering pesticide use and reducing residues on food products. Increasing resistance in livestock decreases the need for prophylactic antibiotics, slowing antimicrobial resistance development.

Such integrative systems optimize resource efficiency and resilience. For example, disease-resistant crop varieties facilitate intercropping and crop rotation, which suppress pathogen cycles and conserve biodiversity. Disease-tolerant animals coupled with improved husbandry practices reduce disease transmission risks and improve animal welfare. These holistic approaches align with the United Nations' sustainable development goals and emerging frameworks for climate-smart and regenerative agriculture.

Challenges and Limitations in Developing Disease-Resistant Organisms

Despite tremendous progress, challenges remain in designing disease-resistant crops and livestock. Pathogens evolve rapidly, potentially overcoming resistance genes and compromising long-term effectiveness. Pyramiding multiple resistance genes can mitigate this risk but increases complexity and regulatory hurdles. Off-target effects, unintended genetic changes, and altered host physiology must be carefully evaluated for safety and sustainability.

Public perceptions and regulatory landscapes pose additional barriers. Acceptance of genetically engineered organisms varies globally, shaped by cultural, political, and economic factors. Transparent communication, robust risk assessments, and participatory approaches involving stakeholders are essential to build trust and ensure responsible deployment. Access and equity considerations must guarantee that smallholder farmers and disadvantaged communities' benefit from advanced technologies without exacerbating inequalities. Balancing yield, resistance, and other agronomic traits requires multidisciplinary collaboration, combining molecular biology, genomics, bioinformatics, breeding, and field sciences to achieve optimal outcomes under diverse environmental conditions.

3.3.3 Synthetic Biology in Food Production and Alternatives

Synthetic biology is not just a set of tools; it is a paradigm shift in food production, offering a pathway to a more sustainable, healthy, and abundant food future. By programming biology, we can create foods that are kinder to the planet, more nutritious, and tailored to human needs, while reducing the ethical concerns associated with industrial animal agriculture. From engineering microbes to produce life-saving nutrients and exquisite flavors to growing real meat without slaughter and crafting dairy from yeast, SynBio is unlocking possibilities once confined to science fiction.

The challenges of regulation, public acceptance, and scaling are significant but not insurmountable. Through transparent dialogue, robust science, and thoughtful policy, these hurdles can be overcome. The culinary revolution driven by synthetic biology is already underway, and its impact will only grow. As we stand on the brink of this new era, the potential to nourish a growing global population while healing the environment is immense. The future of food is being written in genetic code, promising a feast that is not only delicious but also fundamentally better for people and the planet.

Decoding Synthetic Biology: Engineering Life for the Plate

At its core, synthetic biology applies engineering principles to biology. It involves the design, construction, and optimization of biological systems using standardized genetic parts (like BioBricks), genome editing tools (like CRISPR-Cas9), and computational modeling. SynBio allows scientists to move beyond simply observing and manipulating natural systems to creating novel biological functions. Key techniques include DNA synthesis (writing and assembling genetic sequences from scratch), genome editing (precisely modifying existing DNA), metabolic engineering (re-designing metabolic pathways within cells to produce desired compounds), and synthetic gene circuits (programming cells with logic functions, such as turning on production only in response to specific conditions). In the context of food, SynBio is applied to microorganisms (bacteria, yeast, fungi, algae), plant cells, and animal cells. By inserting, deleting, or editing genes, scientists can instruct these biological systems to produce specific molecules–flavors, fragrances, nutrients, proteins, or even structural components–that would be difficult, expensive, or unsustainable to obtain through traditional agriculture or chemical synthesis. This level of control enables the creation of foods with tailored properties, improved functionality, and reduced environmental impact. SynBio is the toolkit that turns cells into miniature, programmable factories for next-generation food production.

Engineering Flavor and Function: Synbio in Food Ingredients

One of the most immediate and impactful applications of SynBio in food is the engineering of microorganisms to produce high-value food ingredients and additives. This approach, often called precision fermentation, uses engineered microbes as cell factories. For instance, vanillin, the primary component of vanilla bean extract, has traditionally been sourced from vanilla orchids, a labor-intensive and environmentally damaging crop, or produced chemically from petrochemicals. SynBio allows companies to insert the genes responsible for vanillin production into yeast (Saccharomyces cerevisiae) or bacteria. These engineered microbes are then fermented in large bioreactors using sustainable feedstocks like sugarcane or corn syrup, producing pure vanillin without the need for synthetic chemicals or vast plantations. This method is more efficient, scalable, and eco-friendly. Similarly, citrus flavors like nootkatone (grapefruit) and valencene (orange) can be produced by yeast engineered with genes from oranges or grapefruits. Beyond flavors, SynBio is used to create nutraceuticals and functional ingredients. For example, omega-3 fatty acids (EPA and DHA), essential for heart and brain health, are commonly derived from fish oil, raising concerns about overfishing and contamination. Engineered yeast can now produce these beneficial oils directly. Vitamins like B12, often derived from animal sources or synthesized chemically, can be produced by engineered bacteria, offering vegan-friendly alternatives. Enzymes used in food processing (e.g., rennet for cheese, amylases for baking) are also produced via SynBio, improving efficiency and purity. Even sweeteners like rebaudioside M (a zero-calorie stevia compound) are being produced by engineered yeast, providing a more sustainable and consistent supply than extracting from stevia leaves. These applications demonstrate SynBio's power to create cleaner, more sustainable, and ethically produced food components.

Cultivating Alternatives: Lab-Grown Meat and Plant-Based Innovations

SynBio is not only enhancing traditional food ingredients but also driving the development of entirely new categories of food, particularly alternative proteins that aim to reduce reliance on animal agriculture. Cultivated meat (also known as lab-grown, cell-based, or in vitro meat) is a prime example. This technology involves taking a small sample of muscle cells from an animal (like a cow or chicken) and placing them in a bioreactor filled with a nutrient-rich serum.

The cells are then stimulated to grow and differentiate into muscle tissue, forming edible meat without slaughtering the animal. While the initial growth media often contained fetal bovine serum (FBS), SynBio is playing a crucial role in making this process more sustainable and scalable. Engineered microbes are used to produce animal-free growth factors and serum replacements–the expensive and ethically problematic components of the growth medium. By programming yeast or bacteria to produce these specific proteins, the cost of cultivated meat production can be drastically reduced, and the process can become completely animal-independent. Furthermore, SynBio helps optimize the cell lines themselves for better growth, texture, and nutritional profile.

Similarly, the booming plant-based meat and dairy sector heavily relies on SynBio. Companies are engineering microbes (like yeast or filamentous fungi) to produce plant proteins (e.g., pea, soy) through fermentation, creating ingredients with improved functionality (better texture, solubility, binding) for plant-based burgers, nuggets, and cheeses. Even more transformative, SynBio is used to create dairy proteins without cows. By inserting the genes for casein and whey proteins (the key components of cow's milk) into yeast, scientists can produce identical proteins via fermentation. These proteins are then used to create milk, cheese, yogurt, and ice cream that are nutritionally identical to dairy but produced without animals, using far less land, water, and energy, and generating fewer greenhouse gas emissions. This is molecular farming for the modern age.

Sustainability on the Menu: Environmental and Health Benefits

The integration of SynBio into food production offers profound environmental and health advantages, addressing many of the critical challenges facing the global food system. Environmental sustainability is a major driver. Traditional animal agriculture is a leading contributor to greenhouse gas emissions, deforestation, water pollution, and biodiversity loss. SynBio-enabled alternatives like cultivated meat and animal-free dairy promise to drastically reduce this footprint. Precision fermentation for ingredients uses significantly less land and water than conventional agriculture and can utilize renewable energy sources. It also eliminates the need for pesticides, herbicides, and antibiotics associated with livestock farming. Furthermore, SynBio can create climate-resilient crops by engineering plants for traits like drought tolerance, salinity resistance, or enhanced nitrogen fixation, reducing the environmental burden of fertilizers. Health benefits are equally compelling. SynBio allows for the nutritional enhancement of foods. For example, scientists can engineer crops to produce higher levels of essential vitamins (like Vitamin A in Golden Rice 2, though facing deployment challenges) or minerals (like iron and zinc to combat malnutrition).

In alternative proteins, cultivated meat can be engineered to be leaner and free from pathogens like Salmonella or E. coli, while plant-based alternatives can be designed to have improved protein quality, fiber content, and reduced saturated fats. SynBio also enables the removal of allergens or the reduction of compounds linked to health issues (e.g., reducing lactose in dairy alternatives or gluten in cereals). The ability to precisely control the composition of food ingredients paves the way for personalized nutrition, where foods are tailored to individual dietary needs and health goals.

Navigating the Challenges: Regulatory Hurdles and Public Perception

Despite its immense promise, the widespread adoption of SynBio in food faces significant challenges, primarily centered around regulation and public acceptance. Regulatory frameworks struggle to keep pace with the rapid innovation in SynBio. Governments worldwide are still developing clear pathways for approving foods produced using genetic engineering, especially for novel applications like cultivated meat and precision fermentation products.

The process vs. product distinction is a key debate: is the safety assessment based on the final product's characteristics or the genetic engineering process used? Agencies like the FDA (in the US) and EFSA (in Europe) are working to establish science-based regulations, but uncertainty and inconsistency across regions can hinder investment and market entry.

Public perception is another major hurdle. While consumer interest in sustainable and ethical foods is high, there is significant skepticism and misunderstanding about genetic engineering and "lab-grown" foods. Terms like "GMO" often carry negative connotations due to historical debates, despite the precision of modern techniques like CRISPR. Concerns about "unnatural" foods, corporate control, and long-term health impacts persist.

Building trust and transparency is crucial. The food industry must engage in clear communication about the benefits, safety measures, and underlying science. Labeling policies also need to be carefully considered to provide consumers with information without unnecessary alarm. Cost and scalability remain economic challenges, particularly for cultivated meat. While SynBio is driving costs down (e.g., for growth factors), achieving price parity with conventional meat will take time and further technological innovation. Scaling up bioreactor capacity and downstream processing also presents engineering challenges.

3.4 MEDICINE AND HEALTHCARE

3.4.1 Synthetic Biology in Drug Discovery

Synthetic biology represents a paradigm shift in the way drugs are discovered, designed, and produced. By harnessing the power of biological systems and engineering principles, scientists can create living platforms capable of producing complex therapeutics that were once unattainable. From microbial biosynthesis of natural products to personalized medicine and synthetic vaccines, the impact of this technology spans every aspect of modern drug development. While challenges related to ethics, safety, and regulation remain, the advantages of synthetic biology—speed, precision, sustainability, and versatility—make it indispensable for the future of healthcare. As technology advances, synthetic biology will continue to expand the boundaries of drug discovery, bringing innovative, affordable, and effective treatments to patients around the world.

Evolution of Synthetic Biology

The concept of synthetic biology is rooted in molecular biology and genetic engineering, but its modern incarnation evolved with the ability to design and synthesize genetic circuits and metabolic pathways from scratch. Initially, drug discovery relied heavily on natural products derived from plants, microbes, and marine organisms. However, many bioactive molecules were difficult to produce in large quantities due to the scarcity of natural sources or complex chemical structures. Synthetic biology revolutionized this process by enabling scientists to reprogram organisms such as Escherichia coli and Saccharomyces cerevisiae to produce valuable drug molecules. The advent of DNA synthesis, genome editing tools like CRISPR-Cas9, and computational modeling accelerated this transformation, allowing researchers to create tailor-made biological systems for drug discovery and production.

Principles of Synthetic Biology in Drug Discovery

At its core, synthetic biology operates on the principle of treating biological components as standardized parts that can be assembled to create new functions. In drug discovery, this involves designing biological circuits that mimic or enhance natural biosynthetic pathways to produce therapeutic compounds. By integrating computational modeling with biological experimentation, scientists can predict how genetic modifications will influence metabolic outputs. This rational design approach minimizes trial-and-error experimentation and increases precision. Synthetic biology also embraces modularity, enabling the reuse of standardized genetic parts to build multiple biosynthetic systems. Through the design-build-test-learn (DBTL) cycle, drug discovery becomes a more efficient and predictable process.

Role of Microbial Engineering

Microorganisms are at the heart of synthetic biology applications in drug discovery. Engineered microbes serve as mini-factories for the production of complex molecules that are difficult or expensive to synthesize chemically. For example, the antimalarial drug artemisinin, originally derived from the plant Artemisia annua, is now produced through yeast engineered with synthetic genes encoding the necessary biosynthetic enzymes.

Similarly, bacteria are designed to produce antibiotics, anticancer agents, and immunosuppressants. Microbial chassis systems such as E. coli, Streptomyces, and Pseudomonas are modified to express non-native pathways, enabling them to produce diverse bioactive compounds. The scalability and controllability of microbial production make it ideal for industrial applications, ensuring a consistent and sustainable supply of pharmaceuticals.

Synthetic Biology Tools and Techniques

The success of synthetic biology in drug discovery depends on a suite of powerful tools and techniques that allow precise genetic manipulation and metabolic pathway optimization. Key among these are:

- CRISPR-Cas9 genome editing for precise insertion, deletion, or replacement of genes.
- DNA synthesis and assembly technologies that enable rapid construction of genetic circuits.
- Metabolic pathway engineering for redirecting cellular resources toward drug production.
- Synthetic promoters and regulatory elements for controlling gene expression levels.
- Automated biofoundries for high-throughput genetic design and screening.
- Machine learning and AI algorithms for predicting biosynthetic outcomes.
- Optogenetic tools for light-controlled gene regulation in drug-producing organisms.
- Cell-free biosynthesis systems for producing compounds outside living cells.
- Microfluidic platforms for rapid testing of metabolic variants.
- Biosensors for detecting and quantifying metabolites during pathway optimization.

Together, these tools empower researchers to design highly efficient biological systems tailored to produce new or improved drug candidates.

Discovery of Novel Natural Products

Natural products remain a cornerstone of drug discovery, with many antibiotics, anticancer agents, and immunosuppressants originating from microbial or plant metabolites. However, traditional discovery methods have exhausted easily accessible natural sources. Synthetic biology revitalizes natural product discovery by allowing researchers to mine and reconstruct biosynthetic gene clusters from uncultured microorganisms.

Through metagenomics and pathway refactoring, genes responsible for natural product biosynthesis can be expressed in laboratory strains to produce previously unknown compounds. This "genome mining" approach expands the chemical diversity available for drug screening. By activating silent or cryptic gene clusters, synthetic biology unlocks new chemical entities that may serve as next-generation therapeutics.

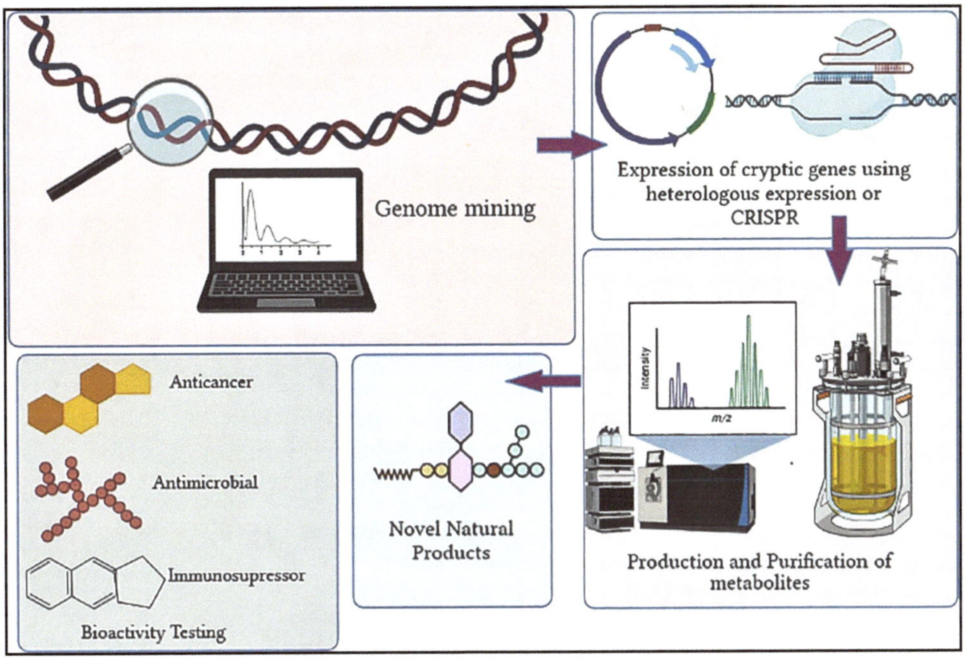

Tapping Into Actinobacterial Genomes for Natural Product Discovery

Pathway Optimization and Rational Design

Once a biosynthetic pathway is introduced into a host organism, optimizing its performance becomes essential. Synthetic biology enables precise control over each step of the pathway, from precursor supply to enzyme activity. Using computational models, scientists can predict how changes in gene expression affect overall yield. Dynamic regulation tools, such as synthetic riboswitches and feedback control loops, allow real-time modulation of metabolic flux. This rational approach ensures that cellular resources are used efficiently, minimizing toxic by-products and maximizing drug yields. Moreover, pathway balancing allows the simultaneous production of multiple intermediates, facilitating combinatorial biosynthesis where new drug analogs can be generated by mixing and matching biosynthetic enzymes.

Synthetic Biology and Personalized Medicine

One of the most exciting frontiers in drug discovery is personalized medicine, where therapies are tailored to individual genetic profiles. Synthetic biology contributes to this field by enabling customized drug production and biosensing systems that can adapt to patient-specific needs. Engineered cells can act as "living therapeutics," sensing disease biomarkers and producing appropriate therapeutic molecules within the body. Synthetic biology also aids in the development of patient-specific organoids and cell lines for drug testing, allowing researchers to predict drug responses before clinical administration. These personalized platforms reduce adverse effects, enhance treatment efficacy, and shorten the timeline from discovery to therapy.

Synthetic Vaccines and Biopharmaceuticals

Synthetic biology has revolutionized vaccine development and biopharmaceutical production. Traditional vaccines often rely on attenuated pathogens or long culture times, but synthetic approaches allow rapid vaccine design based on genetic sequences. The mRNA vaccines for COVID-19 are a prime example, demonstrating how synthetic biology can accelerate response to emerging diseases. By synthesizing viral antigen genes and encapsulating them in lipid nanoparticles, scientists can create vaccines within weeks. Beyond vaccines, synthetic biology facilitates the production of therapeutic proteins, monoclonal antibodies, and biosimilars through optimized expression systems. These biologics are used to treat cancers, autoimmune diseases, and infections with higher specificity and reduced side effects.

Ethical, Environmental, and Regulatory Considerations

While synthetic biology holds tremendous promise, its application in drug discovery raises ethical and safety questions. The creation of genetically modified organisms (GMOs) and synthetic life forms must be regulated to prevent accidental release or misuse. Biosafety protocols and containment systems are essential to ensure environmental protection. Ethical debates also surround the ownership of synthetic genetic material and intellectual property rights related to modified organisms. From a regulatory standpoint, new frameworks are needed to evaluate the safety and efficacy of drugs produced using synthetic biology. International collaboration among regulatory agencies, such as the FDA and EMA, is critical to harmonize standards and ensure public confidence in synthetic biology-derived medicines.

3.4.2 Gene Therapy and Cell Therapy

Gene therapy and cell therapy represent a new paradigm in the treatment of human disease—one that shifts the focus from symptom management to root-cause correction and regeneration. By leveraging the fundamental principles of molecular and cellular biology, these therapies offer durable and often curative solutions for conditions once deemed untreatable. Gene therapy corrects defective genetic instructions, while cell therapy restores or replaces damaged cells, and together they embody the promise of personalized, regenerative healthcare. Although ethical, technical, and regulatory challenges remain, the pace of progress is accelerating rapidly, driven by advances in gene editing, stem cell biology, and bioengineering. As these innovations continue to mature, gene and cell therapies will not only revolutionize clinical practice but also redefine the boundaries of what medicine can achieve, ushering in a new era of transformative and sustainable health solutions for humanity.

Concept of Gene Therapy

Gene therapy involves the deliberate introduction, removal, or alteration of genetic material within a patient's cells to treat or prevent disease. It is based on the understanding that many diseases are caused by mutations or abnormalities in genes that disrupt normal cellular functions. By introducing a functional copy of a defective gene or silencing a harmful gene, gene therapy aims to restore healthy biological activity. The concept emerged in the late twentieth century, driven by advances in molecular genetics and recombinant DNA technology. The first approved gene therapy trials began in the 1990s, but it was only in the last decade that significant clinical success has been achieved, thanks to safer vectors, better gene-editing tools, and more precise delivery systems. Gene therapy can be classified into somatic and germline approaches: somatic gene therapy targets non-reproductive cells, affecting only the treated individual, whereas germline therapy targets reproductive cells, leading to heritable genetic changes. Ethical and safety considerations have largely confined current applications to somatic interventions.

Mechanisms and Delivery Systems in Gene Therapy

A central challenge in gene therapy lies in the efficient and safe delivery of therapeutic genes to the desired target cells. Several vector systems have been developed for this purpose, broadly categorized into viral and non-viral delivery methods. Viral vectors, such as adenoviruses, adeno-associated viruses (AAV), lentiviruses, and retroviruses, are commonly used because of their natural ability to transfer genetic material into host cells. Each vector type has distinct advantages and limitations in terms of immunogenicity, capacity, and duration of expression. Non-viral methods, including liposomes, nanoparticles, and electroporation, provide safer alternatives for transient gene expression and can minimize immune responses.

Once delivered, the therapeutic gene is expressed to produce functional proteins that correct or compensate for defective ones. Modern advancements, particularly the CRISPR-Cas9 gene-editing system, have revolutionized the precision and efficiency of gene therapy. By allowing targeted modification of specific DNA sequences, CRISPR enables direct correction of genetic mutations and fine-tuning of gene expression with unprecedented accuracy.

Types and Applications of Gene Therapy

Gene therapy can be applied to a wide range of medical conditions, from single-gene disorders to complex diseases. In monogenic diseases such as cystic fibrosis, muscular dystrophy, and hemophilia, therapy involves replacing or repairing a single defective gene. In oncology, gene therapy has been used to enhance the immune system's ability to recognize and attack cancer cells by introducing genes encoding tumor-specific antigens or immune-modulating cytokines.

Additionally, gene therapy plays an emerging role in cardiovascular diseases, where it can stimulate angiogenesis or repair damaged heart tissues by delivering growth factor genes. In infectious diseases, such as HIV, gene therapy strategies aim to render host cells resistant to viral infection by editing receptors or blocking viral replication. The versatility of gene therapy lies in its potential to treat the root cause of diseases rather than merely managing symptoms, positioning it as a key pillar of future healthcare innovation.

Concept of Cell Therapy

Cell therapy, also known as cellular therapy, is the transplantation of living cells into a patient to replace damaged tissue or restore lost function. It operates on the principle that cells are the functional units of life and that introducing healthy or engineered cells can repair or regenerate tissues affected by disease or injury. Cell therapy encompasses a broad spectrum of treatments, from simple blood transfusions and bone marrow transplants to sophisticated regenerative and immunomodulatory interventions. Stem cell therapy, in particular, has gained global prominence due to stem cells' unique ability to self-renew and differentiate into various specialized cell types. Depending on their source, cell therapies may use autologous cells derived from the patient or allogeneic cells obtained from a donor. Advances in biotechnology and tissue engineering have made it possible to cultivate, manipulate, and reprogram cells ex vivo before reintroducing them into the body, greatly enhancing therapeutic efficacy and personalization.

Stem Cells and Regenerative Potential

Stem cells are the cornerstone of many cell therapy applications due to their capacity for regeneration and differentiation. Embryonic stem cells (ESCs) are pluripotent, capable of differentiating into almost any cell type in the human body. Adult stem cells, such as hematopoietic stem cells (HSCs) and mesenchymal stem cells (MSCs), are multipotent and can give rise to a limited range of cell types, such as blood, bone, or cartilage. Induced pluripotent stem cells (iPSCs), generated by reprogramming adult somatic cells back into a pluripotent state, represent a major breakthrough in regenerative medicine. iPSCs offer the potential to generate patient-specific cells for transplantation, minimizing immune rejection risks. Stem cell-based therapies are being investigated for treating neurological disorders, spinal cord injuries, diabetes, cardiovascular diseases, and degenerative conditions like osteoarthritis. By enabling tissue regeneration and repair, stem cells hold the promise of restoring organ function and reversing damage previously thought irreversible.

Table: Comparison of Gene Therapy and Cell Therapy

Aspect	Gene Therapy	Cell Therapy
Definition	Introduction or modification of genetic material to correct or alter gene function	Transplantation or modification of living cells to restore or replace tissue function
Core Mechanism	Alters DNA sequences within cells to achieve therapeutic effect	Utilizes functional or engineered cells for repair and regeneration
Primary Tools	Viral vectors, CRISPR-Cas9, plasmids	Stem cells, immune cells, engineered tissues
Applications	Genetic disorders, cancer, cardiovascular and infectious diseases	Regenerative medicine, immunotherapy, tissue repair
Duration of Effect	Potentially long-term or permanent	May require repeated administration depending on cell survival and function

Immunotherapy and Cellular Engineering

A significant breakthrough in cell therapy has come through the development of immune-based treatments such as chimeric antigen receptor T-cell (CAR-T) therapy. In this approach, a patient's own T-cells are genetically modified to express synthetic receptors that recognize specific cancer cell markers. Once reinfused into the patient, these engineered T-cells can identify and destroy tumor cells with high precision. CAR-T therapies have achieved remarkable success in treating certain types of leukemia and lymphoma, offering new hope to patients with otherwise refractory cancers. Beyond CAR-T, natural killer (NK) cells and dendritic cell-based therapies are also being explored for their potential in cancer immunotherapy. The combination of genetic modification and cellular manipulation epitomizes the synergy between gene therapy and cell therapy, demonstrating how genetic engineering can enhance the therapeutic capacity of cells for targeted disease intervention.

Ethical, Regulatory, and Safety Considerations

Despite their enormous potential, both gene and cell therapies raise important ethical and regulatory questions. Gene therapy interventions, particularly those involving germline modifications, pose concerns about heritable changes that could impact future generations. Ethical debates also revolve around access to these advanced therapies, which remain prohibitively expensive for many patients. Similarly, the use of embryonic stem cells in cell therapy raises moral questions related to the destruction of embryos. From a safety standpoint, risks include immune reactions, insertional mutagenesis in gene therapy, and tumorigenic potential in stem cell applications. Regulatory agencies such as the U.S. Food and Drug Administration (FDA) and the European Medicines Agency (EMA) have established stringent frameworks to evaluate the safety, efficacy, and ethical compliance of these therapies. Continued public engagement, transparency, and adherence to ethical standards are essential to maintaining trust and ensuring responsible clinical translation of these powerful technologies.

Current Challenges and Technological Advances

Several challenges hinder the widespread adoption of gene and cell therapies. Manufacturing complexity, high costs, and the need for personalized treatment design limit scalability. Moreover, achieving efficient gene delivery, durable expression, and controlled cell behavior remains technically demanding. However, rapid technological advancements are addressing these limitations. Artificial intelligence and bioinformatics are being used to design optimized gene sequences and predict immune responses. 3D bioprinting and tissue engineering are enhancing cell therapy by creating biomimetic scaffolds that support cell integration and tissue regeneration. Gene-editing technologies continue to evolve, with base editing and prime editing offering even greater precision and safety than CRISPR. As research progresses, these innovations are expected to lower costs, increase accessibility, and expand the range of treatable diseases.

Clinical Applications and Success Stories

The clinical success of gene and cell therapies has validated their therapeutic potential. Gene therapy treatments such as Luxturna for inherited retinal dystrophy and Zolgensma for spinal muscular atrophy have demonstrated long-term efficacy by correcting underlying genetic defects. Similarly, CAR-T cell therapies like Kymriah and Yescarta have revolutionized cancer treatment, achieving remission in patients resistant to conventional therapies. In regenerative medicine, stem cell-based interventions are restoring function in cardiac tissue after heart attacks and improving mobility in patients with spinal cord injuries. These success stories represent milestones in biomedical science, underscoring the real-world impact of combining genetic and cellular engineering to improve human health.

3.4.3 Synthetic Organisms for Diagnostics and Treatment

Synthetic organisms engineered for diagnostics and treatment epitomize the fusion of biology and engineering, delivering revolutionary capabilities for healthcare. By leveraging modular gene circuits, programmable biosensors, and living therapeutic factories, these systems overcome many limitations of traditional diagnostics and therapeutics, offering specificity, real-time responsiveness, and adaptability. While challenges related to safety, control, and clinical translation remain, rapid progress in synthetic biology technologies and design strategies are paving the way for their widespread adoption. As synthetic organisms mature into reliable clinical tools, they hold the promise of transforming disease detection, monitoring, and therapy, ultimately advancing precision medicine and improving patient outcomes on a global scale.

The future of synthetic organisms in medicine promises integration with artificial intelligence, personalized medicine, and next-generation biosensors. AI-guided design will optimize circuit architecture and predict patient-specific responses, customizing diagnostics and therapeutics to individual genetic and environmental profiles. Advances in multi-omics and single-cell technologies will further illuminate disease heterogeneity, enabling tailored living therapeutics that dynamically adapt to cellular states.

Emerging synthetic biology-based implants and wearable devices incorporating living biosensors will provide continuous health monitoring and responsive treatment, merging diagnostics and therapeutics into seamless platforms. Hybrid synthetic systems combining cellular and cell-free components might achieve unprecedented sensitivity, specificity, and operational flexibility.

Multi-modal synthetic organisms capable of interfacing with endogenous immune systems, microbial communities, and host metabolic networks will transcend conventional drug paradigms. These living medicines will offer durable, sustainable, and intelligent healthcare options.

Synthetic Organisms for Advanced Diagnostics

Accurate and timely diagnosis is paramount in disease management, yet conventional diagnostics often suffer from limitations such as prolonged turnaround times, reliance on sophisticated instrumentation, and limited multiplexing capabilities. Synthetic organisms overcome many of these challenges by embodying biosensing capabilities built into living or cell-free biological systems.

Engineered bacteria, yeast, and mammalian cells can be programmed with synthetic gene circuits that detect specific molecular signals—proteins, nucleic acids, metabolites—and translate them into measurable outputs such as fluorescence, color change, or electrical signals. These circuits typically comprise modular components: sensors that recognize target analytes, processors that integrate and interpret signals through logic gates, and reporters that generate observable readouts, enabling assays that are rapid, sensitive, and adaptable to diverse biomarkers.

Paper-based diagnostics incorporating synthetic genes and oligonucleotides provide portable, inexpensive test platforms ideal for point-of-care or field applications. These include toehold switch RNA sensors capable of detecting viral RNA with high specificity and stability at ambient temperatures, exemplified recently by Zika virus detection kits. Cell-based biosensors further extend diagnostic possibilities with living systems embedded in wearable or implantable devices, offering continuous monitoring of physiological states or pathogen presence. This approach enables real-time health surveillance and early disease detection with unprecedented sensitivity and specificity.

Therapeutic Applications of Synthetic Organisms

Beyond diagnostics, synthetic organisms have been engineered for therapeutic interventions, introducing a new class of living medicines. Genetically modified bacteria harboring synthetic gene circuits can home to disease sites such as tumors or inflamed tissues, sense the local microenvironment, and produce therapeutic agents in situ with controlled timing and dosage.

For instance, engineered probiotics have been developed to detect and degrade inflammatory molecules in the gut, offering treatments for conditions like inflammatory bowel disease and colorectal cancer. The control embedded in synthetic circuits grants therapies improved efficacy, reduced side effects, and adaptability to changing disease states.

Similarly, synthetic mammalian cells can be programmed to produce and release drugs in response to physiological cues, functioning as autonomous therapeutic factories. Such designs hold promise for chronic diseases requiring tightly regulated drug delivery, such as diabetes and cancer. Additionally, synthetic bacteriophages and phage-derived systems serve as precision antimicrobials, selectively targeting pathogenic bacteria without disrupting beneficial microbiota, and circumventing antibiotic resistance issues.

Modular Design and Engineering Principles

The success of synthetic organisms in diagnostics and treatment arises from their foundation in modular design and engineering principles. Synthetic biologists use standardized biological parts and well-characterized modules that can be assembled and reconfigured to develop tailored functionalities. These components include promoters, ribosome binding sites, transcription factors, and reporter genes, which collectively enable programmatic control over gene expression. Logic gates and circuits analogous to electronic computing provide the capacity to integrate multiple inputs and execute complex decision-making processes within cells.

Advanced tools such as CRISPR-based regulation, RNA switches, and protein-based sensors contribute layers of control and specificity. The iterative design-build-test-learn cycle underpins ongoing refinement and optimization of synthetic circuits for robustness, sensitivity, and safety. Computational modeling and machine learning further accelerate development by predicting circuit behavior and minimizing unintended interactions.

Cell-Free Synthetic Systems for Diagnostics and Therapeutics

Synthetic biology has progressed beyond living cells to develop cell-free systems, which decouple complex biological reactions from cellular viability constraints. These platforms contain biomolecular machinery—enzymes, ribosomes, nucleotides—housed in reaction mixtures or paper matrices, enabling rapid, scalable, and user-friendly diagnostic assays. Cell-free biosensors can detect nucleic acid sequences, proteins, toxins, and metabolites with high precision, providing results in minutes and requiring minimal instrumentation.

In therapeutics, cell-free platforms facilitate on-demand production of peptides, antibodies, and therapeutic proteins at the point of care, circumventing cold chain requirements and production delays. These systems have inherent safety advantages due to the absence of replicating organisms and are amenable to integration into wearable devices or environmental sensors.

Living Therapeutics: Programmable Bacteria and Mammalian Cells

Genetically engineered bacteria remain at the forefront of living therapeutics, owing to their ease of manipulation, rapid replication, and capacity for colonizing specific body niches. Engineered gut bacteria, for example, can sense biomarkers of metabolic disorders or infections and respond by secreting enzymes, antimicrobial peptides, or signaling molecules that modulate host physiology. Such designs create feedback loops that maintain health or counter disease progression. In oncology, synthetic bacterial strains localize to tumors and deliver cytotoxic agents or immune modulators directly to malignant cells, enhancing therapeutic indices and reducing systemic toxicity. The incorporation of kill-switch mechanisms and environmental sensors prevents unwanted proliferation, ensuring safety and controllability. Similarly, synthetic mammalian cells can be equipped with sensors for hormones, inflammatory markers, or cancer-specific antigens, triggering therapeutic payload release precisely when and where needed. Engineered cell therapies, such as synthetic CAR-T cells with programmable behaviors, exemplify the integration of sensing, logic, and actuation in treatment paradigms.

Challenges in Safety, Control, and Translation

Despite promising advances, deploying synthetic organisms for diagnostics and therapy requires overcoming several key challenges. Ensuring biosafety is paramount; engineered organisms must not persist or propagate uncontrollably in patients or the environment. Synthetic biologists employ genetic containment strategies such as auxotrophy, kill-switches, and synthetic recoding to minimize ecological risks. Immunogenicity and off-target effects must be carefully assessed to avoid adverse reactions. Maintaining robust function in complex and variable biological environments is another hurdle, as synthetic circuits can be sensitive to noise and cellular context. Translation from laboratory prototypes to clinical-grade therapeutics demands rigorous optimization, standardization, regulatory approval, and scalable manufacturing processes.

3.5 ENVIRONMENTAL SUSTAINABILITY

3.5.1 Sustainable Biomanufacturing Practices

Sustainable biomanufacturing is more than a set of practices; it is a transformative vision for the future of industry. By harnessing the inherent capabilities of biology—nature's ultimate innovator—we can create products and processes that are not only efficient and profitable but also harmonious with the environment. From converting agricultural waste into clean energy and biodegradable plastics to engineering microbes for life-saving drugs with minimal ecological impact, the potential is boundless.

The challenges are real, but they are being met with ingenuity, collaboration, and a shared recognition of the urgent need for change. As we stand at the precipice of a new industrial era, sustainable biomanufacturing offers a path to redefine prosperity: one that values planetary health as much as economic growth. Embracing this approach is not merely an option—it is the key to unlocking a thriving, sustainable future for generations to come.

Imperative for Sustainability in Biomanufacturing

The drive towards sustainable biomanufacturing is fueled by compelling environmental, economic, and social imperatives. Environmentally, conventional industrial manufacturing is a major contributor to greenhouse gas emissions, water pollution, and soil degradation. Biomanufacturing, while inherently more eco-friendly than many chemical processes, still faces challenges related to energy consumption, waste generation, and resource use. Embracing sustainability mitigates these impacts by reducing reliance on fossil fuels, curbing emissions, and minimizing toxic byproducts. Economically, sustainable practices enhance efficiency, reduce operational costs through energy and resource savings, and open doors to new markets driven by consumer demand for green products.

They also mitigate risks associated with volatile fossil fuel prices and resource scarcity. Socially, sustainable biomanufacturing promotes healthier workplaces by reducing exposure to hazardous chemicals, supports rural economies through bio-based feedstocks, and addresses food and energy security challenges. Furthermore, it aligns with global initiatives like the UN Sustainable Development Goals (SDGs), particularly those related to climate action, responsible consumption, and industry innovation. Ignoring sustainability is no longer feasible; it is a strategic necessity for long-term industrial resilience and societal well-being.

Foundational Principles of Sustainable Biomanufacturing

Sustainable biomanufacturing rests on several core principles that guide the design and operation of bioprocesses. Renewable resource utilization is paramount, shifting from petroleum-derived feedstocks to biomass such as agricultural residues, non-food crops (e.g., switchgrass, miscanthus), algae, and organic waste streams. This reduces dependence on finite resources and lowers the carbon footprint. Energy efficiency and renewability are critical, emphasizing processes that minimize energy consumption and utilize renewable sources like solar, wind, or biogas.

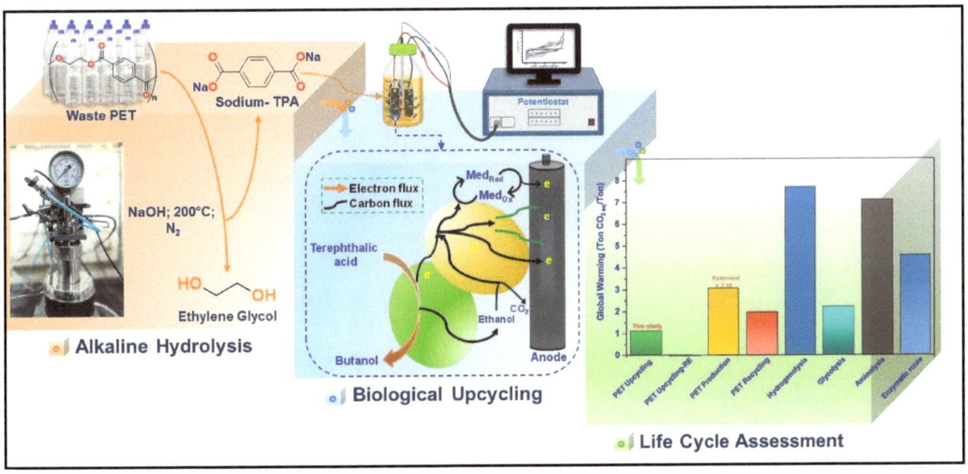

Environmentally Sustainable Biomanufacturing

Waste minimization and valorization embody the circular economy ethos, ensuring that byproducts are not discarded but converted into valuable co-products, such as converting lignin from biomass into biochemicals or using fermentation waste for animal feed. Pollution prevention involves designing processes that avoid the release of harmful substances into air, water, or soil, often through closed-loop systems and advanced treatment technologies. Sustainable design integrates environmental considerations from the outset, encompassing everything from the choice of microbial strains and bioreactor design to downstream processing and product end-of-life. These principles form a framework for creating biomanufacturing systems that are not only productive but also regenerative and restorative.

Enabling Technologies and Practices: Building a Sustainable Framework

The transition to sustainable biomanufacturing is powered by a suite of innovative technologies and practices that enhance efficiency and reduce environmental impact. Advanced feedstock processing is key, particularly the development of lignocellulosic biomass conversion. Techniques like steam explosion, enzymatic hydrolysis, and acid pretreatment break down tough plant materials into fermentable sugars, enabling the use of non-food biomass like corn stover, wheat straw, and wood chips for biofuel and biochemical production, avoiding competition with food crops. Fermentation technology continues to evolve, with solid-state fermentation (SSF) gaining traction for its lower water and energy requirements compared to submerged fermentation, making it ideal for producing enzymes, organic acids, and biofuels. Process intensification strategies, such as continuous fermentation and perfusion systems, improve productivity, reduce downtime, and lower resource consumption by maintaining optimal conditions for longer periods. Energy management is revolutionized by waste heat recovery systems and the integration of biogas production from organic waste via anaerobic digestion, which can power bioreactors and facilities. Water stewardship is enhanced through advanced wastewater treatment using microbial consortia or membrane bioreactors, enabling water recycling and reducing freshwater intake. Green chemistry principles are applied to downstream processing, minimizing the use of toxic solvents through aqueous two-phase systems or supercritical fluid extraction.

Waste as a Resource: The Circular Economy in Action

A cornerstone of sustainable biomanufacturing is the transformation of waste into valuable resources, closing the loop and minimizing environmental burden. Industrial symbiosis models exemplify this, where the waste or byproduct of one process becomes the feedstock for another. For instance, in biorefineries, lignin from cellulose processing can be converted into bioplastics or adhesives, while fermentation CO2 can be captured and used for algae cultivation or chemical synthesis. Nutrient recovery systems extract nitrogen and phosphorus from wastewater for use as fertilizers, reducing reliance on synthetic fertilizers and preventing eutrophication. Anaerobic digestion is a powerful technology for converting organic waste—such as crop residues, food waste, and manure—into biogas (methane) for energy and digestate for soil amendment. Enzyme technology enables the breakdown of complex waste streams into simpler compounds for reuse; for example, ligninolytic enzymes from fungi can degrade persistent pollutants in textile dye waste. Biopolymer production, like polyhydroxyalkanoates (PHAs) or polylactic acid (PLA), uses waste carbon sources (e.g., methane, glycerol) to create biodegradable plastics, addressing the global plastic crisis. These practices not only reduce landfill use and pollution but also create new revenue streams, enhancing economic sustainability.

Realizing Sustainability: Case Studies in Action

The efficacy of sustainable biomanufacturing is demonstrated through diverse real-world applications. POET-DSM's Project LIBERTY in the United States is a pioneering biorefinery that converts corn stover into cellulosic ethanol. Using enzymatic hydrolysis and engineered yeast, it produces 25 million gallons of ethanol annually while generating steam from lignin and recycling water, significantly reducing greenhouse gas emissions compared to gasoline. In the European Union, the PEF (Product Environmental Footprint) initiative has spurred companies like Novozymes to produce industrial enzymes using sustainable practices. Their enzymes, used in detergents, textiles, and biofuels, are produced via fermentation with renewable energy and optimized processes that reduce water and raw material use by up to 50%. Corbion, a leader in sustainable ingredients, produces PLA bioplastics from sugarcane at its facility in Thailand. The process uses renewable energy, recycles water, and ensures that the sugarcane is grown responsibly, with the bioplastic being compostable and carbon-neutral. Pharmaceutical giant Pfizer has implemented green chemistry in its biomanufacturing, using enzymatic catalysis and solvent recovery systems to reduce waste and energy use in producing drugs like atorvastatin. These examples showcase how sustainability drives innovation, efficiency, and environmental stewardship across sectors.

Navigating Challenges: Roadblocks to Sustainable Adoption

Despite its promise, the widespread adoption of sustainable biomanufacturing faces significant challenges. High initial capital costs for advanced bioreactors, pretreatment facilities, and waste valorization systems can be prohibitive, especially for small and medium-sized enterprises. Technological limitations persist, such as the inefficiency of lignocellulosic biomass conversion, high enzyme costs, and the difficulty of scaling novel processes like solid-state fermentation. Regulatory and policy frameworks often lag behind innovation, creating uncertainty for investors and delaying market entry. Inconsistent standards and incentives across regions further complicate matters. Infrastructure gaps, including limited biomass collection and distribution networks, and inadequate renewable energy integration, hinder progress. Consumer acceptance and market competition from established, often cheaper, petrochemical-based products remain hurdles. Additionally, supply chain complexities for sustainable feedstocks can lead to volatility and higher costs. Overcoming these challenges requires concerted efforts: increased public and private investment, supportive policies (e.g., carbon pricing, subsidies for green tech), research breakthroughs, and greater industry collaboration to share best practices and infrastructure.

Horizon Ahead: Trends Shaping

Sustainable biomanufacturing is bright, driven by relentless innovation and a global push for sustainability. Synthetic biology and metabolic engineering are accelerating the development of superior microbial strains with higher yields, substrate flexibility, and tolerance to process stresses, enabling more efficient conversion of diverse feedstocks. AI and machine learning are revolutionizing process design, predictive maintenance, and optimization, reducing energy use and waste by analyzing complex datasets in real-time. Carbon capture, utilization, and storage (CCUS) technologies are being integrated into biomanufacturing, allowing facilities to capture CO2 emissions and convert them into chemicals, fuels, or building materials, turning a liability into an asset. Modular and mobile biorefineries are emerging, bringing processing capabilities closer to feedstock sources, reducing transportation emissions, and supporting rural economies. Nutraceuticals and personalized nutrition produced via sustainable biomanufacturing will rise, with engineered microbes creating health-promoting compounds. Furthermore, policy initiatives like the EU's Green Deal and the US National Biotechnology and Biomanufacturing Initiative are catalyzing investment and setting ambitious targets, signaling a global commitment to a bio-based future. As these trends converge, sustainable biomanufacturing will transition from niche applications to the backbone of a resilient, low-carbon industrial economy.

3.5.2 Waste-to-Value Technologies

Waste-to-value technologies embody the essence of sustainable innovation by transforming one of humanity's greatest challenges—waste—into an opportunity for growth and environmental renewal. Through chemical, biological, and physical processes, waste materials are being converted into fuels, fertilizers, materials, and chemicals that power economies while protecting ecosystems. These technologies promote circularity, reduce carbon emissions, and support cleaner industrial practices. Although challenges related to cost, technology maturity, and regulatory frameworks persist, the momentum toward a circular and resource-efficient economy continues to build. The integration of emerging fields such as artificial intelligence, synthetic biology, and green chemistry will further enhance the efficiency and scalability of waste-to-value systems. Ultimately, by viewing waste as a resource rather than a burden, societies can transition toward a sustainable future where economic development and environmental stewardship coexist harmoniously.

Concept and Importance of Waste-to-Value

The concept of waste-to-value revolves around the idea that waste is not merely a byproduct but a potential feedstock for new processes. By applying advanced technologies, it becomes possible to recover materials, energy, or nutrients from various waste streams, turning liabilities into assets. This approach aligns with the United Nations Sustainable Development Goals (SDGs), particularly those focusing on responsible consumption, clean energy, and climate action. Waste-to-value technologies promote resource efficiency, reduce dependency on raw materials, and support low-carbon industrial growth. They are crucial for industries such as manufacturing, agriculture, energy, and construction, where waste generation is inevitable but can be converted into value-added outputs.

Types of Waste Suitable for Conversion

Different types of waste can serve as raw materials for waste-to-value technologies, depending on their physical and chemical characteristics. Municipal solid waste (MSW), which includes household refuse, plastics, and paper, can be transformed into fuels or building materials. Industrial waste, including metal scraps, sludge, and process residues, often contains valuable metals or chemicals that can be recovered. Agricultural waste such as crop residues and animal manure provides biomass for bioenergy and bio-based products. Electronic waste (e-waste) contains precious metals like gold, silver, and rare earth elements that can be recycled for electronic manufacturing. Even carbon dioxide emissions can be captured and converted into fuels or polymers using catalytic and biological processes. By addressing diverse waste streams, waste-to-value technologies contribute to the broader vision of sustainable resource management.

Principles and Mechanisms

Waste-to-value technologies operate through physical, chemical, and biological processes designed to extract energy or materials from discarded products. The physical processes include sorting, crushing, and thermal treatment to prepare waste for conversion. Chemical processes involve reactions such as pyrolysis, gasification, or transesterification to break down waste into useful molecules. Biological processes, such as anaerobic digestion and fermentation, use microorganisms to convert organic waste into biogas or biofuels. Each process aims to maximize recovery efficiency while minimizing emissions and energy consumption. The selection of a conversion pathway depends on factors like waste composition, desired product, and available technology.

Waste-to-Value Technologies

A range of innovative technologies is used to transform waste into useful resources. The most prominent include:

- Anaerobic digestion, where microbes decompose organic waste to produce biogas and nutrient-rich digestate.
- Gasification, converting carbon-based waste into syngas for electricity or fuel production.
- Pyrolysis, thermally decomposing plastics and biomass into oil, gas, and char without oxygen.
- Fermentation, transforming agricultural residues into bioethanol and biochemicals.
- Composting, turning biodegradable waste into organic fertilizer for agriculture.
- Waste-to-energy (WtE) incineration, generating electricity and heat while reducing landfill volume.
- Catalytic conversion, using catalysts to transform CO_2 or plastics into synthetic fuels and chemicals.
- Hydrothermal liquefaction, converting wet biomass into crude bio-oil under high pressure.
- Metal recovery and recycling, extracting valuable metals from electronic and industrial waste.
- Carbon mineralization, locking carbon dioxide into stable minerals for construction materials.

These technologies differ in energy efficiency, scalability, and end-product value but collectively contribute to reducing waste volumes and greenhouse gas emissions.

Waste-to-Energy Conversion

One of the most widely adopted waste-to-value approaches is waste-to-energy (WtE) conversion, which involves recovering energy from waste materials. Technologies like gasification and incineration convert waste into heat, electricity, or synthetic fuels, replacing fossil fuel-based energy sources. For instance, municipal solid waste can be combusted in high-efficiency incinerators to generate steam that powers turbines. Alternatively, anaerobic digestion of organic waste produces biogas, a renewable energy source composed mainly of methane. WtE facilities significantly reduce landfill dependency while providing renewable energy for urban centers. However, these systems must employ advanced emission control technologies to prevent the release of harmful pollutants.

Waste Valorization in the Chemical Industry

The chemical industry is one of the largest generators of waste but also one of the primary beneficiaries of waste-to-value solutions. Through chemical recycling, polymer waste can be depolymerized back into monomers for the production of new plastics.

Similarly, carbon capture and utilization (CCU) technologies enable the transformation of industrial CO_2 emissions into methanol, formic acid, or polymers, effectively closing the carbon loop. Biological methods also play a role, such as the use of engineered microbes to convert agricultural residues into high-value biochemicals and bioplastics. These valorization processes reduce the industry's carbon footprint while diversifying raw material sources.

Waste-to-Value in Agriculture and Food Systems

Agriculture and food industries produce vast quantities of organic waste, much of which can be valorized through biological processes. Crop residues and manure can be processed via anaerobic digestion to produce biogas and biofertilizers, providing renewable energy and improving soil health. Food waste can be fermented to generate organic acids, enzymes, and animal feed ingredients. The circular utilization of nutrients through composting and biofertilizer production reduces dependency on synthetic fertilizers, lowering greenhouse gas emissions associated with agriculture. Innovations such as insect bioconversion, where larvae convert organic waste into protein-rich feed, illustrate the diverse potential of waste-to-value technologies in agricultural sustainability.

Environmental and Economic Benefits

The transition to waste-to-value systems offers a dual advantage of environmental protection and economic gain. Environmentally, these technologies mitigate pollution, reduce greenhouse gas emissions, and conserve natural resources. Economically, they create new industries and employment opportunities in renewable energy, bio-manufacturing, and recycling sectors. By transforming waste into marketable products, companies can reduce waste management costs while generating additional revenue streams. Moreover, waste-to-value initiatives attract green investments and contribute to national energy security by diversifying the renewable energy mix. In developing economies, decentralized waste-to-value systems can provide rural electrification, clean cooking fuel, and improved sanitation.

Challenges and Barriers

Despite their potential, waste-to-value technologies face several challenges that hinder widespread adoption. The heterogeneity of waste streams complicates processing and requires advanced sorting technologies. High initial capital costs for equipment and infrastructure deter investment, especially in developing regions. Technical inefficiencies, such as incomplete conversion or low yields, can limit profitability. Regulatory and policy gaps in waste management frameworks often impede innovation, while public perception issues—particularly concerning waste-derived fuels or food-related applications—can restrict market acceptance. Finally, environmental safeguards are essential to ensure that conversion processes do not introduce new pollutants or health risks. Overcoming these challenges requires integrated policy support, technological innovation, and international collaboration.

3.6 SYNTHETIC BIOLOGY FOR ENERGY SOLUTIONS

3.6.1 Microbial Fuel Cells

Microbial fuel cells embody the innovative potential of biological systems to address modern energy and environmental challenges. By harnessing the metabolic activity of microorganisms, MFCs provide a clean, renewable, and sustainable means of generating electricity while simultaneously treating organic waste. Their ability to transform waste into energy positions them as a crucial technology in the emerging bioenergy landscape. Although challenges remain—particularly concerning efficiency, scalability, and material costs—ongoing research in microbial engineering, materials science, and reactor optimization is rapidly advancing the field. As these challenges are overcome, microbial fuel cells are likely to play an increasingly important role in decentralized energy systems, environmental monitoring, and waste-to-energy conversion. In the long term, MFCs could form part of a larger network of bioelectrochemical technologies that contribute to global sustainability goals, embodying the fusion of life science and energy innovation in the service of a cleaner and more resilient planet.

Historical Background and Conceptual Foundation

The concept of biological electricity generation is not new. The first observations of electrical activity by microorganisms date back to the early twentieth century when M. C. Potter, in 1911, demonstrated that bacteria could produce electric current during metabolism. However, the lack of sophisticated technology and understanding of microbial electron transfer limited further development. It was not until the late 1990s and early 2000s, with advancements in bioelectrochemistry and molecular microbiology, that MFCs began to emerge as a viable research field. The foundational principle of MFCs is based on the ability of certain microorganisms, known as exoelectrogens or electroactive bacteria, to transfer electrons generated during substrate oxidation to an external electrode rather than to traditional electron acceptors such as oxygen. This electron transfer process establishes an electrical current that can be captured and utilized for power generation. Thus, microbial fuel cells bridge biological and electrochemical systems to produce electricity from renewable biomass sources.

Structure and Working Principle of Microbial Fuel Cells

A typical microbial fuel cell consists of two main chambers: the anode and the cathode, separated by a proton exchange membrane (PEM). The anode chamber is anaerobic, containing microorganisms that oxidize organic substrates such as glucose, acetate, or wastewater compounds. During this process, the microbes generate electrons and protons.

The electrons are transferred to the anode and flow through an external circuit to the cathode, creating an electric current. Meanwhile, protons migrate through the membrane to the cathode chamber, which is aerobic. At the cathode, these protons combine with electrons and oxygen to form water. This continuous electron flow through the circuit constitutes the electricity generated by the microbial fuel cell. The efficiency of this process depends on several factors, including microbial activity, substrate type, electrode material, and system design.

Table: Components and their Functions in Microbial Fuel Cells

Component	Function
Anode Chamber	Hosts electroactive bacteria that oxidize organic substrates under anaerobic conditions, releasing electrons and protons.
Cathode Chamber	Aerobic compartment where electrons and protons combine with oxygen to form water, completing the circuit.
Proton Exchange Membrane (PEM)	Allows proton transfer from anode to cathode while preventing oxygen diffusion into the anode.
Electrodes	Conductive materials that collect and transport electrons; commonly made of carbon-based materials like graphite or carbon cloth.
External Circuit	Pathway through which electrons travel, generating electrical current measurable as power output.

This configuration demonstrates the fundamental bioelectrochemical mechanism underlying MFC operation and highlights how the synergy between biological and electrochemical systems enables sustainable energy conversion.

Microbial Metabolism and Electron Transfer Mechanisms

The core of MFC functionality lies in microbial metabolism. In the anode chamber, microorganisms oxidize organic matter to obtain energy for their growth and maintenance. During this metabolic process, electrons and protons are released. However, for electricity generation, these electrons must be transferred to the anode. Exoelectrogenic bacteria such as Geobacter sulfurreducens, Shewanella oneidensis, and certain species of Pseudomonas are capable of extracellular electron transfer (EET). This electron transfer occurs through three main mechanisms: direct contact, conductive pili (nanowires), and soluble mediators.

In direct contact transfer, bacteria form biofilms on the electrode surface, allowing electron flow via membrane-bound cytochromes. In nanowire-based transfer, bacteria produce conductive protein filaments that transport electrons over distances. Mediated electron transfer involves small redox-active compounds such as flavins or quinones that shuttle electrons between cells and electrodes. The efficiency of these mechanisms determines the power output and stability of microbial fuel cells.

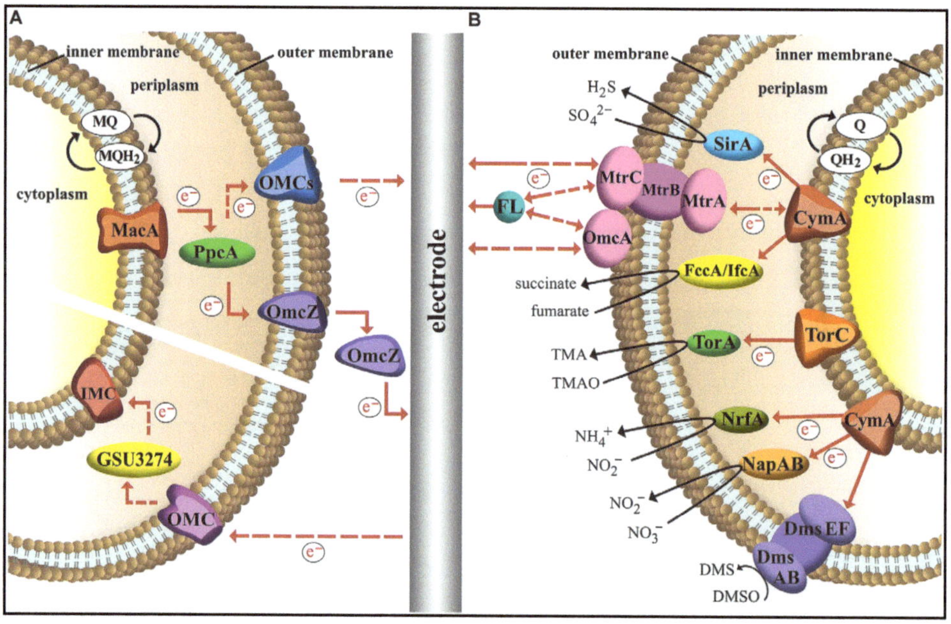

Microbial Electron Transport and Energy Conservation

Types of Microbial Fuel Cells

Microbial fuel cells can be categorized based on design and function. The most common configuration is the dual-chamber MFC, where the anode and cathode chambers are separated by a proton exchange membrane. This design ensures strict anaerobic conditions for the anode but may lead to higher internal resistance. Single-chamber MFCs, on the other hand, eliminate the membrane and use an air cathode, reducing costs and improving oxygen transfer. Variations such as stacked MFCs enhance voltage and current output by connecting multiple units in series or parallel. Specialized MFCs have also been developed for specific purposes, including sediment MFCs for marine environments and photo-MFCs that integrate photosynthetic microorganisms for light-driven bioelectricity production. Each configuration offers distinct advantages in terms of efficiency, scalability, and environmental compatibility, depending on the intended application.

Applications in Wastewater Treatment

One of the most promising applications of microbial fuel cells lies in wastewater treatment. Traditional wastewater treatment methods are energy-intensive and often produce large amounts of sludge. In contrast, MFCs can simultaneously degrade organic pollutants and produce electricity, converting waste into a valuable energy resource. When wastewater serves as the substrate, microorganisms in the anode chamber metabolize organic compounds, reducing chemical oxygen demand (COD) and improving water quality. This dual benefit—pollutant removal and energy recovery—makes MFCs an attractive option for sustainable waste management. Moreover, microbial communities within wastewater are naturally diverse, providing a rich source of electroactive species that enhance performance without requiring external inoculation. Several pilot-scale systems have demonstrated the feasibility of integrating MFCs into existing wastewater treatment plants, potentially offsetting energy costs and reducing environmental impact.

Energy Output and Performance Optimization

Although MFCs offer remarkable sustainability benefits, their electrical output is relatively low compared to conventional energy technologies. The power density of MFCs typically ranges from a few milliwatts to several watts per square meter, depending on system design and substrate type. Optimization efforts focus on improving electrode materials, enhancing microbial electron transfer, and reducing internal resistance. Carbon-based materials such as graphite, carbon cloth, and carbon nanotubes are widely used due to their conductivity and biocompatibility. Modifications with nanomaterials, conductive polymers, and surface treatments further enhance microbial adhesion and electron exchange. Additionally, optimizing operating conditions such as temperature, pH, and substrate concentration can significantly influence microbial activity and power generation. Advances in bioengineering are also enabling the genetic modification of electroactive bacteria to improve electron transfer efficiency, opening new frontiers in synthetic bioelectrochemistry.

Environmental and Economic Benefits

Microbial fuel cells offer a unique combination of environmental and economic advantages. By utilizing organic waste as fuel, they contribute to waste valorization and renewable energy generation simultaneously. The process is carbon-neutral, as the carbon released during microbial metabolism originates from biogenic sources rather than fossil fuels. MFCs also produce minimal greenhouse gas emissions and can operate under mild conditions without the need for expensive catalysts such as platinum. Economically, MFCs could reduce the operational costs of wastewater treatment by generating electricity to offset energy consumption. Although large-scale commercialization is still limited due to cost and performance challenges, continued improvements in materials and reactor design are expected to enhance their competitiveness. The integration of MFCs with other renewable technologies, such as solar or wind power, can further contribute to sustainable energy ecosystems.

Challenges and Limitations

Despite their promise, microbial fuel cells face several technical and practical limitations that hinder widespread adoption. One major challenge is the low power output, which restricts their use in high-energy-demand applications. Scaling up from laboratory prototypes to industrial systems introduces complexities related to mass transfer, biofilm stability, and reactor maintenance. Electrode and membrane costs remain significant barriers, as high-performance materials are often expensive. Furthermore, long-term operation can be affected by biofouling, pH imbalances, and variations in substrate composition. From a biological perspective, maintaining optimal microbial activity and community stability is critical but difficult under fluctuating environmental conditions. Addressing these challenges requires interdisciplinary research combining microbiology, materials science, and environmental engineering to improve both the efficiency and economic feasibility of MFCs.

Microbial Fuel Cells in Environmental Monitoring and Remote Applications

An emerging application of MFCs is in environmental sensing and monitoring. Due to their self-sustaining nature, MFCs can power biosensors that monitor environmental parameters such as dissolved oxygen, organic pollutants, or heavy metals in real time. Sediment-based MFCs, for instance, can operate in marine or riverine environments for extended periods, generating enough power to transmit data wirelessly without external power sources. This capability makes them ideal for use in remote or resource-limited settings where maintenance and battery replacement are challenging. Such applications highlight the versatility of microbial fuel cells not just as energy generators but also as components of smart environmental management systems.

3.6.2 Algae-Based Biofuels

Algae-based biofuels offer a sustainable, high-yielding, and versatile renewable energy alternative with immense potential to complement and eventually replace fossil fuels. The diverse array of algal species, coupled with advances in cultivation, genetic engineering, and conversion technologies, make algae prime candidates for next-generation bioenergy production. While economic and technical challenges persist, continuous innovation in bioprocessing, metabolic engineering, and integrated biorefineries is rapidly enhancing viability. As global energy demands increase and the urgency for climate solutions intensifies, algae biofuels stand poised to play a pivotal role in ushering in a cleaner and more sustainable energy future.

Algae as a Resource for Biofuel Production

Algae, due to their photosynthetic nature, efficiently convert sunlight, carbon dioxide, and nutrients into biomass enriched with lipids, carbohydrates, and proteins. Microalgae are particularly notable for their high lipid content, which can exceed 50% of their dry weight in some strains, making them an excellent feedstock for biodiesel. Macroalgae, or seaweeds, also contribute significantly by supplying carbohydrates convertible into bioethanol or biogas through fermentation or anaerobic digestion. The high growth rates of algae enable multiple harvests annually, allowing greater biofuel yields per hectare than terrestrial crops. Additionally, their ability to sequester carbon dioxide while producing oxygen aligns well with global climate goals.

Cultivation Techniques for Algal Biomass

The cultivation of algae forms the initial critical step in algae-based biofuel production, with two predominant methods: open pond systems and closed photobioreactors. Open ponds are economically attractive for large-scale production due to low construction and operational costs, utilizing natural sunlight and ambient temperatures. However, they are susceptible to contamination, evaporation, and limited control over growth conditions. Closed photobioreactors offer enhanced control of environmental parameters—light, temperature, CO2 supply—and minimize contamination, thereby improving biomass productivity and quality. Innovative hybrid systems attempt to balance cost and productivity by integrating both methods.

Nutrient management is essential, as algae require nitrogen, phosphorus, and trace elements. Wastewater streams rich in nutrients have been utilized as culture media, providing sustainable nutrient recycling while treating waste. Genetic and metabolic engineering have begun to play a role in developing algae strains optimized for high lipid production, robust growth under harsh conditions, and ease of harvesting.

Harvesting and Biomass Processing

Harvesting algal biomass efficiently is critical due to the dilute nature of cultures—typically less than 1% biomass concentration in water. Techniques include centrifugation, filtration, flocculation, and sedimentation. Combinations of these methods are often employed to maximize recovery while minimizing energy inputs.

Upon harvesting, algal biomass requires processing to extract or convert the stored biochemicals into useable biofuels. The diverse biochemical composition necessitates tailored conversion pathways. Lipid-rich microalgae are primarily processed for biodiesel production through lipid extraction followed by transesterification. Carbohydrate-rich species undergo hydrolysis to release fermentable sugars for bioethanol production.

Conversion Technologies for Algae-Derived Biofuels

Algal biomass can be converted into various types of biofuels using multiple biochemical and thermochemical processes. Biochemical methods include transesterification for biodiesel, fermentation to bioethanol, and anaerobic digestion to biogas. Thermochemical conversion encompasses pyrolysis, gasification, and hydrothermal liquefaction (HTL), which transform entire biomass components into bio-crude oils or syngas.

Hydrothermal liquefaction is particularly notable for algae as it processes wet biomass directly under high temperature and pressure, converting proteins, lipids, and carbohydrates into biocrude without the need for drying. This biocrude can subsequently be refined to produce drop-in biofuels compatible with existing fuel infrastructure. Pyrolysis and gasification produce liquid and gaseous fuels but require dried biomass, which adds energy and cost burdens.

Fermentation techniques convert algal carbohydrates to ethanol or hydrogen, while anaerobic digestion produces methane-rich biogas. Each method has distinct advantages and challenges, with ongoing research focused on improving yields, reducing costs, and integrating biorefinery concepts to valorize co-products such as proteins and pigments.

Environmental and Economic Considerations

Algae-based biofuels offer substantial environmental benefits when compared to conventional fossil fuels. They reduce net carbon emissions by capturing CO2 during growth, have minimal land footprint, and avoid the food-versus-fuel dilemmas characteristic of crop-based biofuels. When cultured using wastewater streams or in saline waters, algae cultivation can alleviate environmental pollution and conserve freshwater resources. Economic viability remains a key hurdle for algal biofuels.

High capital and operational expenditure associated with cultivation, harvesting, and downstream processing currently limit competitiveness. Nevertheless, advances in large-scale cultivation technologies, strain improvement via synthetic biology, process integration in biorefineries, and co-production of high-value compounds improve the economic outlook. Policies supporting renewable energy and carbon reduction also foster industry growth.

Advances in Genetic and Metabolic Engineering

Genetic engineering of algal strains is rapidly advancing to boost biofuel yields and process resilience. Modifications target lipid biosynthesis pathways to increase storage and alter fatty acid profiles favorable for fuel properties. Engineering efforts also aim to reduce cellular components like carbohydrates and proteins that compete with lipid accumulation or complicate biomass processing. Synthetic biology approaches enable the construction of optimized metabolic pathways, enhanced carbon fixation, and improved tolerance to environmental stresses such as high light intensity, temperature fluctuations, or nutrient limitation. These engineered strains promise higher productivity and lower processing costs, facilitating commercial-scale operations.

Integration with Biorefinery and Circular Economy Concepts

Algae-based biofuels are increasingly envisioned as components of integrated biorefineries, wherein multiple valuable products are obtained from a single biomass feedstock. Beyond fuels, algae provide proteins for animal feed, carotenoids for nutraceuticals, pigments for cosmetics, and bioactive compounds for pharmaceuticals. This co-production enhances profitability and resource efficiency. Circular economy principles are embedded in algae biofuel production by recycling water, nutrients, and residue streams, minimizing waste and environmental impact. Algae can also remediate wastewater by removing excess nutrients and contaminants, contributing to environmental sustainability.

CHAPTER 4

Key Techniques and Tools in Synthetic Biology

4.1 DNA ASSEMBLY AND ENGINEERING TOOLS

4.1.1 Golden Gate Assembly and Gibson Assembly

Golden Gate Assembly and Gibson Assembly have fundamentally transformed the landscape of molecular biology by providing researchers with powerful, efficient, and flexible tools for DNA construction. These methods have eliminated many traditional barriers to genetic engineering, enabling projects of unprecedented scope and complexity that were previously impossible or impractical. The complementary strengths of each method provide researchers with a comprehensive toolkit for addressing diverse DNA assembly challenges, from simple gene cloning to synthetic genome construction.

As these techniques continue to evolve and improve, they will undoubtedly play crucial roles in addressing global challenges in health, agriculture, energy, and environmental sustainability. The ongoing democratization of these technologies, combined with decreasing DNA synthesis costs and increasing automation, promises to accelerate the pace of discovery and innovation in biotechnology. Mastering these assembly methods has become essential for modern molecular biologists, as they represent fundamental enabling technologies for the bioeconomy of the future.

Fundamentals of DNA Assembly Methods

The evolution of DNA assembly methods reflects the growing complexity of genetic engineering projects and the need for more efficient and flexible cloning strategies. Traditional restriction enzyme cloning, while foundational to molecular biology, suffers from significant limitations including the requirement for unique restriction sites, the introduction of scar sequences between assembled fragments, and the inability to assemble multiple fragments simultaneously. Modern assembly methods overcome these constraints by employing different strategies to join DNA fragments with high efficiency and fidelity. The fundamental challenge in DNA assembly involves creating specific, stable connections between DNA fragments while maintaining the desired sequence order and orientation. Both Golden Gate and Gibson Assembly address this challenge through distinct mechanisms that exploit different enzymatic activities and molecular interactions. These methods share common goals of achieving high assembly efficiency, reducing hands-on time, enabling automation, and producing predictable results that facilitate the design-build-test-learn cycle central to synthetic biology. The success of these assembly methods depends on careful design of DNA fragments with appropriate overlaps or restriction sites, optimization of reaction conditions, and selection of suitable vector systems for propagating assembled constructs.

Golden Gate Assembly: Principles and Mechanism

Golden Gate Assembly harnesses the unique properties of Type IIS restriction enzymes to achieve highly efficient, directional assembly of multiple DNA fragments in a single reaction. Unlike conventional restriction enzymes that cleave within their recognition sequences, Type IIS enzymes such as BsaI, BbsI, and SapI cut outside their recognition sites, generating programmable overhangs that can be designed to create unique, complementary ends for each fragment junction. This remarkable property enables the creation of customized four-base overhangs that serve as address tags, directing the precise order and orientation of fragment assembly. The Golden Gate reaction combines restriction digestion and ligation in a single tube, with the Type IIS enzyme cutting the DNA fragments to generate compatible overhangs while T4 DNA ligase simultaneously joins the complementary ends. The thermodynamic cycling between cutting and ligation temperatures drives the reaction toward the desired product, as correctly assembled constructs lack internal Type IIS sites and become protected from further digestion. The design process involves careful selection of overhang sequences to minimize cross-reactivity and ensure high assembly fidelity, with computational tools available to optimize overhang sets for specific projects. Golden Gate Assembly achieves remarkable efficiency, routinely assembling ten or more fragments with success rates exceeding ninety percent, making it particularly valuable for combinatorial library construction and modular pathway assembly.

Gibson Assembly: Methodology and Applications

Gibson Assembly employs a sophisticated isothermal reaction that combines three enzymatic activities to seamlessly join multiple DNA fragments with overlapping ends. The method utilizes T5 exonuclease to create single-stranded DNA overhangs, DNA polymerase to fill in gaps, and DNA ligase to seal nicks, all working in concert at a constant temperature of fifty degrees Celsius. The process begins with the design of DNA fragments containing fifteen to forty base pair overlapping sequences at their termini, which provide specificity for assembly and serve as templates for polymerase extension. During the reaction, T5 exonuclease removes nucleotides from the five-prime ends of double-stranded DNA, creating three-prime overhangs that anneal to complementary sequences on adjacent fragments. The DNA polymerase fills in gaps created by the exonuclease activity, while DNA ligase seals the remaining nicks to produce covalently closed circular DNA or linear constructs depending on the design. The beauty of Gibson Assembly lies in its simplicity and flexibility, requiring only the mixing of DNA fragments with the enzyme cocktail and incubation for fifteen minutes to one hour. This method excels at assembling large DNA fragments, having been used to construct entire bacterial genomes and large viral genomes, demonstrating its utility for ambitious synthetic biology projects that require the assembly of hundreds of kilobases of DNA.

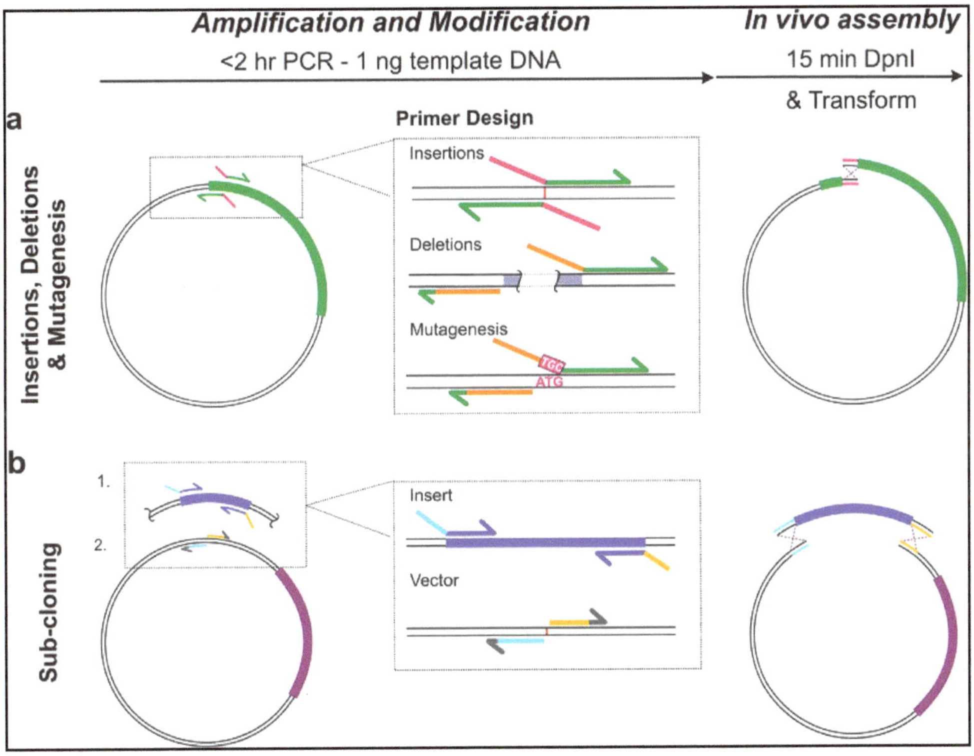

Methodological Overview of Gibson Assembly

Comparative Analysis of Both Methods

While both Golden Gate and Gibson Assembly offer powerful solutions for DNA construction, they differ significantly in their mechanisms, requirements, and optimal use cases. Golden Gate Assembly provides superior specificity through its programmable overhangs, making it ideal for assembling standardized parts and creating combinatorial libraries where precise control over assembly order is critical. The method's ability to generate scarless junctions between fragments makes it particularly valuable for protein engineering applications where additional amino acids could affect function. Gibson Assembly offers greater flexibility in fragment design, as overlapping sequences can be added through PCR primers without modifying the internal sequences of DNA fragments. This flexibility makes Gibson Assembly more suitable for assembling fragments from diverse sources or when working with sequences that cannot accommodate Type IIS restriction sites. Golden Gate Assembly typically shows higher efficiency for assembling many small fragments, while Gibson Assembly excels at joining larger fragments with high fidelity. The choice between methods often depends on specific project requirements, with Golden Gate preferred for standardized, modular assemblies and Gibson chosen for more complex, custom constructions. Many laboratories now employ both methods complementarily, using Golden Gate for initial module construction and Gibson for final assembly of large constructs.

Applications in Synthetic Biology

The widespread adoption of Golden Gate and Gibson Assembly has catalyzed numerous breakthroughs in synthetic biology and biotechnology applications. Metabolic pathway engineering benefits enormously from these assembly methods, enabling rapid construction of multi-gene pathways for producing valuable chemicals, pharmaceuticals, and biofuels. Researchers routinely use Golden Gate Assembly to create libraries of pathway variants with different promoters, ribosome binding sites, and gene orders to optimize production yields. Gibson Assembly has proven invaluable for constructing synthetic chromosomes and genomes, including the landmark synthesis of the first synthetic bacterial genome by the J. Craig Venter Institute. CRISPR gene editing applications leverage both assembly methods to construct guide RNA arrays and donor templates for multiplexed genome engineering. Protein engineering projects utilize Golden Gate Assembly to create fusion proteins and domain-shuffled variants without introducing unwanted linker sequences. The development of biological circuits for biosensing, biocomputing, and therapeutic applications relies heavily on these assembly methods to rapidly prototype and test different circuit architectures. Agricultural biotechnology employs these techniques to stack multiple traits in crop plants, combining herbicide resistance, insect protection, and nutritional enhancements in single transformation events.

Technical Advantages and Current Limitations

Both assembly methods offer substantial advantages over traditional cloning but also face certain technical limitations that researchers must consider when planning experiments. The high efficiency and automation compatibility of both methods reduce labor costs and accelerate research timelines, with robotic systems capable of performing hundreds of assemblies in parallel. The scarless nature of assemblies preserves reading frames and regulatory elements, crucial for maintaining biological function. Golden Gate Assembly's modular approach facilitates the creation of standardized biological parts that can be readily shared and reused across different projects and laboratories. Gibson Assembly's ability to join any sequences with designed overlaps provides unmatched flexibility for complex constructions. However, Golden Gate Assembly requires the absence of internal Type IIS sites within fragments, sometimes necessitating site removal through silent mutations or fragment subdivision. Gibson Assembly can suffer from reduced efficiency when assembling fragments with high GC content or significant secondary structures, and the overlapping regions can sometimes lead to unwanted recombination events. Both methods require careful optimization of fragment ratios and reaction conditions for complex assemblies, and the cost of synthetic DNA and enzymes can be significant for large-scale projects.

Recent Innovations and Technical Improvements

Continuous innovation has enhanced both Golden Gate and Gibson Assembly methods, addressing limitations and expanding their capabilities for increasingly complex applications. Advanced Golden Gate systems now incorporate multiple Type IIS enzymes with different recognition sequences, enabling hierarchical assembly strategies that can construct megabase-scale DNA from standardized parts. Loop assembly techniques extend Golden Gate methodology to enable recursive assembly of repetitive sequences commonly found in plant centromeres and heterochromatin. Enhanced Gibson Assembly protocols incorporate additional enzymes such as RecA recombinase to improve assembly of difficult sequences and reduce errors. Microfluidic platforms have been developed for both methods, enabling assembly reactions in nanoliter volumes and reducing reagent costs while increasing throughput. Error correction strategies integrated into assembly workflows identify and eliminate incorrect assemblies before transformation, improving overall success rates. Computational tools for assembly design have become increasingly sophisticated, incorporating machine learning algorithms to predict assembly success and optimize fragment design. The development of cell-free assembly systems allows researchers to perform and validate assemblies without the constraints of bacterial transformation and selection.

4.1.2 Modular Cloning Systems

Modular cloning systems represent a transformative advancement in molecular biology, enabling scientists to assemble complex genetic constructs with unprecedented speed, accuracy, and standardization. By leveraging type IIS restriction enzymes and hierarchical assembly principles, these systems allow for seamless integration of genetic parts, facilitating innovation across synthetic biology, metabolic engineering, and biotechnology. Although challenges related to standard compatibility and sequence domestication remain, ongoing efforts toward automation, open-source sharing, and digital integration continue to expand their potential. The modular cloning paradigm has redefined how scientists approach genetic design, making the creation of new biological systems as systematic and predictable as engineering circuits or mechanical devices. As synthetic biology continues to mature, modular cloning will remain a cornerstone technology, shaping the next generation of bioengineering solutions and sustainable biotechnological innovations.

Concept of Modular Cloning

Modular cloning (MoClo) is a modern DNA assembly method that enables researchers to build multi-gene constructs using standardized parts. The key idea behind MoClo is modularity: each DNA part, such as a promoter or gene, is treated as a self-contained unit that can be combined with others through compatible ends. Unlike traditional restriction enzyme cloning, modular cloning uses type IIS restriction enzymes, such as BsaI or BpiI, which cut DNA outside of their recognition sites. This allows scientists to design DNA parts with customized overhangs that determine how fragments fit together, enabling seamless assembly without leaving unwanted sequences between modules. The system's hierarchical design means that simple constructs can be progressively assembled into more complex genetic networks, making it ideal for applications in metabolic engineering, synthetic biology, and biotechnology research.

Historical Development and Evolution

The concept of modular cloning evolved from earlier DNA assembly strategies such as Golden Gate Assembly, introduced in the late 2000s. Golden Gate cloning demonstrated that type IIS restriction enzymes could be used to assemble multiple fragments in a single reaction. Building on this principle, Werner and colleagues developed the Modular Cloning (MoClo) system in 2012, which introduced a standardized hierarchical architecture for assembling transcriptional units and multi-gene constructs. This approach inspired the creation of various related systems, such as GoldenBraid, Loop assembly, and Mobius cloning, each introducing unique design features and compatibility standards. Over time, modular cloning systems have become central tools for synthetic biology, providing a robust and flexible framework for assembling genetic constructs across different organisms.

Principles and Mechanisms of Modular Cloning

Modular cloning is based on standardized overhangs and type IIS restriction enzyme digestion-ligation cycles. In this process, each DNA part is flanked by recognition sites for a type IIS enzyme that generates specific overhangs when cut. These overhangs guide the order and orientation of parts during ligation. For example, a promoter fragment may have overhangs that fit only upstream of a coding sequence, ensuring proper transcriptional arrangement. The ligation of multiple fragments into a destination vector occurs in a single reaction cycle using DNA ligase, enabling efficient one-pot assembly of up to several genes. Because the recognition sites are removed during ligation, the resulting construct is “scarless,” meaning no extraneous sequences disrupt the genetic code. This mechanism makes MoClo highly accurate and efficient for assembling large DNA constructs.

Standardization and Hierarchical Assembly

One of the greatest strengths of modular cloning lies in its standardization. The MoClo system defines specific positions for genetic parts within a transcriptional unit: promoters, 5’ untranslated regions (UTRs), coding sequences, terminators, and others. Each part is assigned compatible overhangs to ensure correct order during assembly. This standardization facilitates hierarchical assembly, where smaller modules are combined into larger units.

For example, Level 0 parts (basic genetic elements) are combined into Level 1 transcription units, which can then be assembled into Level 2 multi-gene constructs. This hierarchical design allows researchers to build increasingly complex genetic systems with precision and reproducibility, significantly accelerating the design-build-test cycle in synthetic biology.

Key Components and Tools

Modular cloning systems are composed of several essential components that work together to ensure flexibility and compatibility across projects and laboratories.

- Type IIS restriction enzymes such as BsaI, BsmBI, or BpiI for generating specific overhangs.
- Standardized DNA parts, including promoters, ribosome binding sites, coding sequences, and terminators.
- Vector backbones designed for Level 0, Level 1, and Level 2 assemblies.
- DNA ligase enzymes for covalent linkage of DNA fragments.
- Overhang design rules that specify how fragments are joined.
- Antibiotic selection markers for identifying successful constructs.
- Reporter genes (e.g., GFP or LacZ) for visual confirmation of cloning success.
- Bioinformatics tools for designing modular parts and verifying compatibility.
- Automated liquid handling systems that streamline high-throughput cloning.
- Databases and repositories such as Addgene or MoClo repositories for sharing standardized parts.

These elements together enable modular cloning systems to be reproducible and shareable across laboratories, promoting collaboration and open-source innovation in genetic engineering.

Advantages of Modular Cloning Systems

Modular cloning systems offer several advantages over traditional cloning approaches. Their speed and efficiency allow for simultaneous assembly of multiple fragments in one reaction, drastically reducing time compared to stepwise cloning. The standardization of overhangs and part types enables compatibility between projects and laboratories, fostering a collaborative ecosystem in synthetic biology. MoClo systems are also highly accurate, with low error rates due to the precise recognition of enzyme sites. Moreover, the hierarchical structure allows for reusability of parts, so researchers can mix and match existing elements to create new constructs. These benefits collectively make modular cloning an indispensable tool for metabolic pathway engineering, biosensor construction, and genetic circuit design.

Applications in Synthetic Biology and Biotechnology

Modular cloning has revolutionized the way genetic constructs are designed and implemented in various scientific and industrial fields. In synthetic biology, it allows for the rapid prototyping of genetic circuits that control gene expression in predictable ways. In metabolic engineering, researchers use modular cloning to assemble multi-gene pathways for the production of biofuels, pharmaceuticals, and bioplastics. The system also facilitates functional genomics, enabling the systematic analysis of gene function by creating combinatorial gene libraries. In plant biotechnology, modular cloning systems have been adapted for use in species such as Arabidopsis and Nicotiana benthamiana, allowing for the development of genetically enhanced crops. Furthermore, MoClo has applications in therapeutic research, such as constructing vectors for gene therapy and vaccine development, where precise control over gene expression is critical.

Limitations and Challenges

Despite its numerous advantages, modular cloning systems are not without challenges. The reliance on type IIS restriction sites means that internal recognition sequences must be removed from DNA parts before assembly, a process known as domestication, which can be time-consuming. Additionally, the incompatibility between different modular cloning standards (e.g., MoClo vs. GoldenBraid) can hinder interoperability across platforms. The efficiency of assembly reactions can also vary depending on fragment size and sequence complexity. While automation and computational tools have improved reproducibility, human error in overhang design or vector preparation can still lead to failed assemblies. Moreover, intellectual property constraints and proprietary vector systems can limit open sharing of modular parts. Addressing these issues requires greater standardization, open-access resources, and cross-platform compatibility to fully realize the potential of modular cloning technologies.

Significance in the Context of Synthetic Biology

Modular cloning embodies the engineering philosophy at the heart of synthetic biology—standardization, abstraction, and modularity. By providing a robust and reproducible framework for assembling genetic constructs, it enables researchers to treat biological systems like programmable machines. This engineering mindset transforms biology into a predictable and designable discipline, accelerating innovation in areas ranging from agriculture to medicine. Modular cloning systems make it possible to build synthetic organisms that produce renewable fuels, biodegradable plastics, or therapeutic proteins with high precision. As global challenges such as climate change and antibiotic resistance demand biological solutions, modular cloning stands out as a foundational technology that bridges molecular biology and biotechnology engineering.

4.2 SYNTHETIC PROMOTERS AND REGULATORS

Synthetic promoters and regulators have revolutionized the control of gene expression, providing tools that are precise, tunable, and modular. By engineering DNA sequences and regulatory elements, researchers can achieve predictable transcriptional output, enabling the construction of sophisticated gene circuits, metabolic pathways, and therapeutic interventions. Their applications span biotechnology, medicine, environmental monitoring, and industrial bioproduction, demonstrating their versatility and transformative potential. Despite challenges related to context-dependency, stability, and host interactions, advances in computational modeling, high-throughput screening, and genome editing are overcoming these limitations. As the field continues to evolve, synthetic promoters and regulators will play a central role in designing programmable biological systems, advancing precision medicine, sustainable bioproduction, and functional synthetic biology. These innovations exemplify the power of engineering principles applied to living systems, providing new avenues for scientific discovery and technological advancement.

4.2.1 Fundamentals of Promoters and Gene Regulation

Promoters are DNA sequences located upstream of genes that serve as binding sites for RNA polymerase and associated transcription factors. They are fundamental to the initiation of transcription, dictating the timing, location, and level of gene expression. In natural systems, promoters are regulated by complex networks of activators, repressors, and enhancers, which respond to intracellular and extracellular signals. Synthetic promoters leverage this foundational understanding but are designed to provide greater predictability and modularity. By altering core promoter elements, transcription factor binding sites, or spacing between regulatory motifs, synthetic promoters can be tailored to achieve specific expression profiles, ranging from constitutive high-level expression to finely tuned inducible activity. Synthetic regulators extend this capability by providing mechanisms to control promoter activity dynamically, allowing cells to respond to chemical inducers, environmental conditions, or metabolic states.

4.2.2 Design Principles of Synthetic Promoters

The design of synthetic promoters involves careful consideration of multiple sequence features, including the core promoter region, upstream activating sequences, and transcription factor binding sites. Core promoter elements such as the TATA box, initiator sequences, and downstream promoter elements determine the basal level of transcription. By modifying these sequences, researchers can adjust the strength of transcription initiation. Upstream sequences, including enhancers and operator sites, enable regulation by synthetic transcription factors or small molecules, providing inducible or repressible control.

Computational tools and machine learning approaches are increasingly used to predict promoter strength and behavior based on sequence composition, allowing rational design of promoters with desired characteristics. Additionally, synthetic promoters can be combined into libraries with varying strengths, enabling the fine-tuning of multi-gene pathways and minimizing metabolic burden in engineered organisms.

4.2.3 Synthetic Regulators: Types and Mechanisms

Synthetic regulators encompass a wide range of molecular tools that modulate gene expression. Protein-based regulators include engineered transcription factors that can activate or repress target promoters, often in response to specific ligands or environmental cues. These regulators can be designed using modular DNA-binding domains and customizable effector domains, allowing precise control over transcriptional activity. RNA-based regulators, such as riboswitches and small interfering RNAs, provide additional layers of post-transcriptional control, modulating translation efficiency or mRNA stability. CRISPR-based systems, particularly CRISPR interference (CRISPRi) and CRISPR activation (CRISPRa), offer programmable regulation of endogenous and synthetic genes by targeting promoter regions or regulatory sequences with guide RNAs. These synthetic regulators can be combined with inducible systems to create dynamic and reversible control, enabling sophisticated gene circuit designs for applications in metabolic engineering, synthetic biology, and therapeutic interventions.

4.2.4 Applications in Metabolic Engineering

Synthetic promoters and regulators are particularly valuable in metabolic engineering, where precise control over gene expression is essential to optimize production pathways. By adjusting the transcriptional output of key enzymes, metabolic flux can be redirected toward desired products while minimizing the accumulation of toxic intermediates. For example, in the production of biofuels, pharmaceuticals, or industrial chemicals, synthetic promoters allow fine-tuning of enzyme expression levels to balance precursor supply, cofactor availability, and pathway efficiency. Synthetic regulators provide temporal control, enabling sequential activation of pathway genes to optimize production kinetics. These tools also facilitate the creation of feedback-controlled circuits that respond to intracellular metabolite concentrations, improving yield and stability in engineered microbial strains. By combining synthetic promoters and regulators, metabolic pathways can be optimized with unprecedented precision, enhancing the scalability and economic feasibility of biotechnological processes.

4.2.5 Applications in Synthetic Gene Circuits

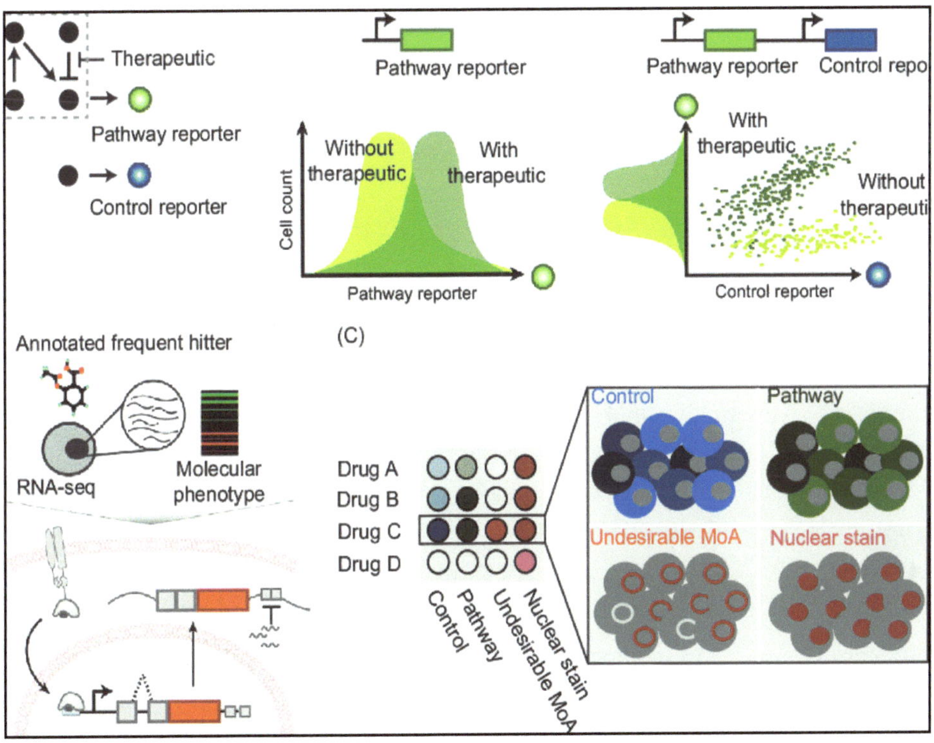

Synthetic Gene Circuits as Tools for Drug Discovery

Synthetic gene circuits, inspired by electrical engineering, rely heavily on synthetic promoters and regulators to achieve predictable behavior. Circuits such as oscillators, toggle switches, logic gates, and memory devices depend on carefully designed promoter-regulator combinations to control gene expression patterns. For example, a synthetic toggle switch requires two mutually repressing promoters and regulators to maintain stable states that can be switched by external inputs. Oscillatory circuits utilize feedback loops with synthetic promoters and regulators to produce rhythmic gene expression.

Logic gates integrate multiple signals using promoters with multiple operator sites, enabling cells to process environmental or chemical inputs in a computational manner. These circuits have applications in biosensing, therapeutic delivery, and biocomputation, demonstrating the versatility of synthetic promoters and regulators in programming cellular behavior.

4.2.6 Environmental and Biosensing Applications

Synthetic promoters and regulators also play a critical role in environmental monitoring and biosensing applications. Engineered microorganisms can be designed to respond to specific chemical or physical stimuli by activating reporter genes under the control of synthetic promoters. For example, bacteria can be programmed to detect heavy metals, toxins, or pollutants in soil or water and produce measurable outputs such as fluorescence, color change, or electrical signals.

Synthetic regulators enable dynamic responses, allowing cells to respond only under specific conditions or thresholds, reducing background noise and improving sensitivity. These systems offer cost-effective, scalable, and deployable solutions for environmental monitoring, complementing conventional analytical methods and providing real-time feedback for ecological management and pollution control.

4.2.7 Therapeutic Applications

In medicine, synthetic promoters and regulators enable precise control of therapeutic gene expression, enhancing the safety and efficacy of gene therapy and cell therapy. Inducible promoters allow the expression of therapeutic proteins or genetic circuits only in response to specific signals, minimizing off-target effects and toxicity. Synthetic regulators can be used to modulate immune cell activity, control apoptosis, or regulate metabolic pathways in engineered cells.

For example, in chimeric antigen receptor (CAR) T-cell therapy, synthetic promoters can be employed to control the expression of CAR constructs, reducing the risk of cytokine release syndrome and improving patient safety. Similarly, RNA-based regulators or CRISPR-based transcriptional modulators can be incorporated into therapeutic cells to provide reversible and tunable control over gene expression, expanding the possibilities for personalized medicine and precision therapeutics.

4.2.8 Challenges in Designing Synthetic Promoters and Regulators

Despite their advantages, the design and implementation of synthetic promoters and regulators present several challenges. One major issue is context-dependency, where promoter strength and regulator activity can vary depending on the host organism, genomic location, or environmental conditions. Unintended interactions with native transcriptional machinery can lead to unpredictable behavior or metabolic burden. Achieving fine-tuned control over multiple genes simultaneously remains complex, particularly in eukaryotic systems with elaborate regulatory networks. Additionally, the introduction of synthetic elements may trigger host immune responses or stress responses, impacting cell viability and stability. Overcoming these challenges requires iterative design-build-test cycles, computational modeling, and high-throughput screening to identify robust and reliable promoter-regulator combinations.

4.2.9 Advances in Computational and High-Throughput Approaches

Recent technological advances have greatly facilitated the design of synthetic promoters and regulators. Computational models enable the prediction of promoter strength, regulatory interactions, and circuit behavior, reducing experimental trial and error. Machine learning algorithms are increasingly used to analyze large datasets of promoter sequences and gene expression profiles, identifying motifs and features that influence transcriptional output. High-throughput screening platforms allow rapid testing of thousands of synthetic promoter variants and regulatory configurations, enabling the selection of optimal designs for specific applications. These tools accelerate the engineering of genetic systems, making it possible to construct complex synthetic pathways and gene circuits with predictable performance in diverse organisms.

4.3 SYNTHETIC CIRCUITS: FROM BIOLOGICAL PATHWAYS TO GENE NETWORKS

Synthetic circuits exemplify the power of engineering biology to decipher, emulate, and extend natural gene regulatory networks. From foundational genetic switches to intricate, integrated gene networks, synthetic circuits enable cells to process information dynamically, make decisions, and execute complex biological functions with programmable precision. Their modular construction, versatile behaviors, and expanding applications reflect a deepening understanding of biological design principles and a growing toolkit for innovation in biotechnology and medicine. Though challenges related to complexity, stability, and predictability persist, ongoing multidisciplinary advances foreshadow transformative impacts on bioproduction, healthcare, and environmental sustainability by harnessing the programmability of life itself.

4.3.1 Foundations and Components of Synthetic Circuits

Synthetic circuits build on well-characterized biological components—promoters, transcription factors, ribosome binding sites, and regulatory RNAs—that are assembled modularly to generate desired input-output relationships. Early synthetic gene circuits, such as the repressilator, showcased the capacity for engineered oscillations driven by transcriptional repression, while bistable switches exemplified cellular memory through feedback loops.

These foundational motifs mirror natural regulatory elements but are designed with orthogonality in mind, minimizing interference with host cellular processes. Increasingly sophisticated components have expanded the synthetic toolkit, including CRISPR-Cas-based regulators that enable RNA-guided gene expression control, RNA aptamers interacting with small molecules for environmental sensing, and engineered protein-protein interactions that modulate signaling cascades. Together, these components enable transcriptional, post-transcriptional, translational, and even post-translational control layers, allowing circuits to function robustly in diverse biological contexts.

4.3.2 Design Principles and Circuit Architectures

The architecture of synthetic circuits is typically defined by three functional modules: sensors, processors, and actuators. Sensors detect specific biochemical or physical signals and translate them into transcriptional or translational outputs. Processors, implemented via logic gates, feedback loops, or integrative modules, interpret these input signals to determine the appropriate cellular response. Finally, actuators convert processed signals into functional outputs such as enzyme production, fluorescent reporter expression, or metabolic changes. One hallmark of advanced synthetic circuits is their ability to implement Boolean logic, allowing cells to perform computations akin to electronic devices; these logic gates include AND, OR, NOT, and more complex combinational structures that enable conditional decision-making based on multiple inputs. Feedback mechanisms, both positive and negative, confer bistability, oscillations, and noise resistance, imparting robustness and dynamic control often critical for therapeutic or industrial applications.

4.3.3 Integration with Native Gene Networks

Initially, synthetic circuits were designed to operate as autonomous entities, independent from endogenous cellular machinery. However, contemporary approaches increasingly emphasize integration with host regulatory networks to harness cellular resources effectively and achieve context-dependent control. The interplay between synthetic and natural networks requires careful tuning to avoid deleterious crosstalk, burden, or instability. Advances in orthogonal gene expression systems—derived from non-host organisms or synthetically recoded genetic elements—aid in minimizing host interference. Moreover, dynamic interfaces allow synthetic circuits to modulate native pathways, reprogram cellular states, or respond adaptively to environmental or developmental cues, enhancing their functional versatility. Such hybrid designs offer refined control suitable for therapeutic cell engineering, metabolic regulation, and complex biosensing.

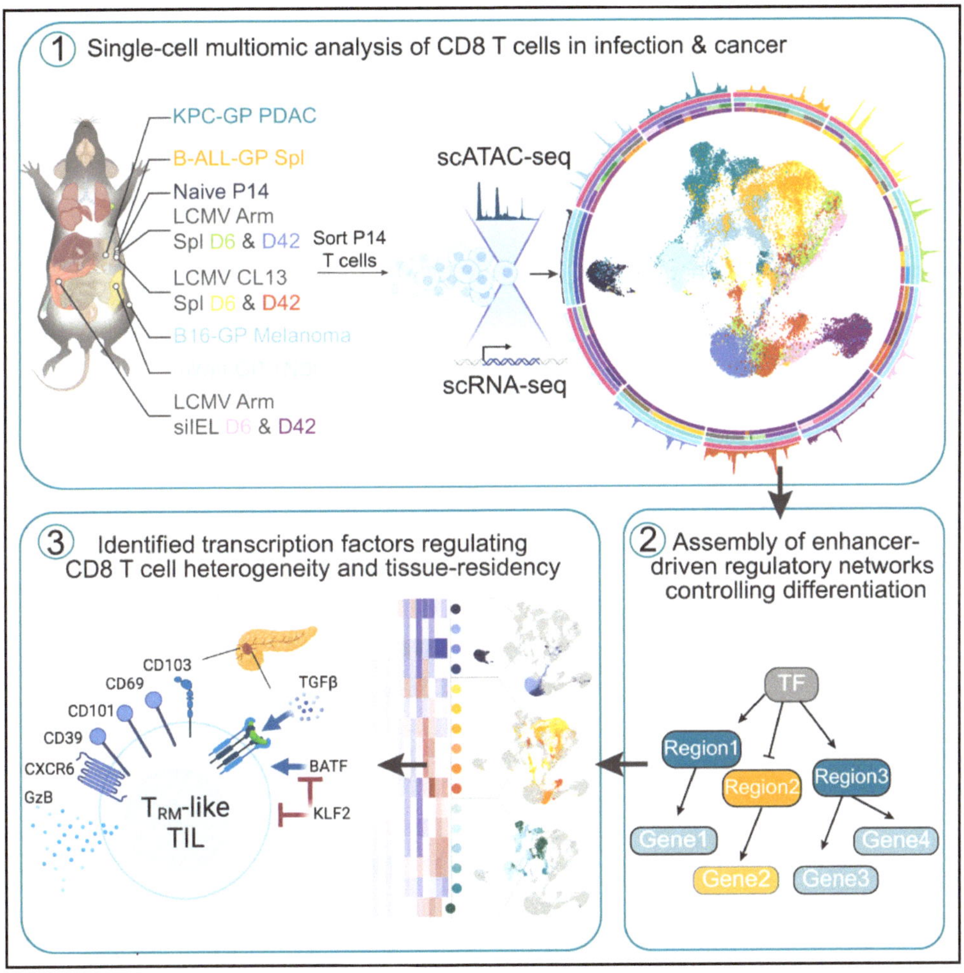

Genomic Networks in Native Context

4.3.4 Functional Behaviors: Memory, Oscillations, and Logic

Synthetic gene networks demonstrate a rich repertoire of dynamic behaviors essential for biological computation and control. Bistable switches enable cells to record past stimuli or toggle between phenotypic states, serving as cellular memory devices useful in differentiation or therapeutic contexts. Oscillatory circuits generate rhythmic gene expression or metabolic activity, emulating biological clocks and enabling time-based regulation of cellular functions. Multi-input logic circuits empower cells to execute complex decisions, responding only when specified combinations of signals are present, which is critical in distinguishing nuanced physiological conditions in biosensing or targeted therapy. Signal amplification and noise filtering are also engineered features enhancing sensitivity and specificity in cellular responses, improving the reliability of synthetic circuits in noisy biological environments.

4.3.5 Applications in Biotechnology and Medicine

The practical applications of synthetic circuits span biosensing, bioproduction, cellular therapy, and environmental engineering. In biotechnological settings, synthetic circuits regulate metabolic pathways achieving high-yield biosynthesis of chemicals, pharmaceuticals, and biofuels by dynamically adjusting enzyme expression in response to metabolic states. In medical biotechnology, engineered immune cells armed with synthetic gene circuits execute logic-gated attack on tumors, secreting therapeutic molecules only in the presence of cancer-specific markers, reducing off-target effects and improving safety. Synthetic circuits also enable diagnostic cells that detect disease biomarkers in vivo, reporting infection or pathological states through measurable outputs. Furthermore, circuits controlling differentiation promote regenerative medicine by guiding stem cell fate decisions precisely in temporal and spatial dimensions.

4.3.6 Challenges in Design and Implementation

Despite significant advances, constructing functional synthetic circuits faces challenges stemming from biological complexity and variability. Context-dependence and host-cell heterogeneity often cause circuit behavior to deviate from idealized designs. Unintended interactions with endogenous pathways, metabolic load, and cellular stress further complicate predictability and stability. Circuit components may suffer from mutation, epigenetic silencing, or degradation, reducing performance over time. Engineering orthogonal parts with minimal dependence on cellular context remains an ongoing endeavor. Computational modeling and machine learning help to predict circuit dynamics and optimize parameter spaces, but experimental iterations remain critical. Advances in high-throughput screening, directed evolution, and genome editing tools contribute to refining and robustifying synthetic circuits.

4.4 CRISPR AND GENE EDITING FOR SYNTHETIC BIOLOGY APPLICATIONS

CRISPR-Cas gene editing has transformed synthetic biology from a discipline focused on theory to one capable of practical, real-world applications. By providing a precise, efficient, and versatile platform for genetic manipulation, CRISPR has accelerated the design and construction of biological systems that address critical challenges in health, agriculture, energy, and the environment. From engineering microbes to produce life-saving drugs and sustainable materials to creating crops that feed a growing population and therapies that cure genetic diseases, the impact of CRISPR is profound and far-reaching. While challenges related to safety, delivery, ethics, and regulation must be navigated carefully, the ongoing innovations in CRISPR technology continue to push the boundaries of what is possible. As we harness the power of this molecular scalpel, we are not just editing genes; we are rewriting the future of biology, paving the way for a more sustainable, healthy, and technologically advanced world. The CRISPR revolution is well underway, and its potential to reshape life as we know it is only beginning to be realized.

4.4.1 CRISPR: A Molecular Scalpel

CRISPR-Cas (Clustered Regularly Interspaced Short Palindromic Repeats and CRISPR-associated proteins) is an adaptive immune system found in bacteria and archaea, which protects them against invading viruses (bacteriophages) and plasmids. This natural system has been harnessed and adapted for genome editing in a wide range of organisms. The core components of the most commonly used CRISPR system, CRISPR-Cas9, include a guide RNA (gRNA) and the Cas9 enzyme. The gRNA is a synthetic RNA molecule designed to be complementary to a specific DNA sequence within the genome. It directs the Cas9 enzyme to the exact location where the edit is desired.

Cas9 acts as a molecular scalpel: upon binding to the DNA at the target site (guided by the gRNA), it cuts both strands of the DNA double helix, creating a double-strand break (DSB). This break triggers the cell's natural DNA repair mechanisms. The two primary repair pathways are non-homologous end joining (NHEJ), which often results in small insertions or deletions (indels) that can disrupt gene function (knockout), and homology-directed repair (HDR), which uses a provided DNA template to repair the break with high fidelity, allowing for precise insertions or modifications (knock-in). The simplicity and programmability of the gRNA component make CRISPR exceptionally versatile, as only the gRNA sequence needs to be changed to target a different genomic location.

4.4.2 Precision Engineering: How CRISPR Enables Genetic Editing

CRISPR-Cas9 has revolutionized genetic editing by providing a straightforward, efficient, and relatively inexpensive method for making precise changes to DNA. Its applications extend far beyond simple knockouts and knock-ins. Base editing is a refinement that allows for direct, irreversible conversion of one base pair to another without requiring a DSB or a DNA template. This is achieved by fusing a catalytically impaired Cas9 (nickase) to a base-modifying enzyme, such as a cytidine deaminase or adenosine deaminase. Base editors can correct point mutations associated with genetic diseases or introduce specific mutations for research purposes. Prime editing represents another leap forward, offering even greater precision and flexibility. It uses a Cas9 nickase fused to a reverse transcriptase and a specialized prime editing guide RNA (pegRNA). The pegRNA both specifies the target site and contains the desired edit sequence. Prime editing can perform all 12 possible base-to-base conversions, small insertions, and small deletions with high efficiency and minimal byproducts, making it ideal for correcting complex mutations. Epigenome editing utilizes catalytically dead Cas9 (dCas9) fused to epigenetic modifiers, such as DNA methyltransferases or histone acetyltransferases. This allows researchers to turn gene expression on or off by altering epigenetic marks without changing the underlying DNA sequence, providing powerful insights into gene regulation. Furthermore, multiplexed editing is a hallmark of CRISPR, enabling the simultaneous targeting of multiple genes in a single experiment by introducing multiple gRNAs. This capability is crucial for synthetic biology, where complex genetic circuits often require the coordinated manipulation of several genes.

4.4.3 Building Biological Systems: CRISPR in Synthetic Biology Applications

CRISPR-Cas9 has become an indispensable tool for synthetic biologists, enabling the rapid prototyping, construction, and optimization of genetic parts and systems. Its precision and multiplexing capabilities allow for the creation of increasingly sophisticated biological devices.

Metabolic Engineering and Biomanufacturing

One of the most impactful applications is in engineering microbial cell factories for the sustainable production of chemicals, fuels, and materials. CRISPR enables the precise knockout of competing metabolic pathways, the knock-in of heterologous genes from other organisms to introduce novel functions, and the fine-tuning of gene expression levels. For example, scientists have used CRISPR to engineer E. coli and yeast strains to produce biofuels like isobutanol and isoprenoids, bioplastics like PHA, and high-value pharmaceuticals such as artemisinin (an antimalarial drug) and insulin. By multiplexing edits, complex pathways can be constructed and optimized in a single step, significantly accelerating the design-build-test-learn cycle.

Sustainable Solutions: Engineering Microbes for a Greener Future

CRISPR is pivotal in developing solutions for environmental sustainability. Engineered microbes can convert renewable feedstocks (e.g., lignocellulosic biomass, CO2) into valuable products. CRISPR has been used to enhance the efficiency of enzymes in cellulose degradation, enabling better conversion of plant waste into fermentable sugars for biofuel production. It has also been applied to cyanobacteria and algae to improve their photosynthetic efficiency and lipid production for biodiesel. Furthermore, CRISPR-edited microbes are being developed for bioremediation, such as bacteria that can degrade persistent pollutants like plastics and chlorinated solvents.

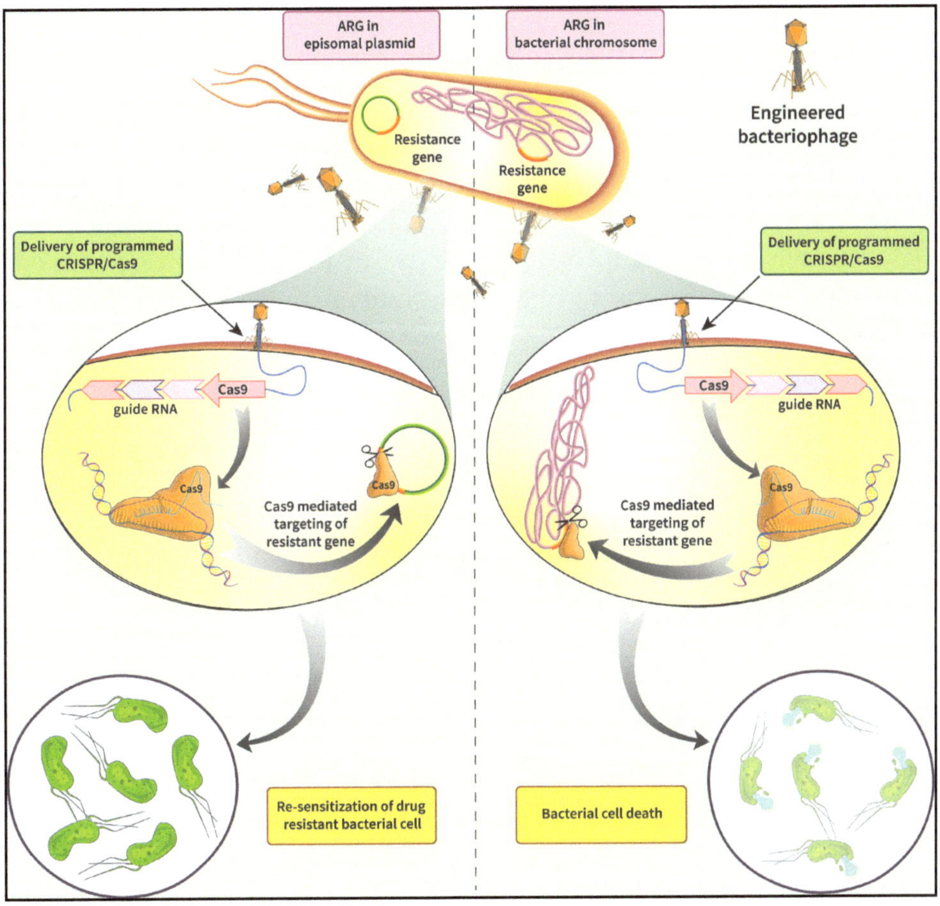

CRISPR-Cas Systems and Anti-CRISPR Proteins in Bacterial Antibiotic Resistance

Genetic Circuits and Biosensors: Programming Cellular Logic

Synthetic biologists design genetic circuits—networks of genes and regulatory elements—that perform logical operations within cells, akin to electronic circuits. CRISPR-Cas systems, particularly using dCas9 for transcriptional repression (CRISPRi) or activation (CRISPRa), provide a powerful platform for building these circuits. dCas9 can be targeted to specific promoter regions to block or activate transcription without cutting DNA, allowing for tunable and reversible control. This has enabled the construction of biosensors that detect environmental toxins, pathogens, or disease biomarkers and respond by producing a signal (e.g., fluorescence) or a therapeutic agent. Genetic circuits for population control, oscillators, and logic gates have been implemented using CRISPR-based regulation.

Revolutionizing Agriculture: CRISPR in Crop Improvement

Agriculture faces immense challenges from climate change, pests, and diseases. CRISPR-Cas9 is accelerating crop breeding by allowing precise edits to plant genomes. Unlike traditional genetic modification, CRISPR can create mutations that mimic natural variations, often without introducing foreign DNA, resulting in non-transgenic plants that are more readily accepted.

Examples include creating disease-resistant varieties (e.g., wheat resistant to powdery mildew, rice resistant to bacterial blight), developing crops with improved nutritional profiles (e.g., high-oleic soybeans, vitamin-A-enriched rice), and engineering traits for climate resilience, such as drought tolerance and salinity resistance. CRISPR is also used to edit livestock genomes for disease resistance (e.g., pigs resistant to PRRS virus) and hornless dairy cows, improving animal welfare.

Biomedical Frontiers: Gene Therapy and Disease Modeling

In medicine, CRISPR holds transformative potential for gene therapy—curing genetic diseases by correcting mutations at their source. Clinical trials are underway for conditions like sickle cell disease, beta-thalassemia, and certain inherited blindness disorders, using CRISPR to edit patient cells ex vivo or in vivo. CRISPR is also revolutionizing disease modeling and drug discovery. Researchers can create cellular and animal models with specific mutations to study disease mechanisms and test therapeutics. CRISPR screens (both loss-of-function and gain-of-function) allow for high-throughput identification of genes involved in diseases, revealing new drug targets.

4.4.4 CRISPR: Advantages Driving Adoption

CRISPR-Cas9 has rapidly become the dominant gene-editing technology due to several key advantages:

- **Unprecedented Precision:** It allows targeting of specific DNA sequences with high accuracy, minimizing off-target effects compared to older methods like ZFNs and TALENs.
- **High Efficiency:** The simplicity of the system enables efficient editing in a wide variety of cell types and organisms, including previously recalcitrant ones.
- **Multiplexing Capability:** Multiple genes can be edited simultaneously in a single experiment, which is essential for complex synthetic biology projects.
- **Cost-Effectiveness:** The design and synthesis of gRNAs are relatively inexpensive, making CRISPR accessible to most laboratories.

- **Speed and Versatility:** The system can be adapted quickly for new targets, and the development of base and prime editing has expanded its range of applications.
- **Minimal Footprint:** Edits can be made without leaving behind foreign DNA, facilitating the generation of transgene-free organisms.

These advantages have democratized genetic engineering, accelerating research and development across the life sciences.

4.4.5 Navigating the Challenges: Hurdles and Ethical Debates

Despite its power, CRISPR faces significant challenges and ethical concerns:

- **Off-Target Effects:** Cas9 can sometimes cut DNA at unintended sites, leading to mutations that could have harmful consequences, especially in therapeutic applications. Ongoing research focuses on improving specificity through engineered Cas variants (e.g., high-fidelity Cas9) and optimized gRNA design.
- **Delivery Difficulties:** Efficiently delivering CRISPR components (Cas9 and gRNA) into target cells, particularly in vivo, remains a major hurdle. Viral vectors (e.g., AAV) are common but have size limitations and immunogenicity risks. Non-viral delivery methods (e.g., nanoparticles) are being developed but need improvement.
- **HDR Efficiency:** Precise edits via HDR are inefficient in many cell types, especially non-dividing cells, compared to error-prone NHEJ. Strategies to bias repair towards HDR or use base/prime editing are helping to overcome this.
- **Regulatory and Biosafety Issues:** The regulatory landscape for CRISPR-edited organisms varies globally, creating uncertainty. In agriculture, edited crops face differing regulations; some countries exempt them from GMO rules if no foreign DNA is present, while others regulate them strictly. For human therapy, rigorous safety testing and long-term monitoring are essential.
- **Ethical Concerns:** The most profound ethical debates surround human germline editing—editing sperm, eggs, or embryos to make heritable changes. This raises concerns about unintended consequences for future generations, eugenics, and social inequality. The 2018 case of He Jiankui creating CRISPR-edited babies sparked international condemnation and calls for moratoria. Most countries ban germline editing for reproduction, but discussions continue on its potential for preventing severe genetic diseases.

4.4.6 Next Frontier: Innovations and Future Directions

CRISPR technology is rapidly evolving, with continuous innovations expanding its capabilities:

- **Advanced Editors:** Base and prime editors are being refined for higher efficiency, fewer byproducts, and broader targeting. Dual-function editors that combine base editing with transcriptional regulation are emerging.
- **Epigenome Editing:** Tools for precise epigenome editing are becoming more sophisticated, enabling targeted control of gene expression without altering DNA sequence, with applications in development and disease.
- **CRISPR Activation and Repression:** CRISPRi and CRISPRa are being enhanced with more potent activators/repressors and used to build complex regulatory networks.
- **AI and Machine Learning:** Integration of AI for gRNA design, predicting off-target sites, and optimizing editing outcomes is improving accuracy and efficiency.
- **In Vivo Delivery:** Advances in lipid nanoparticles (LNPs) and other non-viral delivery systems are making in vivo editing more feasible for therapeutic applications.
- **Synthetic Gene Drives:** CRISPR-based gene drives, which bias inheritance to spread edits through populations, are being explored for controlling pests and vector-borne diseases but raise ecological concerns.
- **Cell-Free Systems:** CRISPR components are being used in cell-free systems for rapid prototyping of genetic parts and biosensors.

These innovations promise to overcome current limitations and unlock new applications, from curing genetic diseases to creating climate-resilient crops and sustainable biomaterials.

4.5 COMPUTATIONAL TOOLS AND MODELING FOR SYNTHETIC BIOLOGY

4.5.1 System Biology and Modeling Approaches

Systems biology represents a transformative approach to understanding life by integrating experimental data with computational modeling to study biological systems as a whole. By combining principles of dynamics, robustness, and modularity, systems biology enables the prediction of system behavior, identification of regulatory nodes, and optimization of biological functions. Its modeling approaches—from deterministic and stochastic models to agent-based and multi-scale simulations—provide versatile tools to address diverse biological questions. Applications in medicine, biotechnology, and synthetic biology demonstrate the field's capacity to accelerate drug discovery, enhance personalized therapies, and optimize bioproduction systems. Despite challenges in data integration, computational complexity, and model validation, ongoing advances in omics technologies, artificial intelligence, and high-throughput experimentation continue to propel systems biology forward. As this field matures, it will remain indispensable for decoding the complexity of life, guiding experimental design, and translating systems-level insights into practical solutions for health, industry, and the environment.

Concept and Scope of Systems Biology

Systems biology is defined as the study of biological systems as a whole, emphasizing the relationships and interactions between components rather than isolated functions. The primary goal is to understand how molecular networks, cellular pathways, and physiological systems work together to maintain homeostasis or respond to environmental changes. The scope of systems biology extends from molecular networks at the cellular level to organ systems and even population-level interactions. By employing both experimental and computational methods, systems biology seeks to identify patterns, regulatory mechanisms, and emergent properties that cannot be discerned from studying individual elements alone. This holistic perspective is critical for deciphering the complexity of diseases, developing predictive models, and designing novel therapeutic interventions.

Importance of Systems Biology

The importance of systems biology lies in its ability to integrate vast amounts of heterogeneous biological data into coherent models that can predict system behavior. In medicine, systems biology provides insights into complex diseases such as cancer, diabetes, and neurodegenerative disorders by mapping interactions between genes, proteins, and metabolites. It enables the identification of key regulatory nodes, potential drug targets, and biomarkers for early diagnosis. In biotechnology, systems biology assists in optimizing metabolic pathways for the production of biofuels, pharmaceuticals, and industrial enzymes. Environmental and ecological studies also benefit, as systems-level analysis helps understand microbial communities, biogeochemical cycles, and ecosystem dynamics. Overall, systems biology transforms biology from a descriptive science into a predictive and engineering-driven discipline.

Fundamental Principles

Several fundamental principles underlie systems biology. Integration is central, combining experimental data from genomics, transcriptomics, proteomics, and metabolomics into unified networks. Dynamics emphasizes that biological systems are constantly changing in response to internal and external stimuli, necessitating time-dependent modeling.

Nonlinearity recognizes that interactions among biological components often lead to emergent behaviors that are not directly proportional to individual effects. Robustness is another key principle, highlighting the capacity of systems to maintain function despite perturbations or environmental fluctuations. Finally, modularity facilitates the decomposition of complex networks into functional units, making modeling and analysis more tractable. These principles collectively guide the development of computational models and experimental designs in systems biology.

Systems Biology Data Sources

Systems biology relies heavily on high-throughput experimental data. Modern technologies allow researchers to generate massive datasets capturing different layers of biological information. Genomic sequencing provides comprehensive maps of DNA sequences and structural variations. Transcriptomics quantifies gene expression patterns under varying conditions, revealing regulatory dynamics.

Proteomics measures protein abundance, post-translational modifications, and interactions, while metabolomics characterizes small-molecule metabolites within cells and tissues. Interactomics maps physical and functional interactions between proteins, nucleic acids, and other molecules. In addition, phenomics and imaging data provide insight into cellular and tissue-level behaviors. The`integration of these diverse datasets is essential for building accurate, predictive models of biological systems.

Modeling Approaches in Systems Biology

Computational modeling is at the core of systems biology, providing a framework to simulate and predict the behavior of complex systems. Several modeling approaches are commonly used:

- Deterministic models based on ordinary differential equations (ODEs) describe continuous changes in molecular concentrations over time.
- Stochastic models account for random fluctuations and noise in biological processes, which are critical at low molecule numbers.
- Boolean networks simplify gene regulatory or signaling networks by representing components as binary states (on/off).
- Bayesian networks provide probabilistic models to infer causal relationships among genes or proteins.

- Agent-based models simulate interactions of individual agents (cells or molecules) to observe emergent behavior at the population level.
- Constraint-based models, such as flux balance analysis (FBA), optimize metabolic network fluxes based on stoichiometric constraints.
- Hybrid models integrate deterministic and stochastic approaches to capture multiple scales of biological processes.
- Spatial models account for compartmentalization and localization effects within cells or tissues.
- Machine learning-based models detect patterns and predict system behavior using data-driven approaches.
- Multi-scale models combine molecular, cellular, and tissue-level dynamics for comprehensive system simulation.

Each modeling approach has specific advantages and limitations, and the choice depends on the system under study, data availability, and desired predictive accuracy.

Applications in Medicine

Systems biology has revolutionized the field of medicine by enabling the identification of disease mechanisms and potential therapeutic targets. For example, in cancer research, systems-level models help elucidate signaling pathways responsible for uncontrolled proliferation and resistance to therapy. Predictive models can identify key nodes in these pathways that serve as drug targets or combination therapy candidates. In personalized medicine, systems biology integrates genomic, transcriptomic, and proteomic data to tailor treatments to individual patients, improving efficacy and reducing adverse effects. Additionally, systems pharmacology uses network models to predict drug interactions, off-target effects, and optimal dosing strategies. The application of systems biology in understanding infectious diseases, metabolic disorders, and neurodegeneration is also expanding rapidly, highlighting its critical role in advancing healthcare.

Applications in Biotechnology and Synthetic Biology

In biotechnology, systems biology provides the foundation for rational design and optimization of metabolic pathways. By mapping cellular networks, researchers can identify bottlenecks and regulatory constraints that limit production of biofuels, pharmaceuticals, or industrial enzymes. Systems-level modeling enables the prediction of gene knockouts, overexpression targets, or pathway rerouting strategies to maximize yield. In synthetic biology, systems biology informs the design of synthetic genetic circuits, biosensors, and engineered microbes. High-throughput modeling and simulation reduce trial-and-error experimentation, accelerating the development of efficient and robust bioproduction systems. Moreover, systems biology supports the integration of synthetic circuits into host organisms while minimizing unintended interactions and metabolic burden.

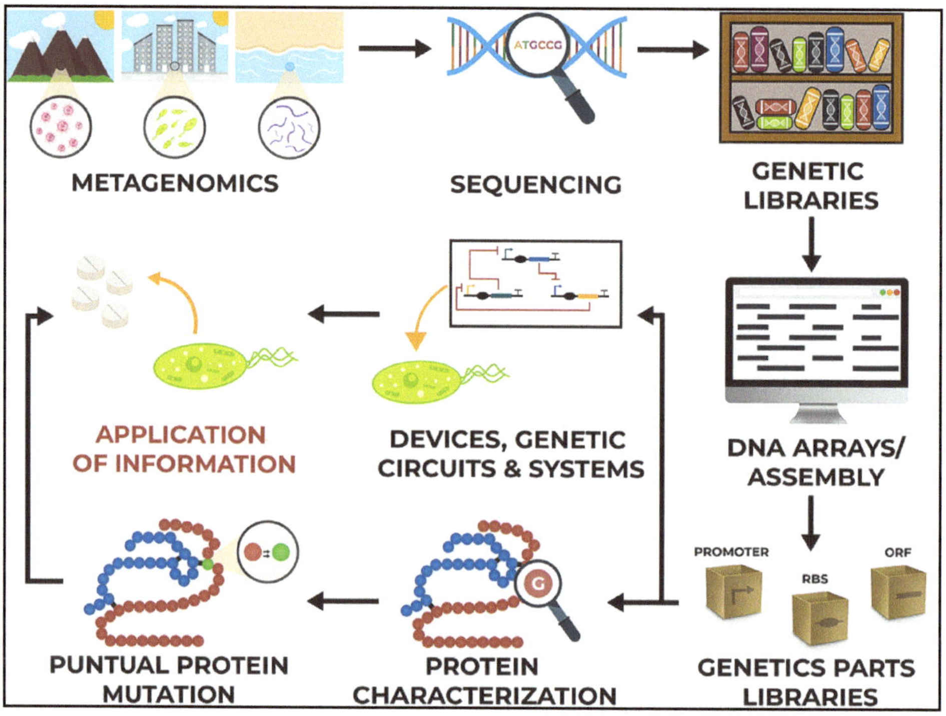

Developments in Synthetic Biology and Biotechnology

Challenges and Limitations

Despite its promise, systems biology faces several challenges. Biological systems are inherently complex and dynamic, making accurate modeling difficult. High-dimensional datasets often contain noise, missing values, or inconsistencies, complicating data integration. Parameter estimation in computational models is challenging due to limited experimental measurements and inherent variability. Multi-scale modeling requires extensive computational resources and sophisticated algorithms to capture dynamics across molecular, cellular, and tissue levels. Furthermore, standardization of data formats, modeling frameworks, and annotation practices remains incomplete, hindering reproducibility and collaboration. Finally, interpreting model predictions and translating them into experimental or clinical interventions requires careful validation, as computational models are approximations of real biological systems.

4.5.2 Predictive Design and Simulation Software

Predictive design and simulation software has revolutionized the way scientists and engineers approach complex problems, enabling virtual experimentation, optimization, and performance evaluation across diverse fields. By combining computational modeling, visualization, and algorithmic analysis, these tools allow accurate predictions of system behavior, reducing the need for costly and time-consuming physical prototypes. Applications span mechanical and civil engineering, biotechnology, electronics, environmental modeling, and energy systems, demonstrating the versatility and transformative potential of predictive software. Integration with machine learning, AI, and high-performance computing further enhances predictive capabilities, enabling intelligent, data-driven design and decision-making. Despite challenges related to data quality, computational demands, and model validation, continued advancements are expanding the scope, reliability, and accessibility of predictive design tools. As industries and research sectors increasingly embrace virtual modeling and simulation, predictive software will continue to drive innovation, optimize performance, and reduce risk, cementing its role as a critical enabler of modern science and technology.

Foundations of Predictive Design

At its core, predictive design relies on understanding the principles and behavior of a system and translating them into computational models. These models are mathematical representations of physical, chemical, or biological processes, capturing essential variables and interactions that govern system performance. For example, in mechanical engineering, predictive design may model stress distribution in a structural component, fluid flow in a pipeline, or thermal behavior in an engine. In biological systems, simulation software can model metabolic pathways, gene regulatory networks, or drug interactions. Accurate predictive design depends on high-quality data, robust algorithms, and the ability to account for real-world variability. The objective is to anticipate system responses to different inputs, environmental conditions, and constraints, thereby enabling informed decisions about design, optimization, and implementation.

Types of Simulation Software

Predictive simulation software can be broadly categorized based on the type of analysis performed and the systems they model. Finite element analysis (FEA) software is widely used in mechanical and structural engineering to simulate stress, strain, vibration, and deformation in materials and components. Computational fluid dynamics (CFD) software models fluid flow, heat transfer, and mass transport, providing insights into aerodynamics, hydraulics, and chemical mixing processes. Multiphysics simulation platforms integrate different physical phenomena, such as thermal, mechanical, and electrical interactions, into a unified model, allowing comprehensive evaluation of complex systems. In electronics and electrical engineering, circuit simulation software predicts signal behavior, power distribution, and electromagnetic interference. In the realm of biology and biotechnology, predictive software simulates metabolic pathways, synthetic genetic circuits, and cellular behavior. These diverse simulation tools provide domain-specific capabilities while sharing the common goal of virtual experimentation and predictive analysis.

Principles of Predictive Modeling

Predictive modeling within simulation software relies on several fundamental principles. First, the system of interest is abstracted into a mathematical model, often expressed as a set of equations that describe the relationships between variables. These equations can include linear or nonlinear dynamics, differential equations, stochastic processes, or agent-based models. Second, initial conditions and boundary conditions are defined to reflect the real-world environment in which the system operates. Third, computational algorithms solve these equations iteratively, simulating system behavior over time or under varying input parameters. High-fidelity simulations may also incorporate uncertainty analysis, sensitivity analysis, and optimization algorithms to assess the robustness and reliability of predictions. The integration of artificial intelligence and machine learning into predictive design further enhances accuracy by allowing models to learn from historical data and adapt to complex, nonlinear behaviors that traditional methods may not capture effectively.

Applications in Engineering and Product Design

In engineering and product development, predictive design software is essential for optimizing performance, safety, and cost-effectiveness. For example, automotive engineers use FEA and CFD to simulate crashworthiness, aerodynamics, and thermal management in vehicles, reducing the need for multiple physical prototypes. Aerospace engineers employ simulation to model fluid dynamics, structural loads, and propulsion efficiency in aircraft and spacecraft. Civil engineers use predictive software to design earthquake-resistant buildings, optimize load-bearing structures, and evaluate the impact of environmental factors on infrastructure. Predictive modeling allows designers to explore alternative materials, geometries, and configurations virtually, accelerating the innovation cycle while minimizing risks and expenses. By providing quantitative insights, simulation software ensures that final products meet regulatory standards, performance criteria, and customer expectations.

Applications in Biotechnology and Life Sciences

In biotechnology, predictive design and simulation software has transformed the way researchers approach drug development, genetic engineering, and synthetic biology. Computational models can simulate metabolic networks, predict enzyme kinetics, and analyze gene regulatory circuits, enabling the rational design of engineered organisms. Drug discovery benefits from predictive modeling by simulating molecular interactions, protein-ligand binding, and pharmacokinetics, reducing the need for extensive laboratory testing. Synthetic biology relies on simulation software to predict the behavior of genetic constructs, assess circuit stability, and optimize metabolic pathways for biofuel production, pharmaceuticals, or biomaterials. These tools allow researchers to anticipate potential bottlenecks, identify optimal design strategies, and improve the efficiency and success rate of experimental workflows. The integration of predictive software in life sciences accelerates translational research and supports the development of innovative therapies and bioproducts.

Environmental and Energy Applications

Predictive design and simulation software plays a critical role in environmental engineering and renewable energy development. Environmental models simulate pollutant dispersion, water quality, climate patterns, and ecosystem dynamics, providing data-driven insights for policy and management decisions. In renewable energy, simulation software evaluates the performance of solar panels, wind turbines, hydropower systems, and bioenergy facilities under varying environmental conditions. Predictive modeling enables optimization of energy generation, efficiency improvements, and the design of resilient energy infrastructure. By integrating environmental, economic, and operational factors, simulation tools facilitate sustainable design strategies and support decision-making for complex, multidisciplinary challenges.

Optimization and Performance Evaluation

A key advantage of predictive design software is its ability to optimize system performance before physical implementation. Optimization involves adjusting design parameters to achieve desired objectives, such as maximizing efficiency, minimizing cost, or reducing environmental impact. For example, in aerospace engineering, optimization algorithms can identify the optimal wing shape and material composition to balance lift, drag, and structural integrity. In biotechnology, metabolic pathway simulations can determine the combination of enzyme expression levels that maximizes product yield while minimizing byproduct formation. Simulation software allows for parametric studies, scenario analysis, and sensitivity testing, helping designers understand trade-offs and identify robust solutions. By providing quantitative evaluations, predictive tools reduce uncertainty, improve decision-making, and enhance the overall quality of design outcomes.

Integration with Machine Learning and Artificial Intelligence

The integration of machine learning (ML) and artificial intelligence (AI) has significantly enhanced the predictive capabilities of simulation software. ML algorithms can analyze vast datasets to identify patterns, correlations, and predictive features that traditional models might overlook. In predictive design, AI-driven software can generate optimized designs autonomously, predict system failures, and recommend modifications based on performance criteria.

For instance, in structural engineering, AI models can forecast fatigue failure or corrosion rates, while in biotechnology, machine learning can predict enzyme efficiency or gene expression patterns. This synergy between AI and simulation not only accelerates the design process but also improves accuracy, adaptability, and predictive power, making modern simulation tools more intelligent and versatile.

User Interfaces and Visualization

Effective predictive design relies not only on computational power but also on intuitive user interfaces and visualization capabilities. Modern simulation software provides interactive graphical interfaces that allow users to define geometries, set parameters, and monitor simulations in real time. Visualization tools enable the interpretation of complex datasets through 2D and 3D representations, animations, and heat maps, facilitating the understanding of system behavior. These features are particularly valuable in multidisciplinary projects where collaboration among engineers, biologists, and decision-makers is essential. By providing clear and interpretable outputs, predictive design software bridges the gap between complex mathematical models and actionable insights, enabling informed decisions across diverse applications.

Table: Examples of Predictive Design and Simulation Software

Domain	Software Examples	Primary Applications
Mechanical Engineering	ANSYS, SolidWorks Simulation	Stress analysis, thermal modeling, product optimization
Fluid Dynamics	COMSOL Multiphysics, Fluent	Aerodynamics, hydrodynamics, heat transfer
Biotechnology	COPASI, CellDesigner, MATLAB SimBiology	Metabolic pathway modeling, gene circuit simulation, drug discovery
Environmental Modeling	STELLA, OpenFOAM	Pollution dispersion, climate modeling, water resource management
Electronics	SPICE, Multisim	Circuit design, signal integrity, power distribution

Challenges and Limitations

Despite their transformative potential, predictive design and simulation software face several challenges. Accurate simulations require high-quality data, precise model parameters, and realistic assumptions; errors or oversimplifications can lead to unreliable predictions. Computational demands may be high, particularly for multiphysics or large-scale systems, necessitating access to high-performance computing resources. Model validation remains critical, as simulation results must be corroborated against experimental or empirical data. Additionally, the complexity of software tools and steep learning curves can limit accessibility for non-specialist users. Addressing these challenges requires continued advancements in algorithm efficiency, user-friendly interfaces, data integration, and model validation protocols to ensure reliability and applicability in real-world scenarios.

4.6 SYNTHETIC BIOLOGY IN MICROBIAL ENGINEERING

4.6.1 Engineering Microbes for Production and Sensing

Engineering microbes for production and sensing exemplifies the powerful synergy of biology, engineering, and computation, enabling living systems to perform tailored biosynthesis and dynamic environmental monitoring. By reprogramming metabolic pathways and incorporating sophisticated biosensing circuits, engineered microbes achieve efficiency, adaptability, and precision in diverse applications spanning medicine, industry, agriculture, and environmental science. Although challenges persist in stability, scalability, and safety, continual technological innovations and interdisciplinary integration poise microbial engineering to play a pivotal role in sustainable biomanufacturing and smart biosensing solutions, ultimately addressing global challenges related to health, energy, and environment.

Metabolic Engineering for Enhanced Production

Central to microbial production is metabolic engineering, which involves manipulating native and heterologous metabolic pathways to channel resources towards desired products. This approach necessitates a detailed understanding of host metabolism, enabling genetic modifications such as gene overexpression, knockouts, or the introduction of novel biosynthetic modules.

Techniques such as CRISPR-Cas9 genome editing, synthetic promoters, and ribosome binding site tuning offer precision tools to control gene expression and enzymatic activity. Complementary adaptive laboratory evolution efforts select for strains with improved growth and product formation under industry-relevant stresses, such as tolerance to toxic intermediates or substrate impurities. Balancing precursor supply, cofactor availability, and energy consumption allows microbial cells to achieve high titers, rates, and yields needed for commercial viability.

Biosensing Microbes and Whole-Cell Biosensors

Parallel to production, engineered microbes equipped with biosensing capabilities form living biosensors that detect and report on chemical signals, environmental pollutants, disease biomarkers, or metabolic states. Biosensor designs leverage natural sensory proteins or synthetic gene circuits that recognize target molecules and trigger reporter outputs like fluorescence, luminescence, or metabolic changes. Two-component regulatory systems, riboswitches, and transcription factor-based sensors constitute versatile modules that can be repurposed or engineered to broaden sensing abilities. Microbial whole-cell biosensors (MWCBs) find applications in environmental monitoring—detecting contaminants in water and soil—industrial process control, and clinical diagnostics. Multiplexed biosensing and signal integration enable microbes to process combinatorial biochemical information, thereby enhancing selectivity and robustness.

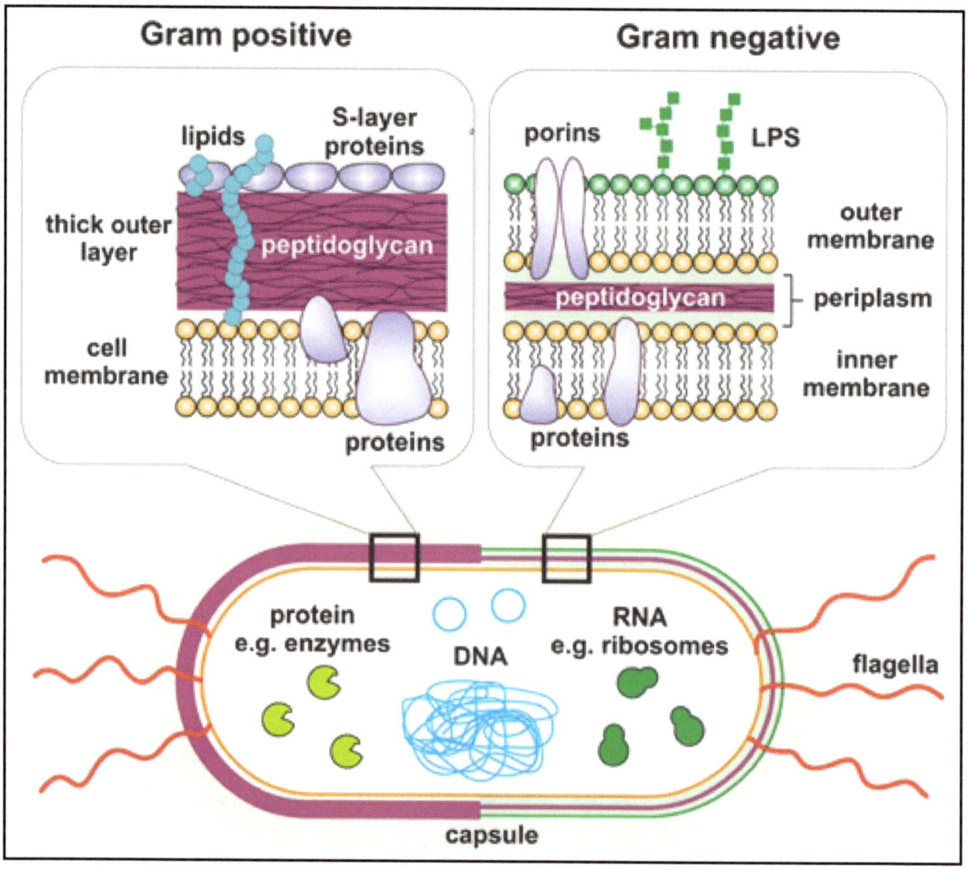

Biosensors for Whole-Cell Bacterial Detection

Integration of Production and Sensing Functions

The frontier of microbial engineering integrates production and sensing into closed-loop systems that autonomously regulate metabolic flux based on intracellular or extracellular cues. Feedback-controlled gene circuits dynamically adjust enzyme expression levels or pathway activity in response to accumulated intermediates or environmental stresses, reducing toxic buildup and metabolic burden. This adaptive regulation increases process stability, decreases resource waste, and improves product quality. Moreover, synthetic quorum sensing systems confer population-level coordination, synchronizing biosynthesis or biosensing activities across microbial communities. Distributed metabolic specialization and communication in microbial consortia exemplify strategies to overcome limitations of single strains by leveraging division of labor and community resilience.

Applications across Sectors

Engineered microbes for production and sensing have application breadth encompassing pharmaceuticals, biofuels, agriculture, environmental remediation, and food industries. Therapeutic compounds like antibiotics, anticancer agents, and insulin analogs are produced at scale in engineered bacteria or yeast, benefiting from optimized biosynthetic routes and real-time sensing of pathway metrics. Biofuel production utilizes microbes converting renewable feedstocks into ethanol, biodiesel precursors, or biogas, with sensing circuits monitoring culture health and substrate utilization. In agriculture, biosensing microbes detect soil nutrient status or pathogen presence, and biosynthetic strains produce biopesticides or growth-promoting compounds on demand. Environmental cleanup leverages microbial biosensors for pollutant detection coupled with biodegradation pathways, enabling smart remediation. Food fermentation and flavor production employ engineered strains that coordinate metabolite synthesis with environmental conditions, improving product consistency and safety.

Challenges in Microbial Engineering

Despite advancements, engineering microbes for dual production and sensing roles faces several challenges. Cellular complexity and physiological variability introduce unpredictability in circuit function and metabolic efficiencies. Off-target effects, metabolic burden, and genetic instability can compromise long-term performance. Achieving orthogonality between synthetic modules and native host pathways requires extensive characterization and minimization of crosstalk. Scaling microbial systems from controlled laboratory environments to industrial reactors involves hurdles including mass transfer limitations, stress tolerance, and phenotypic heterogeneity. Biosafety concerns necessitate stringent containment and kill-switch mechanisms to prevent accidental environmental release or horizontal gene transfer. Addressing these challenges demands iterative design-build-test-learn cycles, coupling experimental insights with computational modeling and machine learning.

4.6.2 Synthetic Yeasts and Bacteria

Synthetic yeasts and bacteria are far more than scientific curiosities; they are the vanguard of a new biological age. By reprogramming the fundamental code of life, scientists are unlocking unprecedented capabilities to heal, feed, and sustain humanity while protecting the planet. From producing life-saving medicines and sustainable materials to cleaning up pollution and creating climate-resilient crops, the applications are limited only by human ingenuity.

The journey is not without challenges—technical, ethical, and societal—but the potential rewards are immense. As we continue to refine our tools, deepen our understanding, and engage in thoughtful dialogue about the implications, synthetic biology holds the promise of a future where biology is not just studied but harnessed as a powerful technology for the betterment of all. The age of synthetic life is here, and it is poised to redefine our world in ways we are only beginning to imagine.

Synthetic Yeasts and Bacteria

Synthetic yeasts and bacteria are microorganisms whose genomes have been intentionally altered using genetic engineering techniques to endow them with new or enhanced functions. This goes beyond traditional genetic modification, which might involve inserting a single foreign gene.

Synthetic biology aims for a more comprehensive redesign, often involving the synthesis of entire genetic circuits, pathways, or even chromosomes. Yeasts, particularly Saccharomyces cerevisiae (baker's yeast), and bacteria, like Escherichia coli (E. coli), are the primary workhorses due to their well-understood genetics, rapid growth, ease of manipulation, and generally recognized safety. "Synthetic" in this context can mean different levels of engineering:

- **Genetically Modified (GM):** Organisms with specific genes added, deleted, or altered to achieve a desired trait (e.g., bacteria producing human insulin).
- **Metabolically Engineered:** Organisms whose metabolic pathways have been re-wired to optimize the production of a target molecule (e.g., yeast producing biofuels).
- **Genome-Reduced or Streamlined:** Organisms with non-essential genes removed to create simplified chassis for easier engineering and more efficient production (e.g., the synthetic yeast genome project, Sc2.0).
- **Xenobiology (XNA):** Organisms engineered to use alternative genetic polymers (Xeno-Nucleic Acids) instead of DNA/RNA, creating biological systems with expanded functionality and biocontainment.

These engineered microbes are not just tools; they are foundational platforms for innovation across science and industry.

Engineering Microbes: The Toolkit of Synthetic Biology

The creation of synthetic yeasts and bacteria relies on a sophisticated toolkit of genetic engineering and synthetic biology techniques, continuously refined for precision and power.

- **CRISPR-Cas Genome Editing:** This revolutionary technology, adapted from a bacterial immune system, allows for precise, targeted modifications to DNA. Using a guide RNA (gRNA) to direct the Cas enzyme (e.g., Cas9) to a specific genomic location, scientists can make cuts to knock out genes, insert new sequences (knock-ins), or make base edits. CRISPR enables multiplexed editing (targeting multiple genes simultaneously), making it indispensable for complex synthetic biology projects like building metabolic pathways or genetic circuits in yeasts and bacteria.
- **Metabolic Engineering:** This involves the deliberate re-design of an organism's metabolic network. Using computational models and genetic tools, scientists map out existing pathways, identify bottlenecks, and then use gene knockouts, knock-ins, and expression tuning (e.g., promoter engineering) to redirect metabolic flux towards the desired product. This is key for maximizing yields in biomanufacturing.
- **DNA Synthesis and Assembly:** The ability to chemically synthesize DNA oligonucleotides and assemble them into larger fragments, genes, or even entire genomes is fundamental. Techniques like Golden Gate Assembly and Gibson Assembly allow for the seamless construction of complex DNA constructs from smaller parts, enabling the building of synthetic chromosomes or genetic devices.
- **Synthetic Gene Circuits:** Inspired by electronics, these are networks of genes and regulatory elements designed to perform logical operations within the cell. Components like promoters, ribosome binding sites (RBS), and terminators are standardized (e.g., BioBricks) and assembled to create circuits that sense inputs (e.g., chemicals, light) and produce outputs (e.g., fluorescence, enzymes). CRISPR-based systems (CRISPRi/a) are often used for flexible transcriptional control.
- **Genome-Scale Engineering:** Projects like the Synthetic Yeast Genome Project (Sc2.0) aim to design and build a fully synthetic eukaryotic genome. This involves synthesizing chromosomes with streamlined sequences, removing repetitive and transposable elements, and incorporating genetic "firewalls" for biocontainment. Similar efforts are underway for bacteria.

Applications Across Industries: From Medicine to Manufacturing

The impact of synthetic yeasts and bacteria is vast and growing, transforming numerous sectors:

- **Medicine and Pharmaceuticals:** Engineered microbes are revolutionizing drug production and therapy. *E. coli* and yeast are used to produce recombinant proteins like human insulin, growth hormones, monoclonal antibodies, and vaccines (e.g., hepatitis B surface antigen). Synthetic biology enables the creation of microbes that can detect disease biomarkers in the body and produce therapeutic molecules in response (e.g., engineered bacteria for cancer therapy or gut microbiome engineering). CRISPR-edited microbes are being developed for gene therapy and as diagnostic tools.
- **Agriculture and Food Production:** Synthetic microbes are enhancing crop resilience and food quality. Engineered bacteria can act as biofertilizers, fixing nitrogen or solubilizing phosphorus to reduce synthetic fertilizer use. Yeasts are used to produce alternative proteins (e.g., egg white proteins without chickens) and food ingredients (e.g., flavors, enzymes). CRISPR is used to edit crops directly for traits like disease resistance, drought tolerance, and improved nutrition, though often via engineered microbes in the production process.
- **Industrial Biomanufacturing:** This is perhaps the most mature application. Synthetic yeasts and bacteria serve as efficient, sustainable cell factories for producing a wide range of chemicals, materials, and fuels. Examples include:
- **Biofuels:** *E. coli* and yeast engineered to produce ethanol, butanol, and advanced biofuels from renewable sugars or waste biomass.
- **Bioplastics:** Bacteria like *Cupriavidus necator* produce polyhydroxyalkanoates (PHAs), and yeast is engineered for polylactic acid (PLA) precursors.
- **Bulk Chemicals:** Microbes produce organic acids (e.g., lactic, succinic, citric acid), solvents, and platform chemicals traditionally derived from petroleum.
- **Specialty Chemicals:** Flavors, fragrances, dyes, and high-value compounds like artemisinin (antimalarial) are produced using engineered yeast.
- **Environmental Remediation:** Synthetic microbes are designed to tackle pollution. Engineered bacteria can degrade persistent environmental contaminants like plastic waste (e.g., PET), chlorinated solvents, oil

spills, and heavy metals. Algal systems, often engineered, are used for carbon capture and wastewater treatment. These solutions offer more sustainable and targeted approaches than traditional methods.

- **Biosensors and Biocomputing:** Engineered yeasts and bacteria can be programmed to detect specific molecules (toxins, pathogens, explosives) and produce a visible signal (e.g., fluorescence, color change). This enables rapid, portable, and cost-effective diagnostics for environmental monitoring, food safety, and medical testing. More complex genetic circuits are being explored for biocomputing applications.

Real-World Success: Case Studies in Synthetic Microbes

Several high-profile examples illustrate the power and potential of synthetic yeasts and bacteria:

- **Synthetic Artemisinin:** Scientists at Amyris and the Institute for OneWorld Health engineered yeast (*S. cerevisiae*) to produce artemisinic acid, a precursor to artemisinin, a life-saving antimalarial drug. This provided a stable, scalable, and cost-effective source, addressing shortages and price volatility of plant-derived artemisinin.
- **The Sc2.0 Synthetic Yeast Genome Project:** An international consortium is synthesizing a 16-chromosome yeast genome, redesigned for simplicity, stability, and enhanced functionality. This "synthetic yeast" will serve as a superior chassis for industrial applications, enabling more efficient production of biofuels, chemicals, and pharmaceuticals.
- **Engineered Bacteria for Plastic Degradation:** Researchers have engineered *Ideonella sakaiensis* bacteria, which naturally degrade PET plastic, to enhance their efficiency. Similarly, *E. coli* has been engineered to convert PET waste into vanillin, demonstrating upcycling potential.
- **NitroFix: Synthetic Nitrogen-Fixing Bacteria:** Projects aim to engineer bacteria like *E. coli* or *Klebsiella* to fix atmospheric nitrogen into ammonia, a key fertilizer component, reducing reliance on energy-intensive industrial fertilizer production.
- **COVID-19 Diagnostics and Vaccines:** Synthetic biology played a crucial role during the pandemic. Engineered *E. coli* was used to produce components for rapid diagnostic tests, and yeast was used in the production of some viral vector vaccines (e.g., AstraZeneca's ChAdOx1).

Challenges and Ethical Considerations

Despite the promise, significant challenges and ethical questions surround the creation and deployment of synthetic yeasts and bacteria:

- **Technical Hurdles:** Engineering biological systems is inherently complex. Challenges include:
- **Off-Target Effects:** Unintended mutations from gene editing (e.g., CRISPR) can have unpredictable consequences.
- **Delivery and Stability:** Getting engineered components into cells efficiently and ensuring genetic stability over generations in industrial settings.
- **Scale-Up Difficulties:** Translating lab success to cost-effective, large-scale industrial fermentation.
- **Unpredictable Emergent Properties:** Synthetic genetic circuits or pathways might interact in unforeseen ways within the host cell.
- **Biosafety and Biosecurity:** Ensuring that synthetic organisms do not pose risks to human health or the environment is paramount. Key concerns include:
- **Environmental Release:** Accidental or deliberate release of engineered microbes could disrupt ecosystems if they outcompete native species or transfer genes horizontally.
- **Biocontainment:** Strategies like xenobiology (using XNA), auxotrophy (dependence on supplied nutrients), and "kill switches" are being developed to prevent survival outside controlled environments.
- **Dual Use:** The technology could be misused to create harmful biological agents, necessitating strict oversight and responsible research practices.

Ethical and Social Issues

- **"Playing God" Concerns:** Some view the creation of synthetic life as ethically problematic, raising questions about the moral status of engineered organisms.
- **Environmental Ethics:** The potential impact of releasing synthetic organisms into the environment requires careful assessment.
- **Socioeconomic Impacts:** The benefits of synthetic biology might not be evenly distributed, potentially exacerbating global inequalities. For example, patented technologies could concentrate benefits in wealthy nations/corporations.
- **Public Perception and Acceptance:** Fear of "GMOs" and "lab-grown" organisms persists, fueled by misinformation. Transparent communication and education are crucial for public trust.

CHAPTER 5

Challenges and Limitations in Synthetic Biology

5.1 TECHNICAL AND SCIENTIFIC CHALLENGES

5.1.1 Scalability and Production Efficiency

Scalability and production efficiency are fundamental principles that determine the success of industrial and biotechnological processes. Scalability ensures that laboratory or pilot-scale processes can be reliably transitioned to commercial production without loss of performance, while production efficiency optimizes the use of resources, time, and energy to maximize output and profitability. Factors such as equipment design, process kinetics, raw material quality, and process control significantly influence scalability, whereas strategies like lean manufacturing, automation, and process integration enhance efficiency.

Modern technologies, including computational modeling, AI, continuous production systems, and advanced sensors, support both objectives across industries ranging from pharmaceuticals to renewable energy. Although challenges persist—such as process variability, resource limitations, and regulatory constraints—ongoing innovation and integration of digital, modular, and sustainable approaches offer promising solutions. By prioritizing scalability and production efficiency, industries can achieve reliable, cost-effective, and environmentally responsible operations, meeting global demand while maintaining quality and competitiveness.

Importance of Scalability

Scalability is a key factor in determining whether a process developed in a laboratory or pilot plant can be successfully transitioned to commercial-scale production. Many processes that function efficiently at a small scale may encounter challenges when scaled up, such as changes in reaction kinetics, mass transfer limitations, or equipment constraints. In pharmaceuticals, for example, small-scale synthesis of a drug may produce high yields in the lab, but scaling to industrial reactors can introduce variations in temperature, mixing, and flow rates that affect product quality. Similarly, in biotechnology, microbial fermentation or cell culture systems may behave differently when volume increases, requiring careful optimization of aeration, nutrient supply, and agitation. Scalability ensures that processes remain robust and reproducible, reducing the risk of costly delays, regulatory complications, and production failures.

Production Efficiency Defined

Production efficiency refers to the ability to produce the desired output with minimal waste, time, energy, and cost. High efficiency is achieved when every step of the production process is optimized, resources are used effectively, and bottlenecks are minimized. Efficiency can be evaluated in terms of yield, throughput, energy consumption, material utilization, and labor productivity.

In industrial settings, production efficiency directly impacts profitability, sustainability, and competitiveness. Efficient production reduces raw material requirements, lowers operational costs, and minimizes environmental footprint. It also allows companies to respond more rapidly to market demand, scale operations as needed, and maintain consistent product quality.

Factors Affecting Scalability

Several factors influence the scalability of a process or system. Equipment design is critical, as reactors, fermenters, and assembly lines must accommodate increased volume while maintaining uniform conditions. Process kinetics can change at larger scales due to heat transfer limitations, diffusion rates, or changes in mixing patterns. Raw material availability and quality can impact scalability, particularly when sourcing unique or high-purity components.

Process control systems, including monitoring sensors and feedback loops, must be capable of handling increased complexity. Regulatory compliance, especially in pharmaceutical and food industries, adds further constraints, requiring scaled processes to meet stringent safety and quality standards. Finally, economic considerations, such as cost per unit, labor, and energy efficiency, determine whether scaling is commercially viable.

Strategies to Improve Scalability

Effective strategies are essential to overcome challenges in scaling production processes. The first approach is modular design, which allows equipment or process units to be replicated or expanded without redesigning the entire system. Process simulation and modeling help predict behavior at larger scales, enabling preemptive optimization of parameters such as temperature, mixing, or flow. Standardization of procedures and materials ensures consistency across batches and reduces variability during scale-up. Pilot studies and stepwise scale-up enable gradual transition from laboratory to commercial scale, allowing identification and mitigation of potential issues. Automation and advanced process control maintain precision and stability during scaling. Continuous production systems, rather than batch operations, can also improve scalability by maintaining steady-state conditions and reducing downtime.

Enhancing Production Efficiency

Improving production efficiency requires a combination of technical, operational, and managerial strategies. Lean manufacturing principles eliminate waste in materials, energy, and time while streamlining workflows. Six Sigma and quality management approaches reduce variability and defects, ensuring high yields and consistent quality.

Energy recovery and optimization reduce operational costs and environmental impact, particularly in chemical and bioprocess industries. Process integration, such as combining reaction and separation steps, reduces processing time and equipment needs. Supply chain optimization ensures timely delivery of raw materials and reduces storage or inventory costs. Predictive maintenance of equipment minimizes downtime and enhances throughput. Additionally, employee training and skill development improve operational efficiency by reducing errors and improving responsiveness to process fluctuations.

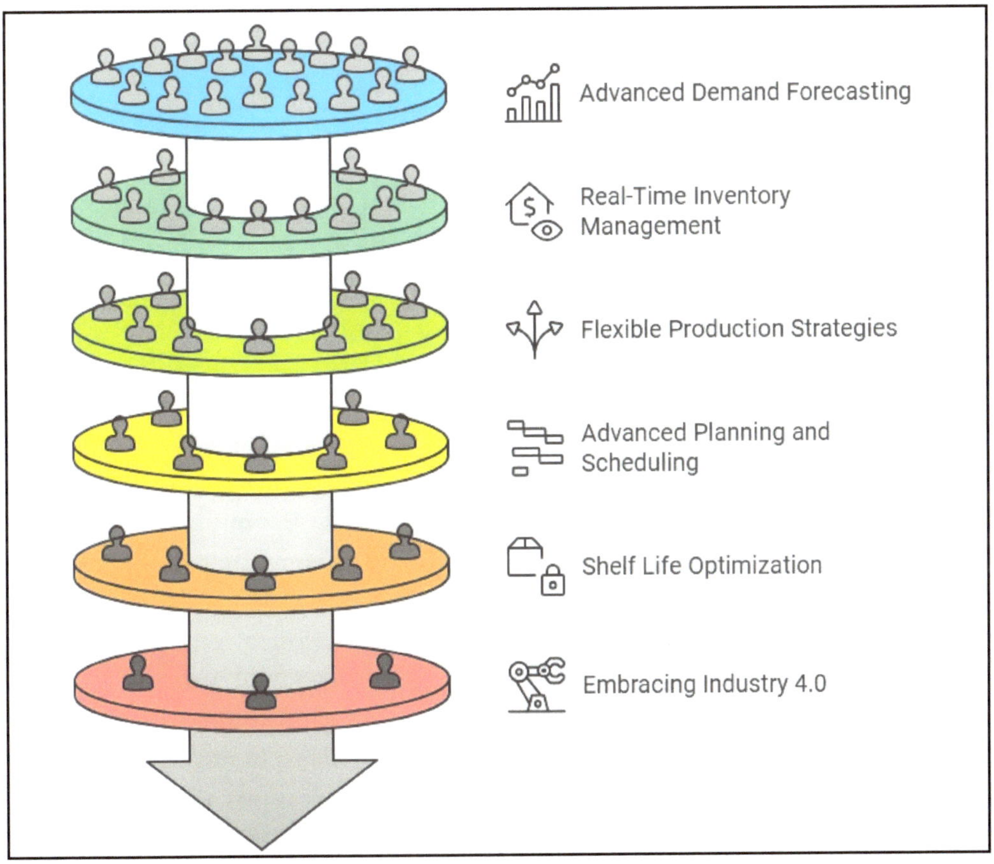

Enhancing Production Efficiency

Technologies Supporting Scalability and Efficiency

Modern technologies play a critical role in enhancing scalability and production efficiency.

- Computer-aided process modeling simulates scale-up scenarios and predicts potential bottlenecks.
- Automation and robotics improve precision, consistency, and throughput in production lines.
- Real-time monitoring sensors track critical parameters such as temperature, pressure, pH, and flow.
- Advanced control systems enable feedback loops for maintaining optimal operating conditions.
- High-throughput screening accelerates the identification of optimal conditions or formulations.
- Continuous flow reactors allow stable, large-scale production with minimal downtime.
- Data analytics and AI optimize process parameters, predict equipment failures, and improve decision-making.
- 3D printing and modular equipment enable rapid prototyping and flexible scaling of production units.
- Bioreactor design innovations enhance oxygen transfer, mixing, and nutrient delivery in cell or microbial cultures.
- Sustainable technologies, such as waste-to-value integration and energy-efficient processes, improve overall resource utilization.

These tools collectively support the reliable, efficient, and scalable production of goods across industries.

Applications across Industries

Scalability and production efficiency are crucial in multiple sectors. In pharmaceutical manufacturing, scale-up strategies ensure that drugs developed in the lab can be produced at industrial scale without compromising potency or purity. In biotechnology, fermentation-based production of enzymes, vaccines, or biopharmaceuticals relies on optimized bioreactors and process control to scale efficiently. Chemical industries benefit from continuous flow processes, automation, and energy-efficient designs to maximize yield and minimize costs. In food and beverage manufacturing, production lines must be scaled to meet fluctuating demand while maintaining quality and safety standards. In renewable energy, particularly biofuel production, scalability determines the feasibility of converting biomass into energy at commercial volumes. Across all sectors, balancing scalability and efficiency ensures that products are delivered reliably, affordably, and sustainably.

Challenges in Achieving High Efficiency and Scalability

Despite technological advances, several challenges persist. Scale-up often introduces unforeseen variations in reaction kinetics or process dynamics, requiring iterative adjustments. Large-scale equipment may not replicate laboratory conditions exactly, leading to inefficiencies or reduced yields. Resource limitations, including availability of high-quality raw materials or energy, can constrain production. Regulatory compliance is particularly challenging in pharmaceuticals and food production, where scaled processes must meet stringent safety and quality standards. Capital costs for scaling equipment or infrastructure can be prohibitive. Furthermore, maintaining process efficiency over time requires continuous monitoring, optimization, and maintenance, which can be resource-intensive. Overcoming these challenges demands careful planning, technological innovation, and skilled operational management.

5.1.2 Controlling Biological Systems and Ensuring Stability

Controlling biological systems and ensuring stability is a fundamental challenge in modern biology and engineering, essential for both natural and engineered systems. Biological stability arises from complex networks of feedback, feedforward, and adaptive regulatory mechanisms, which can be harnessed through synthetic biology and control theory. Mathematical modeling, computational simulations, and synthetic regulatory elements provide powerful tools for designing predictable and robust biological circuits. Applications span metabolic engineering, synthetic therapeutics, tissue engineering, and environmental management, highlighting the broad impact of controlled and stable biological systems. Despite inherent challenges such as noise, context-dependency, and environmental variability, advances in computational modeling, synthetic regulatory tools, and monitoring technologies are enabling increasingly sophisticated strategies to maintain stability. The integration of these approaches promises to transform biotechnology, medicine, and environmental engineering by enabling the precise, reliable, and safe manipulation of living systems. By combining predictive design, synthetic control, and robust monitoring, researchers can ensure that biological systems operate within desired parameters, maintaining stability while achieving functional objectives, and ultimately advancing the frontier of engineered life.

Foundations of Biological Control

Biological systems are regulated through a combination of feedback loops, signal transduction pathways, and gene regulatory networks. Negative feedback mechanisms, such as enzyme inhibition or transcriptional repression, stabilize systems by counteracting deviations from equilibrium. Positive feedback mechanisms, in contrast, can amplify responses and create bistable or switch-like behavior, which is useful for decision-making processes in cells but may compromise stability if uncontrolled. Synthetic biology aims to harness these natural principles to engineer controllable systems with predictable behaviors. By designing regulatory modules that mimic natural feedback loops, researchers can stabilize gene expression, maintain metabolite levels, or modulate cellular responses to environmental signals. Control strategies often involve adjusting the expression of key genes, modulating protein activity, or introducing synthetic regulatory elements that confer tunable responses. The ultimate goal is to achieve robust system behavior that remains functional under a wide range of conditions.

Mathematical Modeling in Biological Systems

Mathematical modeling plays a critical role in understanding and controlling biological systems. Models describe the dynamics of system components and their interactions using differential equations, stochastic models, or agent-based simulations. Deterministic models capture average behavior over time, whereas stochastic models account for the probabilistic nature of molecular interactions, particularly in low-copy-number scenarios such as gene expression. These models allow the prediction of system responses to perturbations, identification of critical nodes or control points, and assessment of stability margins. Computational simulations based on these models help in testing different control strategies virtually before experimental implementation. For example, in metabolic engineering, modeling can identify bottleneck enzymes, predict the impact of gene knockouts, and suggest interventions that stabilize product yield. In synthetic gene circuits, modeling informs the design of oscillators, toggle switches, and feedback controllers to ensure predictable temporal dynamics.

Feedback Control Mechanisms

Feedback control is a fundamental principle in maintaining stability in biological systems. Negative feedback reduces variability by damping deviations from the desired state. For instance, in metabolic pathways, end-product inhibition ensures that metabolite concentrations remain within optimal ranges. Synthetic biologists employ negative feedback to regulate gene expression levels, using transcriptional repressors, RNA-based regulators, or protein degradation tags to prevent overexpression. Positive feedback, while potentially destabilizing, is useful for generating bistability, memory, and switch-like responses. Combining positive and negative feedback loops enables complex behaviors such as oscillations or adaptive responses, as seen in circadian clocks and stress-response networks. Feedforward loops add an additional layer of control by anticipating changes in input signals and adjusting system output preemptively, enhancing robustness. Understanding the interplay of these feedback mechanisms is essential for designing interventions that stabilize biological systems without disrupting their natural functionality.

Synthetic Control Elements

Synthetic biology has enabled the development of engineered control elements that offer precise regulation of biological systems. Synthetic promoters and transcription factors allow tunable gene expression in response to chemical, optical, or environmental inputs. Riboswitches and RNA-based regulators provide post-transcriptional control, modulating translation efficiency or mRNA stability. CRISPR-based systems, including CRISPR interference (CRISPRi) and CRISPR activation (CRISPRa), offer programmable regulation of endogenous and synthetic genes. These synthetic control elements can be combined into regulatory circuits that implement feedback, feedforward, and adaptive control strategies. For example, synthetic gene circuits have been designed to maintain protein concentrations within narrow ranges, reduce noise in gene expression, or produce oscillatory behavior. By integrating multiple control elements, researchers can create modular, predictable, and robust systems capable of functioning reliably even in fluctuating environments.

Ensuring Stability in Gene Circuits

Gene circuits, both natural and synthetic, are subject to intrinsic and extrinsic noise. Intrinsic noise arises from stochastic fluctuations in gene transcription and translation, while extrinsic noise results from environmental variations or changes in cellular resources. Ensuring stability in gene circuits involves designing architectures that mitigate these fluctuations. Negative feedback loops reduce intrinsic noise, while network motifs such as incoherent feedforward loops buffer against external disturbances. Modular design principles, insulation of circuit components, and the use of orthogonal regulatory elements minimize unintended interactions with host cellular processes. Additionally, careful selection of promoters, ribosome binding sites, and degradation tags allows fine-tuning of expression levels to achieve desired stability. Computational modeling and high-throughput screening are often employed to identify optimal designs before experimental implementation, increasing the likelihood of robust system behavior.

Applications in Metabolic Engineering

Controlling biological systems and ensuring stability is particularly important in metabolic engineering, where engineered microorganisms are tasked with producing biofuels, pharmaceuticals, or industrial chemicals. Metabolic pathways are highly interconnected, and perturbations in one pathway can lead to accumulation of toxic intermediates, depletion of essential cofactors, or metabolic imbalances. By implementing synthetic control elements and feedback regulation, researchers can stabilize metabolite concentrations, optimize flux through desired pathways, and maintain cell viability. Dynamic control strategies, such as inducible promoters or metabolite-responsive regulators, allow temporal tuning of enzyme expression to maximize product yield while minimizing stress. These approaches enhance the efficiency and predictability of microbial bioproduction, reducing variability and improving scalability in industrial applications.

Applications in Synthetic Biology and Therapeutics

Synthetic biology applications, including living therapeutics and engineered tissue constructs, require robust control and stability to function safely and effectively. Engineered cells must maintain precise gene expression levels to deliver therapeutic molecules, sense disease biomarkers, or respond to environmental signals. Synthetic promoters, riboswitches, and CRISPR-based regulators enable controlled expression of therapeutic genes, reducing the risk of overexpression, cytotoxicity, or unintended interactions. Feedback circuits and stability-enhancing motifs ensure that cells respond consistently despite fluctuations in the host environment. In tissue engineering, maintaining homeostasis within cell populations is essential for proper tissue formation, differentiation, and function. By implementing controlled regulatory circuits, engineered tissues can achieve stable growth, maturation, and functionality, enhancing their utility in regenerative medicine and therapeutic applications.

Environmental and Ecological Stability

Controlling biological systems is not limited to engineered cells; it extends to ecological and environmental contexts. Microbial consortia, bioreactors, and environmental remediation systems require stable community composition and metabolic activity to perform reliably. Perturbations such as nutrient fluctuations, temperature changes, or invasion by unwanted species can destabilize these systems. Strategies for ensuring stability include engineered feedback regulation, quorum sensing-mediated communication, and spatial organization of microbial populations. Computational modeling and monitoring tools allow prediction of community dynamics and identification of interventions that maintain functionality. These approaches are particularly relevant for applications such as wastewater treatment, soil remediation, and sustainable bioenergy production, where consistent system performance is critical for environmental and economic outcomes.

Challenges in Controlling Biological Systems

Despite advances in synthetic biology and control theory, controlling biological systems and ensuring stability remains challenging. Biological networks are inherently nonlinear, interconnected, and context-dependent, making predictions difficult. Noise, both intrinsic and extrinsic, can introduce variability in system behavior. Environmental changes, host physiology, and metabolic burden further complicate control strategies.

Additionally, interactions between synthetic circuits and host cellular machinery can lead to unintended feedback, cross-talk, or metabolic competition. Scaling laboratory designs to industrial or therapeutic contexts introduces additional complexity, as larger populations or heterogeneous environments may behave differently than small-scale prototypes. Addressing these challenges requires a combination of robust design principles, predictive modeling, real-time monitoring, and iterative optimization.

5.2 ETHICAL AND SOCIAL IMPLICATIONS

5.2.1 Genetic Privacy and Safety Concerns

Genetic privacy and safety concerns represent critical challenges intersecting biotechnology, medicine, ethics, and national security. The uniquely identifiable nature of genetic information demands careful stewardship to prevent unauthorized access, misuse, and discrimination. Parallelly, synthetic biology's transformative potential is shadowed by biosecurity risks, necessitating stringent safeguards against malicious or accidental harm.

A holistic approach combining technological, legal, ethical, and societal measures is essential to protect individuals' genetic privacy while enabling scientific progress. Transparency, inclusive governance, and international cooperation underpin responsible innovation in the genomic and synthetic biology era. Navigating this complex landscape is vital to securing trust and maximizing the benefits of transformative genetic technologies for humanity.

Nature of Genetic Information and Privacy Risks

Genetic data is fundamentally identifiable because it contains long sequences of nucleotides unique to individuals and their relatives. The sharing of genetic data, especially within open-access databases or through consumer services, can expose sensitive personal information to unintended third parties. Techniques such as familial searching have revealed the potential to identify individuals indirectly through relatives who have uploaded their genetic profiles, as exemplified by forensic uses in solving cold cases. Risks such as re-identification attacks arise when anonymized or aggregated genomic data is cross-referenced with auxiliary datasets, challenging traditional notions of anonymity. Moreover, genetic data's permanence differentiates it from other personal data—once compromised, it cannot be changed or revoked, highlighting the urgency of protective policies and secure data governance frameworks.

Ethical and Legal Concerns in Genetic Data Handling

The ethical dimensions surrounding genetic privacy encompass issues of informed consent, data ownership, and equitable access. Individuals often have limited control over how their genetic information is used once submitted to research, clinical, or commercial entities. The possibility of genetic discrimination by employers or insurers based on predispositions uncovered by genetic testing remains a pervasive fear despite legislative efforts like the Genetic Information Nondiscrimination Act (GINA) in the United States. Moreover, the commercialization and patenting of genetic discoveries provoke debates about the privatization of genetic resources versus open scientific collaboration. Disparities in who benefits from genetic technologies also pose justice concerns, with marginalized communities frequently underrepresented in genetic studies yet disproportionately vulnerable to exploitation or exclusion from advances.

Biosecurity and Dual-Use Risks of Synthetic Biology

Synthetic biology's capacity to manipulate genetic material raises legitimate biosecurity concerns. The increasing accessibility and affordability of DNA synthesis enable unprecedented ease in creating or modifying organisms, presenting risks of accidental release or deliberate misuse. While the creation of novel super-pathogens is technically challenging and currently deemed unlikely, the synthesis of known highly pathogenic agents or toxins remains a worrisome scenario.

'Biohacker' culture and democratization of genetic tools broaden the dissemination of potentially hazardous knowledge beyond traditional safety frameworks. This dual-use dilemma necessitates vigilant governance, screening of DNA synthesis orders, and international cooperation to balance enabling innovation while preventing bioterrorism. Frameworks such as the International Genetically Engineered Machine (iGEM) Foundation's safety policies advocate community-based responsibility and shared norms to mitigate these risks.

Technological Safeguards and Data Security Strategies

To address the multifaceted privacy and safety challenges, technical strategies alongside policy measures are imperative. Techniques such as homomorphic encryption, differential privacy, and generation of synthetic genomic datasets serve to protect sensitive information while facilitating research and analytics. Rigorous cybersecurity protocols combat data breaches targeting genomic databases and protect intellectual property. Genomic data controllers and companies increasingly adopt comprehensive compliance programs adhering to regulations like GDPR and HIPAA, emphasizing consent management, transparency, and data minimization. In synthetic biology, biocontainment tools—genetic safeguards like kill switches, auxotrophy, and synthetic amino acid dependency—limit the survival and proliferation of engineered organisms outside controlled settings, ensuring biosafety. Continuous risk assessment, real-time monitoring, and adaptive response frameworks bolster resilience against deliberate or accidental threats.

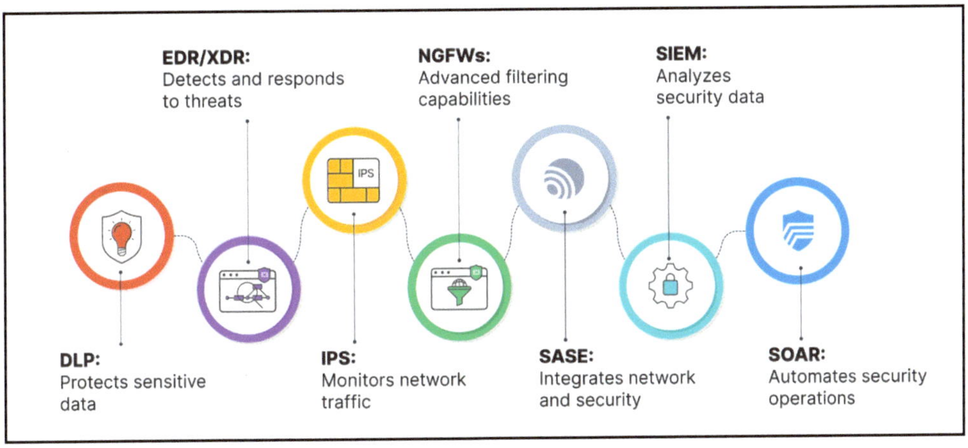

Safeguarding Data through Technology and Security Strategies

Public Perceptions, Societal Implications, and Governance

Public attitudes toward genetic data collection and synthetic biology influence acceptance, trust, and policy development. Misinformation, high-profile breaches, and ethical controversies contribute to skepticism and concern, underscoring the need for transparent engagement, education, and dialogue. Inclusive governance mechanisms that incorporate diverse stakeholder perspectives promote balanced decision-making, respecting cultural values and human rights. International coordination is critical due to the borderless nature of genomic information and biosecurity risks, fostering harmonized standards and cooperation. Ethical guidelines emphasize responsible stewardship, equitable access, and democratic deliberation. As synthetic biology matures, anticipatory governance frameworks must evolve alongside technological advances to address emerging privacy and safety challenges proactively.

5.2.2 Dual-Use Technologies and Biosecurity

The management of dual-use technologies and biosecurity represents one of the defining challenges of the twenty-first century, requiring unprecedented cooperation between scientific, security, and policy communities across national boundaries. As biotechnology continues to advance at an exponential pace, the dual-use dilemma will only become more complex, with emerging technologies presenting both tremendous opportunities for human benefit and novel risks that must be carefully managed. The COVID-19 pandemic has starkly illustrated both the critical importance of biological research for addressing global health threats and the potential consequences of biological events, whether natural or otherwise.

Moving forward, effective governance of dual-use technologies will require flexible, adaptive frameworks that can evolve with technological change while maintaining their legitimacy and effectiveness.

Success in this endeavor demands ongoing dialogue between all stakeholders, commitment to both scientific progress and security, and recognition that perfect security is neither achievable nor desirable if it comes at the cost of beneficial research. The ultimate goal must be to create a global biosecurity culture that enables the life sciences to fulfill their potential for improving human welfare while minimizing risks of misuse or accidents, ensuring that biotechnology remains a force for good in addressing humanity's greatest challenges.

Nature and Scope of Dual-Use Research

The dual-use dilemma in biological research stems from the fundamental reality that the same scientific knowledge and techniques used to understand and combat diseases can potentially be misused to cause harm. This duality is inherent in many areas of life sciences research, where understanding pathogen biology, host-pathogen interactions, and immune system mechanisms necessarily involves working with potentially dangerous materials and generating information that could theoretically be misappropriated. The scope of dual-use concerns extends far beyond traditional pathogens to encompass synthetic biology, neuroscience, genomics, and even agricultural research.

For instance, research aimed at understanding viral transmission to develop better vaccines inherently produces knowledge about how to potentially enhance transmissibility. Similarly, advances in gene synthesis technology that enable rapid vaccine development also lower barriers to recreating dangerous pathogens from published sequences. The democratization of biotechnology through decreasing costs and increasing accessibility of sophisticated equipment and techniques amplifies these concerns, as capabilities once confined to well-funded government and academic laboratories become available to smaller institutions and even individual researchers. This expanding scope requires continuous reassessment of what constitutes dual-use research and how to manage its risks appropriately.

Historical Context and Evolution

The recognition of dual-use potential in biological research has evolved significantly over the past century, shaped by historical events and technological advances that demonstrated both the promise and peril of biotechnology. The development of biological weapons programs during World War II and the Cold War first highlighted how biological knowledge could be weaponized, leading to the establishment of the Biological Weapons Convention in 1972. However, the dual-use debate intensified dramatically following the 2001 anthrax attacks in the

United States, which demonstrated that sophisticated biological agents could be produced and deployed by non-state actors. Several controversial publications in the early 2000s, including research on mousepox virus enhancement and the reconstruction of the 1918 influenza virus, sparked intense discussions about the responsibilities of scientists and publishers in managing dual-use information. The H5N1 influenza transmission studies of 2011-2012 marked a watershed moment, triggering global debates about gain-of-function research and leading to new policies and oversight mechanisms. More recently, the COVID-19 pandemic has further complicated these discussions, highlighting both the critical importance of virus research for pandemic preparedness and the potential risks associated with certain types of experiments.

Categories of Dual-Use Biotechnologies

The landscape of dual-use biotechnologies encompasses diverse research areas and applications, each presenting unique benefits and risks that require tailored management approaches. Pathogen research, including studies on virulence factors, host range, transmissibility, and drug resistance, remains a primary concern as this knowledge could potentially be used to enhance the harmful characteristics of biological agents. Synthetic biology and gene synthesis technologies enable the construction of novel organisms and the recreation of extinct pathogens, raising concerns about both intentional misuse and unintended consequences. Genome editing tools like CRISPR could theoretically be used to create more dangerous pathogens or to develop gene drives with unpredictable ecological impacts. Neuroscience research, particularly studies on neurotoxins and incapacitating agents, presents dual-use concerns related to both therapeutic applications and potential misuse for harm. Agricultural biotechnology, including research on plant and animal pathogens, could be exploited for economic warfare or food security threats. Aerobiology and delivery system research, while essential for understanding disease transmission and developing medical devices, also has obvious weapons potential. Even seemingly benign research in areas such as fermentation technology and bioprocessing could be repurposed for large-scale production of harmful agents.

Biosecurity Frameworks and Risk Management

Comprehensive biosecurity frameworks have been developed at institutional, national, and international levels to manage the risks associated with dual-use research while preserving scientific freedom and innovation. These frameworks typically incorporate multiple layers of oversight, including institutional biosafety committees, national regulatory agencies, and international agreements. Physical security measures protect biological materials and facilities from unauthorized access, while personnel reliability programs ensure that individuals with access to dangerous agents are properly vetted and trained. Information security protocols aim to prevent sensitive knowledge from being acquired by those who might misuse it, though determining what information should be restricted remains contentious. Laboratory biosafety practices reduce the risk of accidental releases, with facilities classified into biosafety levels based on the agents they handle and the corresponding containment measures required. Culture of responsibility initiatives promote awareness among researchers about dual-use implications and encourage proactive consideration of potential misuse. Risk assessment methodologies evaluate both the potential benefits and risks of specific research projects, though quantifying these factors remains challenging. These frameworks must continuously evolve to address emerging technologies and changing threat landscapes while maintaining their effectiveness and legitimacy within the scientific community.

International Cooperation and Governance Challenges

The global nature of biological research and the transboundary character of biological threats necessitate international cooperation in managing dual-use technologies, yet achieving effective governance remains fraught with challenges. Different countries have varying approaches to biosecurity oversight, reflecting diverse regulatory traditions, threat perceptions, and scientific capabilities. The absence of universally accepted standards for dual-use research oversight creates potential vulnerabilities, as research deemed too risky in one jurisdiction might be conducted in another with less stringent controls. International agreements such as the Biological Weapons Convention provide important normative frameworks but lack robust verification mechanisms and struggle to keep pace with technological advances. Scientific collaboration across borders, while essential for addressing global health challenges, complicates oversight and raises questions about information sharing and technology transfer. Export control regimes attempt to prevent sensitive materials and knowledge from reaching problematic actors but must balance security concerns with the need for legitimate scientific exchange. Capacity building in developing countries presents particular challenges, as improving biological research capabilities for public health purposes inherently creates dual-use potential. Cultural differences in risk perception and scientific governance further complicate efforts to establish common approaches to managing dual-use research.

Emerging Technologies and Future Biosecurity Concerns

Rapid advances in biotechnology continuously expand the dual-use landscape, presenting novel biosecurity challenges that existing governance frameworks struggle to address adequately. Artificial intelligence and machine learning applications in biological research accelerate discovery but could also be used to design more effective biological weapons or identify vulnerabilities in public health systems. The convergence of biology with other technologies, such as nanotechnology and robotics, creates new possibilities for both beneficial applications and potential misuse. Cloud laboratories and automated research platforms democratize access to sophisticated capabilities while potentially reducing human oversight of experiments. Digital sequence information and bioinformatics tools enable virtual experimentation and design of organisms without physical access to biological materials, complicating traditional biosecurity approaches. Do-it-yourself biology communities and biohacker spaces operate outside traditional institutional oversight structures, raising questions about how to engage these groups in biosecurity discussions. Advances in gene drives and environmental modification technologies present unprecedented possibilities for large-scale ecological interventions with difficult-to-predict consequences. The increasing feasibility of de-extinction and creation of novel organisms pushes the boundaries of current regulatory frameworks and ethical considerations.

Balancing Innovation with Security

Finding the appropriate balance between promoting beneficial research and managing security risks represents the central challenge in dual-use governance, requiring careful consideration of multiple factors and stakeholder perspectives. Overly restrictive measures could impede legitimate research, delay medical breakthroughs, and disadvantage countries or institutions that comply with strict regulations while others do not. Conversely, insufficient oversight could enable catastrophic events through either intentional misuse or accidental release. Proportionate governance approaches attempt to match oversight intensity to actual risk levels, though determining these levels remains contentious and often uncertain. Stakeholder engagement involving scientists, security professionals, ethicists, and public representatives helps ensure that governance measures are both effective and acceptable to those they affect. Transparency in decision-making processes builds trust and legitimacy, though some security-related information necessarily remains confidential. Educational initiatives raise awareness about dual-use issues without providing recipes for misuse, requiring careful consideration of content and audience. Adaptive governance mechanisms that can respond to emerging technologies and changing threat landscapes while maintaining consistency and predictability represent an ongoing challenge. International harmonization efforts seek to prevent regulatory arbitrage while respecting sovereignty and cultural differences in risk tolerance.

5.2.3 Intellectual Property and Access to Synthetic Biology Innovations

Intellectual property and access are central to the development, dissemination, and societal impact of synthetic biology innovations. While IP frameworks provide crucial incentives for research investment and commercialization, they can also create barriers to equitable access, knowledge sharing, and global collaboration. Effective management of intellectual property requires a combination of legal protection, licensing strategies, public policy interventions, and ethical considerations.

Open science initiatives, non-exclusive licensing, and collaborative patent pools provide practical models for balancing private innovation incentives with public access. Public funding and global technology transfer programs further support equitable participation in synthetic biology research and applications. Moving forward, adaptive and inclusive IP strategies will be essential to ensure that synthetic biology fulfills its potential to address global challenges in medicine, agriculture, industry, and environmental sustainability. By fostering an ecosystem that promotes innovation while safeguarding equitable access, society can fully benefit from the transformative promise of synthetic biology.

Concept of Intellectual Property in Synthetic Biology

Intellectual property in synthetic biology encompasses patents, trade secrets, copyrights, and, in some cases, plant variety protections. Patents are the most commonly applied form, protecting novel DNA sequences, engineered organisms, metabolic pathways, or devices used for gene synthesis.

By granting exclusive rights, patents provide a legal mechanism to recoup investment costs and encourage private sector research. Trade secrets protect proprietary processes, algorithms, or design tools that are not disclosed publicly. Copyrights may cover software tools, bioinformatics platforms, or educational materials. Plant variety protections can extend to genetically engineered crops developed using synthetic biology. Collectively, these mechanisms create a legal and economic structure to support research and commercialization.

Importance of IP in Advancing Innovation

Intellectual property rights play a central role in accelerating synthetic biology research and commercialization. Developing synthetic biology applications, such as engineered microbes for biofuels, gene therapies, or disease-resistant crops, often requires significant investment in laboratories, personnel, and regulatory compliance.

Patents and other IP protections incentivize companies and institutions to fund these efforts by providing the prospect of market exclusivity. IP frameworks also facilitate technology transfer, licensing, and partnerships, enabling knowledge and innovation to move from academic research to industrial application. Moreover, clear ownership of inventions reduces disputes and encourages collaborations across disciplines and institutions. By balancing protection with dissemination, IP fosters both innovation and economic growth.

Access Challenges and Equity Issues

Despite the benefits of IP, synthetic biology also faces challenges related to access and equity. High costs associated with patented technologies can limit availability to researchers, small enterprises, and low-income countries. Exclusive licensing arrangements may restrict use in developing regions, particularly for medical therapies, vaccines, or agricultural innovations. The concentration of patents in the hands of a few large corporations can create "patent thickets", where overlapping claims hinder downstream innovation or complicate research. Publicly funded projects may also be constrained by IP policies, limiting open sharing of genetic parts, standardized sequences, or bioinformatics tools. These access barriers can slow innovation in areas where societal impact is greatest, such as neglected diseases or sustainable agriculture.

Legal and Regulatory Frameworks

IP in synthetic biology is governed by national and international legal frameworks. Patent laws differ across countries, with criteria including novelty, inventive step, and industrial applicability. The World Trade Organization's Agreement on Trade-Related Aspects of Intellectual Property Rights (TRIPS) sets minimum standards for patent protection globally, influencing access to synthetic biology innovations.

Additionally, regulatory agencies evaluate the safety, ethical, and environmental implications of synthetic biology products before commercialization. Ethical frameworks and biosafety guidelines further influence the dissemination and licensing of synthetic biology inventions. Navigating this complex landscape requires collaboration between scientists, policymakers, and legal experts to balance innovation incentives with public interest and equitable access.

Licensing Models and Open Science

To address access challenges, various licensing models have emerged in synthetic biology. Exclusive licensing grants rights to a single entity, maximizing commercialization potential but potentially restricting broader access. Non-exclusive licensing allows multiple users to exploit the technology simultaneously, enhancing research collaboration and dissemination. Patent pools and cross-licensing arrangements consolidate rights to reduce litigation risk and facilitate joint development of complex synthetic systems. Open science initiatives promote open-source biology, where genetic parts, software tools, and protocols are freely shared within the community. Examples include the BioBricks Foundation and the iGEM Registry, which provide standardized parts under open-access agreements.

These models encourage innovation while maintaining broader access, particularly for educational and research purposes.

Ethical Considerations in IP and Access

The intersection of IP and synthetic biology raises important ethical questions. Prioritizing exclusivity may limit access to life-saving therapies or sustainable technologies in low-resource settings. There is also the risk of monopolization, where corporate control over essential genetic tools stifles scientific creativity and innovation. Ethical frameworks emphasize the responsibility to share knowledge, especially when public funds have supported research. In environmental applications, restricted access could delay deployment of engineered organisms to address climate change or pollution. Transparency, fairness, and global equity are therefore central considerations in designing IP policies that balance private incentives with societal benefit.

Role of Public Funding and Policy

Public funding plays a significant role in shaping access to synthetic biology innovations. Government agencies and non-profit organizations invest in research that may not be immediately commercially profitable, such as vaccines for neglected diseases or bio-remediation technologies. Policies linked to public funding often require that resulting innovations be accessible, licensable, or shared under open-access terms. For instance, universities may adopt technology transfer policies that encourage licensing to small enterprises or non-profit organizations. Additionally, national and international strategies, including open data mandates and public-private partnerships, facilitate wider dissemination of synthetic biology knowledge while still enabling commercialization. Such policies help balance innovation incentives with the public interest.

Global Implications and Technology Transfer

Synthetic biology is inherently global, with research and applications spanning continents. Intellectual property arrangements influence the transfer of technology between high-income and low- or middle-income countries. Technology transfer agreements can enable local manufacturing of synthetic biology-based therapeutics, agricultural inputs, or industrial enzymes. However, access is often limited by patent restrictions, costs, or technical complexity. International collaborations, capacity-building programs, and IP licensing strategies are essential to ensure that synthetic biology innovations benefit global populations. Without such mechanisms, disparities in access could exacerbate global inequalities in healthcare, food security, and environmental management.

- Intellectual property rights incentivize innovation in synthetic biology by granting exclusive use and commercialization rights.

- Patents protect engineered DNA sequences, organisms, metabolic pathways, and devices.
- Trade secrets safeguard proprietary processes and software tools used in synthetic biology.
- Exclusive licensing can hinder broad access to life-saving or sustainable technologies.
- Non-exclusive licenses and patent pools promote collaboration and wider dissemination.
- Open science initiatives, such as BioBricks and iGEM, provide freely shareable standardized parts.
- Ethical considerations emphasize fairness, transparency, and societal benefit in IP policies.
- Public funding policies often require accessible or licensable outcomes from supported research.
- Global technology transfer is essential to ensure equitable access in low- and middle-income countries.
- Adaptive licensing, collaborative networks, and policy reforms are needed for future sustainable innovation.

5.3 REGULATORY CHALLENGES AND POLICY FRAMEWORKS

5.3.1 Regulation of Synthetic Biology in Different Regions

The regulation of synthetic biology is a complex, multifaceted process shaped by scientific, ethical, environmental, and societal considerations. Different regions have developed diverse approaches reflecting their legal systems, technological priorities, and cultural values. In the United States, regulation is adaptive and risk-based, emphasizing case-by-case evaluation.

The European Union follows a precautionary approach with extensive risk assessment, public consultation, and harmonized standards. Asian countries demonstrate diversity, balancing biosafety, industrial promotion, and ethical oversight according to national priorities. International coordination, ethical engagement, and biosafety measures are critical for maintaining public trust, facilitating global collaboration, and mitigating risks.

As synthetic biology continues to advance rapidly, regulatory frameworks must evolve to accommodate novel technologies, emerging applications, and the dynamic interplay between innovation and societal responsibility. By aligning regulation with scientific progress, ethical standards, and global harmonization, it is possible to foster a safe, responsible, and innovative synthetic biology ecosystem that delivers benefits across medicine, agriculture, industry, and the environment.

Foundations of Synthetic Biology Regulation

Regulation of synthetic biology is rooted in the broader frameworks governing biotechnology, genetic engineering, and biosafety. Most regulatory systems focus on controlling risks associated with research, development, and commercialization of genetically modified organisms (GMOs), synthetic genes, and engineered microorganisms. Core principles of regulation include risk assessment, risk management, containment, monitoring, and public engagement.

Risk assessment evaluates the potential hazards posed by synthetic biology applications to human health, the environment, and society. Risk management involves implementing safety protocols, containment measures, and approval processes to mitigate identified risks. Biosafety guidelines, often based on international standards such as those developed by the World Health Organization (WHO) and the Organisation for Economic Co-operation and Development (OECD), provide a foundation for national regulations. These frameworks aim to balance innovation with precaution, ensuring that synthetic biology advances without compromising safety, ethics, or societal trust.

Regulatory Frameworks in the United States

In the United States, the regulation of synthetic biology is shaped by a combination of existing federal agencies and legislative frameworks. The primary agencies involved include the Food and Drug Administration (FDA), the Environmental Protection Agency (EPA), and the United States Department of Agriculture (USDA). The FDA oversees products that involve synthetic biology for human consumption, pharmaceuticals, and therapeutic applications, focusing on safety, efficacy, and labeling standards. The EPA regulates the environmental release of genetically engineered microorganisms under the Toxic Substances Control Act and the Federal Insecticide, Fungicide, and Rodenticide Act. The USDA monitors genetically modified crops, plants, and agricultural organisms under the Plant Protection Act. In addition to these agency-specific regulations, the U.S. approach emphasizes case-by-case evaluation rather than creating new, synthetic biology-specific legislation. This approach allows regulators to adapt existing frameworks for novel technologies while ensuring oversight of potential risks. Public engagement and ethical review are also encouraged to address societal concerns, particularly for emerging applications in human therapeutics and environmental release.

Regulation in the European Union

The European Union (EU) maintains one of the most comprehensive and precautionary regulatory frameworks for biotechnology, including synthetic biology. EU legislation emphasizes environmental protection, food safety, and public health. The European Food Safety Authority (EFSA) assesses genetically modified organisms, synthetic biology-derived products, and their potential impact on ecosystems and human health. Directive 2001/18/EC governs the deliberate release of GMOs into the environment, while Regulation (EC) No 1829/2003 addresses genetically modified food and feed. The EU's approach is often described as precautionary, requiring extensive risk assessment, public consultation, and approval before commercialization. Synthetic biology products, particularly those involving novel genetic constructs or microorganisms, fall under these existing frameworks. The EU also promotes harmonization among member states, ensuring consistent evaluation procedures, labeling requirements, and monitoring protocols. Ethical considerations, societal acceptance, and environmental sustainability are integral to regulatory decision-making, reflecting the EU's emphasis on balancing innovation with precaution and public trust.

Regulatory Approaches in Asia

Asia presents a diverse regulatory landscape for synthetic biology, reflecting the varying levels of technological development, economic priorities, and societal attitudes toward biotechnology. In Japan, the regulatory system emphasizes biosafety, environmental protection, and ethical oversight. The Ministry of Health, Labour and Welfare (MHLW) and the Ministry of Agriculture, Forestry and Fisheries (MAFF) oversee synthetic biology applications related to human health and agriculture, while the Ministry of the Environment monitors environmental releases. Japan employs a combination of guidelines, ministerial notifications, and case-by-case approval for novel biotechnologies. In China, regulatory frameworks are evolving rapidly to accommodate the growth of synthetic biology research and industrial applications. The Ministry of Science and Technology, the Ministry of Agriculture and Rural Affairs, and the National Medical Products Administration share responsibilities for oversight. China emphasizes risk assessment, containment, and monitoring while promoting industrial development and innovation. Other Asian countries, including South Korea, Singapore, and India, have developed guidelines that focus on biosafety, laboratory practices, and commercialization approval, often adapting international standards to local contexts. These regulatory approaches reflect both the opportunities and challenges of fostering innovation while ensuring responsible conduct of synthetic biology research.

Biosafety and Biosecurity Considerations

Biosafety and biosecurity are central concerns in regulating synthetic biology worldwide. Biosafety addresses the safe handling, containment, and disposal of synthetic biological materials to prevent accidental exposure or environmental release. Laboratories are categorized according to biosafety levels (BSL), ranging from BSL-1 for low-risk organisms to BSL-4 for high-risk pathogens, with regulatory oversight ensuring compliance with containment standards. Biosecurity focuses on preventing the misuse of synthetic biology for harmful purposes, including bioterrorism and dual-use research of concern (DURC). Governments and international organizations have implemented guidelines for monitoring and controlling access to synthetic genes, DNA synthesis services, and laboratory equipment. Risk assessment frameworks, personnel training, and regulatory oversight combine to mitigate both biosafety and biosecurity threats, ensuring that synthetic biology progresses responsibly without endangering public health or national security.

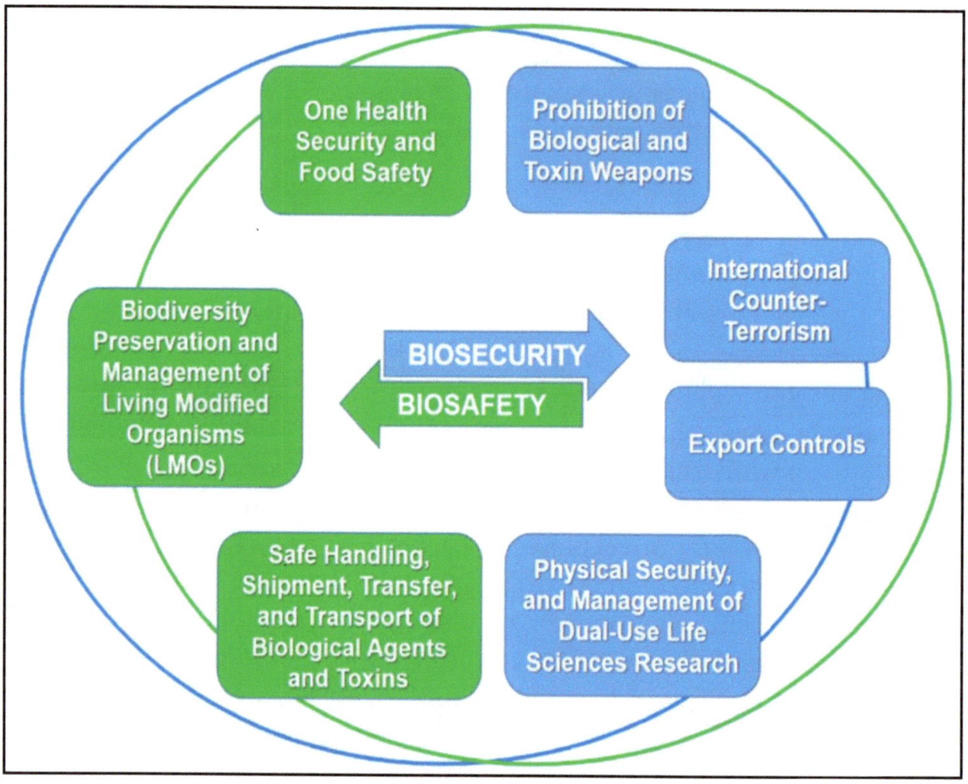

Biosecurity Practices and Biosafety Issues

Ethical and Societal Implications

Regulation of synthetic biology is not limited to technical and safety considerations; it also addresses ethical, social, and societal dimensions. Ethical concerns include the manipulation of life forms, creation of novel organisms, and potential impacts on natural ecosystems. Public perception, societal acceptance, and cultural values significantly influence regulatory policies in different regions. In Europe, ethical review and public engagement are formal components of the regulatory process, ensuring that societal concerns shape approval decisions. In contrast, countries such as the United States emphasize risk-based assessment but still encourage transparency and public consultation. Ethical frameworks aim to balance scientific innovation with social responsibility, ensuring that synthetic biology serves the public good while minimizing moral and ecological concerns. These considerations are increasingly incorporated into international guidelines and regulatory harmonization efforts, reflecting the global dimension of ethical responsibility in synthetic biology.

International Coordination and Standards

Given the global nature of synthetic biology research and commercialization, international coordination is critical. Organizations such as the World Health Organization, the Organisation for Economic Co-operation and Development (OECD), and the International Genetically Engineered Machine (iGEM) Foundation provide guidelines, best practices, and standardization efforts. International agreements emphasize risk assessment, biosafety, ethical conduct, and information sharing. Harmonized standards facilitate collaboration, ensure safe transport and trade of synthetic biology products, and prevent regulatory gaps that could lead to safety or security risks. International coordination also addresses emerging challenges such as synthetic gene synthesis, genome editing, and dual-use research, promoting consistency while allowing flexibility for regional adaptation. Global cooperation supports responsible innovation and reduces the likelihood of conflicting regulations that could hinder scientific progress or market access.

Emerging Trends in Regulation

Regulation of synthetic biology is evolving rapidly to address new technologies such as genome editing, cell-free systems, and synthetic gene drives. Regulatory agencies are increasingly adopting adaptive governance models that allow timely evaluation of novel applications while incorporating stakeholder input. Risk-based frameworks, scenario analysis, and post-market monitoring are becoming standard practices. Public engagement, transparency, and ethical review are emphasized to build trust and societal acceptance. International harmonization efforts continue to expand, promoting shared standards and facilitating cross-border research collaboration and commercialization. Emerging trends also include integrating computational risk modeling, digital tracking of genetic materials, and enhanced oversight of dual-use research to balance innovation with safety and security.

Commercialization and Regulatory Challenges

The commercialization of synthetic biology products presents unique regulatory challenges. Novel therapies, genetically engineered crops, biofuels, and industrial enzymes must navigate complex approval processes to ensure safety, efficacy, and environmental compliance. Companies face long timelines, high costs, and uncertainty due to the rapidly evolving regulatory landscape. Differences between regions, including approval requirements, labeling standards, and risk assessment methodologies, create additional barriers for international market access. Moreover, rapid technological innovation can outpace existing regulations, requiring continuous updates and adaptive governance. Regulatory authorities increasingly recognize the need for flexible, science-based frameworks that accommodate innovation while ensuring public safety, environmental protection, and ethical standards.

Table: Synthetic Biology Regulation by Region

Region	Regulatory Focus	Key Agencies	Approach
United States	Human health, environmental safety, agriculture	FDA, EPA, USDA	Case-by-case, risk-based, adaptive
European Union	Environmental protection, food safety, ethics	EFSA, Member State authorities	Precautionary, harmonized, extensive risk assessment
Japan	Biosafety, ethical oversight	MHLW, MAFF, Ministry of Environment	Guidelines, ministerial approval, case evaluation
China	Industrial development, biosafety	Ministry of Science and Technology, Ministry of Agriculture, National Medical Products Administration	Risk assessment, containment, promotion of innovation
Singapore & South Korea	Laboratory biosafety, commercialization	National biosafety committees, research ministries	Guidelines, case-by-case approval, international standards adaptation

5.3.2 Safety Standards and Risk Assessment in Synthetic Biology

Safety standards and risk assessment constitute pillars for sustainable growth and societal acceptance of synthetic biology. The complex interplay of technical risks, bioethical principles, regulatory frameworks, and societal values necessitates comprehensive, adaptive, and transparent governance systems. Protecting human health and environmental integrity while enabling transformative innovations requires ongoing vigilance, multidisciplinary collaboration, and international harmonization. As synthetic biology continues to expand its scope and impact, robust safety culture and risk governance will be indispensable for harnessing its promise safely and ethically, ensuring beneficial outcomes for current and future generations.

Biosafety Concerns and Containment Strategies

Central to safety standards in synthetic biology is the prevention of unintended release or exposure to genetically engineered organisms, which may pose risks to humans, animals, or ecosystems. Biosafety guidelines categorize organisms based on their pathogenicity, ecological persistence, and gene transfer potential, informing containment measures ranging from physical barriers to genetic safeguards.

Physical containment involves dedicated laboratory facilities—biosafety levels 1 through 4—designed to prevent direct contact or environmental escape. Complementing this are genetic containment strategies engineered into organisms, such as kill switches that induce programmed cell death under specific conditions, dependency on synthetic amino acids or nutrients absent in natural settings, and genetic barriers to horizontal gene transfer. These biological safeguards serve as intrinsic safety nets, enhancing robustness beyond external containment.

Systematic Risk Assessment Frameworks

Risk assessment in synthetic biology employs systematic methodologies to evaluate hazards, exposure likelihood, and potential consequences associated with engineered biological entities and their applications. Risk analyses encompass the characterization of the biological system, including its genetic modifications, stability, replication potential, and ecological fitness. Environmental risk considers gene flow, invasiveness, and impacts on biodiversity and ecosystem functions. Human health risk evaluation examines toxicity, allergenicity, pathogenicity, and potential for antibiotic resistance spread. Risk assessments are iterative, relying on experimental data, computational models, and expert judgment to refine understanding of uncertainties and inform risk management strategies. These frameworks support regulatory decision-making, guide research activities, and shape product development pathways.

Regulatory Landscapes and International Standards

Regulatory oversight of synthetic biology integrates existing frameworks developed for genetically modified organisms (GMOs), pharmaceuticals, and hazardous biological agents while adapting to novel features specific to synthetic systems. Agencies such as the U.S. Food and Drug Administration (FDA), Environmental Protection Agency (EPA), European Food Safety Authority (EFSA), and international bodies provide guidelines, risk assessment criteria, and approval pathways to ensure product safety and efficacy.

Regulations address not only laboratory research but also field trials, commercial production, and distribution. Harmonization efforts, led by organizations such as the Organisation for Economic Co-operation and Development (OECD) and the World Health Organization (WHO), promote consistent standards to facilitate safe innovation and trade while preventing regulatory gaps exploitable by malicious actors. Diverse national policies reflect differing societal values, risk tolerance, and technological capacities, underscoring the need for global dialogue and cooperation.

Ethical and Social Dimensions in Safety Governance

Beyond technical risk management, safety standards in synthetic biology encompass ethical and social considerations that influence governance frameworks. Questions about ecological stewardship, responsible innovation, equitable access, and public engagement shape policies and regulatory approaches. Transparent communication of risks and benefits fosters informed public discourse and trust. Ethical governance emphasizes precautionary principles, considering long-term and indirect effects absent in immediate risk assessments. The inclusion of stakeholders ranging from scientists and policymakers to indigenous communities and civil society ensures diverse perspectives are integrated, mitigating risks related to social acceptability and distributional justice. Anticipatory governance mechanisms incorporate foresight and adaptive regulation to keep pace with rapid technological advances and emerging challenges.

Role of Standardization and Best Practices

Standardization efforts in synthetic biology enhance safety by promoting consistency, reproducibility, and reliability across research and industrial applications. The creation and adoption of standardized biological parts, genetic elements, and assembly methods reduce unpredictability and facilitate quality control. Safety-related best practices include documented protocols for genetic construct characterization, strain validation, and environmental impact testing. International consortia such as the Synthetic Biology Open Language (SBOL) community support data sharing and standard descriptions of biological designs, enabling interoperability and collective safety assessment. Establishing comprehensive registries of engineered organisms and genetic parts with annotated safety and hazard information further promotes transparency and informed risk evaluation.

Emerging Challenges in Safety and Risk Assessment

Rapid advancement in synthetic biology introduces unprecedented challenges in safety regulation and risk assessment. Synthetic organisms with enhanced evolutionary capacities, autonomous behaviors, or synthetic genetic codes challenge traditional containment and predictability assumptions. The convergence of synthetic biology with artificial intelligence and automation accelerates the design-build-test-learn cycle, complicating oversight but also offering opportunities for improved predictive modeling and risk mitigation. Novel therapeutic modalities using engineered cells, gene circuits, and microbiomes require specific biosafety considerations due to their direct interaction with humans. Additionally, globalization and democratization of DNA synthesis technologies raise concerns regarding biosecurity and oversight beyond traditional institutional boundaries. Managing these evolving risks necessitates continuous adaptation of safety frameworks, research into emerging hazards, and international cooperation.

Technological Innovations Supporting Safety

Advances in computational biology, bioinformatics, and machine learning contribute significantly to risk assessment by enabling the simulation of engineered organism behavior, off-target effects, and ecological interactions. Synthetic biology platforms increasingly integrate safety by design, embedding control elements and fail-safe mechanisms from early development stages. Automated DNA synthesis providers implement sequence screening protocols against databases of hazardous sequences to prevent synthesis of dangerous genes or organisms. Real-time environmental monitoring coupled with biosensing and diagnostic technologies facilitates early detection of unintended releases or adverse effects. These technological enablers augment human oversight and regulatory capacities, promoting proactive rather than reactive safety management.

5.4 ENVIRONMENTAL CONCERNS AND ECOLOGICAL IMPACT

5.4.1 Bioengineered Organisms in Natural Ecosystems

The integration of bioengineered organisms into natural ecosystems represents both tremendous opportunity and significant responsibility as humanity develops unprecedented capabilities to reshape the biological world. The potential benefits for addressing environmental degradation, supporting conservation, and adapting to climate change must be weighed against risks of unintended consequences and the ethical implications of deliberately altering natural systems. Success in deploying bioengineered organisms for environmental benefit requires not only technical innovation but also robust risk assessment, effective governance, and meaningful public engagement.

As these technologies continue to advance, society must grapple with fundamental questions about the appropriate level of human intervention in natural ecosystems and our obligations to future generations who will inherit the consequences of today's decisions. The path forward demands interdisciplinary collaboration between ecologists, biotechnologists, social scientists, and ethicists to ensure that bioengineered organisms serve as tools for environmental stewardship rather than sources of new problems. Ultimately, the role of bioengineered organisms in natural ecosystems will be determined by our collective wisdom in balancing innovation with precaution, human benefit with ecological integrity, and present needs with long-term sustainability.

Types and Applications of Bioengineered Organisms

The diversity of bioengineered organisms designed for environmental release reflects the broad range of ecological challenges they aim to address, from agricultural productivity to biodiversity conservation. Genetically modified crops represent the most widespread category of engineered organisms in natural ecosystems, with billions of acres planted globally with varieties expressing traits such as herbicide tolerance, insect resistance, and drought adaptation. These agricultural applications extend beyond crop plants to include bioengineered insects for pest control, such as sterile insect technique programs and more recently, insects carrying gene drives designed to suppress or modify wild populations.

Bioremediation organisms engineered to degrade pollutants or accumulate heavy metals offer promising approaches for cleaning contaminated environments, with modified bacteria and plants showing enhanced abilities to process industrial chemicals, petroleum products, and other environmental contaminants. Conservation applications include organisms designed to restore endangered species through genetic rescue, combat invasive species, or enhance ecosystem resilience to climate change. Modified algae and cyanobacteria are being developed for applications ranging from carbon sequestration to the production of biofuels in outdoor cultivation systems. Even bioengineered probiotics and beneficial microbes designed to protect crops or improve soil health represent intentional introductions of modified organisms into agricultural ecosystems.

Ecological Interactions and Complexity

The behavior of bioengineered organisms in natural ecosystems cannot be fully understood without considering the complex web of ecological interactions that govern biological communities and ecosystem functioning. When released into the environment, engineered organisms become subject to the same ecological and evolutionary forces as their wild counterparts, including competition, predation, parasitism, and mutualism.

These interactions can significantly influence the persistence, spread, and impact of bioengineered organisms in ways that laboratory studies cannot fully predict. Gene flow between engineered organisms and wild relatives represents a primary pathway for the spread of engineered traits, with the potential for creating hybrid populations that possess novel combinations of characteristics. The fitness of bioengineered organisms in natural environments depends on multiple factors including the metabolic cost of engineered traits, environmental conditions, and interactions with other species. Ecological networks can amplify or dampen the effects of engineered organisms through cascading impacts on food webs, nutrient cycling, and ecosystem services. Evolutionary pressures may select for or against engineered traits, potentially leading to unexpected adaptations or the loss of desired characteristics over time. The temporal and spatial scales of ecological processes mean that the full impacts of releasing bioengineered organisms may not become apparent for years or decades after introduction.

Environmental Benefits and Conservation Potential

Bioengineered organisms offer unprecedented opportunities to address environmental degradation and support conservation efforts in ways that traditional approaches cannot achieve. Climate change mitigation strategies increasingly consider engineered organisms for enhanced carbon sequestration, with modified plants and microorganisms designed to capture and store atmospheric carbon more efficiently than natural systems. Bioremediation applications demonstrate remarkable success in cleaning polluted environments, with engineered bacteria breaking down oil spills, degrading plastic waste, and neutralizing toxic chemicals that would otherwise persist for decades. Conservation genetics applications use bioengineering to introduce genetic diversity into endangered populations, potentially saving species from extinction caused by inbreeding depression or lack of adaptive variation. Ecosystem restoration projects employ engineered organisms to accelerate succession, improve soil quality, or reestablish critical ecological functions in degraded habitats. Agricultural sustainability benefits from engineered organisms that reduce the need for chemical pesticides and fertilizers, decreasing agricultural runoff and its associated environmental impacts. Disease control in wildlife populations through engineered vaccines or vector modification could protect endangered species and maintain ecosystem health. The potential for engineered organisms to provide ecosystem services more efficiently than natural systems open new possibilities for sustainable land management and environmental protection.

Risks and Unintended Consequences

The release of bioengineered organisms into natural ecosystems carries inherent risks that stem from both the complexity of ecological systems and the limitations of our ability to predict long-term outcomes. Invasiveness remains a primary concern, as organisms engineered for enhanced fitness or survival could potentially outcompete native species and disrupt ecological communities. Horizontal gene transfer might spread engineered traits to non-target organisms, creating new ecological players with unpredictable characteristics and impacts. Ecosystem disruption could occur through direct effects on species interactions or indirect effects on nutrient cycling, energy flow, and habitat structure. The irreversibility of some releases, particularly those involving gene drives or self-sustaining populations, means that mistakes cannot be easily corrected once organisms are established in the wild. Evolutionary responses by both the engineered organisms and the species they interact with could lead to arms races or unexpected adaptations that undermine the intended benefits or create new problems. Non-target effects on beneficial species such as pollinators or natural enemies of pests could have cascading impacts throughout ecological communities. The potential for bioengineered organisms to affect human health through allergenicity, toxicity, or pathogenicity requires careful assessment, particularly for organisms that might enter food chains or human environments.

Monitoring and Containment Strategies

Monitoring and Containment Cycle

Effective monitoring and containment of bioengineered organisms in natural ecosystems requires sophisticated approaches that combine molecular tools, ecological surveillance, and adaptive management strategies. Molecular markers and tracking systems enable researchers to monitor the presence, abundance, and spread of engineered organisms in the environment, though detecting rare organisms across large landscapes remains technically challenging. Biological containment mechanisms such as genetic kill switches, metabolic dependencies, or reproductive barriers aim to limit the persistence and spread of engineered organisms, though their long-term reliability under selective pressure remains uncertain. Physical containment strategies including buffer zones, barrier crops, and isolation distances provide additional layers of protection but cannot guarantee complete prevention of spread for mobile organisms. Ecological monitoring programs assess both direct impacts on target species or processes and indirect effects on broader ecosystem functioning, requiring long-term data collection and analysis. Early detection systems for unintended spread or unexpected effects enable rapid response before problems become irreversible. Post-release surveillance must continue for extended periods to capture delayed or slow-developing impacts that might not be immediately apparent. Adaptive management frameworks allow for modification of release strategies based on monitoring results, though this requires clear decision criteria and response protocols.

Regulatory Frameworks and Governance

The governance of bioengineered organisms in natural ecosystems involves complex regulatory frameworks that must balance innovation, environmental protection, and public concerns while dealing with scientific uncertainty and evolving technology. National regulations vary significantly in their approach to environmental release, from precautionary principles that require extensive testing before approval to more permissive systems that regulate based on product characteristics rather than production methods. International agreements such as the Cartagena Protocol on Biosafety provide frameworks for transboundary movement and risk assessment but struggle to keep pace with technological advances. Risk assessment procedures evaluate potential environmental impacts through tiered testing from laboratory to confined field trials to commercial release, though standardized methods for assessing ecological risks remain underdeveloped. Public participation in decision-making varies widely across jurisdictions, with some countries mandating extensive consultation while others limit public input to formal comment periods. Liability and compensation mechanisms for potential damages from bioengineered organisms remain contentious and poorly developed in most regulatory systems. The challenge of regulating gene drives and other self-sustaining technologies that could spread across national boundaries requires new forms of international cooperation and governance. Regulatory uncertainty and inconsistency across jurisdictions create challenges for researchers and developers while potentially undermining public trust in oversight systems.

Case Studies and Real-World Examples

Examining specific cases of bioengineered organisms released into natural ecosystems provides valuable insights into both successes and challenges of environmental applications. The widespread adoption of Bt crops expressing insecticidal proteins has demonstrated both benefits in reduced pesticide use and concerns about resistance evolution in target pests and impacts on non-target species. The release of genetically modified mosquitoes in several countries to combat dengue and other diseases has shown promising results while raising questions about community consent and long-term ecological effects. Bioremediation projects using engineered bacteria to clean up oil spills and industrial contamination have achieved significant success in controlled settings but face challenges in open environmental applications. The development of American chestnut trees engineered for blight resistance represents a potential conservation success story but also highlights regulatory and social challenges in releasing engineered organisms for ecosystem restoration. Gene drive mosquitoes tested in contained facilities demonstrate the potential for population suppression but have not yet been released due to ecological and ethical concerns. These real-world examples illustrate the complex interplay between technical feasibility, ecological impact, regulatory approval, and social acceptance that determines the success of environmental applications.

5.4.2 Environmental Risk Assessment and Mitigation Strategies

Environmental risk assessment and mitigation strategies are fundamental to sustainable development, public health, and ecosystem protection. ERA provides a systematic methodology for identifying hazards, evaluating exposure, and quantifying potential environmental impacts. Mitigation strategies, encompassing preventive, corrective, and adaptive measures, reduce the likelihood and severity of environmental damage while promoting sustainability. Modern tools, including life cycle assessment, GIS, ecological modeling, quantitative risk assessment, and remote sensing, support informed decision-making and effective management. Regulatory frameworks and international conventions ensure compliance and promote responsible environmental stewardship. Despite challenges related to uncertainty, resource limitations, and climate change, ongoing innovation in monitoring, modeling, and participatory approaches enhances the effectiveness of risk assessment and mitigation. By integrating scientific, technological, and policy measures, societies can protect natural resources, maintain ecosystem services, and promote sustainable development while addressing the environmental challenges of the 21st century.

Concept of Environmental Risk Assessment

Environmental risk assessment is a structured process used to estimate the likelihood and consequences of environmental hazards. It integrates scientific data, predictive modeling, and regulatory frameworks to evaluate potential harm to ecosystems, human populations, and natural resources. ERA typically involves hazard identification, dose-response assessment, exposure assessment, and risk characterization. Hazard identification determines which agents or activities could cause adverse effects, while dose-response assessment examines the relationship between exposure levels and the severity of effects. Exposure assessment evaluates how organisms, including humans, come into contact with hazards, and risk characterization synthesizes this information to quantify the probability and severity of environmental damage. By providing a scientific basis for management decisions, ERA helps prevent environmental degradation and protect public health.

Importance of Environmental Risk Assessment

The importance of ERA lies in its ability to anticipate and prevent ecological and health crises. By identifying potential hazards before they cause significant damage, ERA allows governments, industries, and communities to implement proactive measures. In industrial contexts, ERA ensures compliance with environmental regulations and helps avoid costly accidents or fines. In agriculture, assessing the risks of pesticides, fertilizers, and genetically modified organisms informs safer application methods and reduces environmental contamination. ERA is also essential for infrastructure development, such as dams, highways, and urban projects, by evaluating potential impacts on water, soil, air, and biodiversity. Overall, ERA enhances sustainability, reduces liability, and promotes informed decision-making in both public and private sectors.

Phases of Environmental Risk Assessment

Environmental risk assessment is generally conducted in a series of structured phases. Problem formulation defines the scope, objectives, and key questions of the assessment, including the identification of sensitive species or ecosystems. Hazard identification involves cataloging potential environmental threats, including chemical pollutants, biological agents, or physical disturbances. Dose-response or effect analysis evaluates the intensity and frequency of adverse effects associated with different levels of exposure. Exposure assessment quantifies the extent, duration, and pathways of environmental exposure. Finally, risk characterization integrates hazard and exposure data to estimate the likelihood and magnitude of adverse effects. Iterative monitoring and re-assessment are essential, as environmental conditions, regulatory standards, and scientific understanding evolve over time.

Types of Environmental Risks

Environmental risks can be categorized into several types based on their sources and impacts. Chemical risks include pollutants such as heavy metals, pesticides, industrial solvents, and greenhouse gases. Biological risks involve invasive species, pathogens, or genetically modified organisms that may disrupt ecosystems. Physical risks encompass habitat destruction, soil erosion, deforestation, and natural disasters exacerbated by human activity. Socioeconomic risks arise when environmental degradation affects livelihoods, public health, or community well-being. Cumulative and synergistic risks result from the combined effects of multiple stressors, which may amplify harm beyond individual contributions. Effective risk assessment requires considering all relevant types to develop comprehensive mitigation strategies.

Mitigation Strategies

Mitigation strategies aim to reduce the likelihood, magnitude, or impact of environmental risks. They can be categorized into preventive, corrective, and adaptive measures. Preventive strategies involve avoiding or minimizing environmental hazards before they occur, such as implementing cleaner production technologies or using eco-friendly materials. Corrective measures address environmental damage that has already occurred, including soil remediation, reforestation, or water treatment. Adaptive strategies enhance the resilience of ecosystems and communities to cope with environmental stressors, such as floodplain restoration or climate-resilient agriculture. Successful mitigation strategies often require an integrated approach combining regulatory enforcement, technological innovation, stakeholder engagement, and continuous monitoring.

Tools and Approaches for Risk Assessment

Various tools and approaches facilitate environmental risk assessment and mitigation.

- Life Cycle Assessment (LCA) evaluates environmental impacts of products or processes from resource extraction to disposal.
- Geographic Information Systems (GIS) map and analyze spatial patterns of environmental hazards and vulnerable areas.
- Ecological modeling simulates interactions among species, habitats, and pollutants to predict ecological responses.
- Quantitative risk assessment (QRA) uses statistical and probabilistic methods to estimate the likelihood of adverse outcomes.
- Environmental impact assessment (EIA) examines potential consequences of proposed projects on ecosystems and communities.
- Environmental monitoring programs track pollutant levels, biodiversity changes, and ecosystem health over time.
- Remote sensing technologies provide large-scale data on land use, vegetation cover, and environmental degradation.
- Toxicological studies evaluate the effects of chemical or biological agents on organisms at different exposure levels.

- Decision support systems (DSS) integrate multiple data sources to guide policy and management interventions.
- Stakeholder consultation and participatory assessment involve communities and industries in risk identification and mitigation planning.

These tools enable data-driven assessment, predictive modeling, and informed decision-making in environmental management.

Regulatory Frameworks and Policies

Environmental risk assessment and mitigation are supported by national and international regulations and policies. In many countries, environmental protection agencies enforce standards for air and water quality, waste management, and chemical usage. Environmental impact assessment (EIA) legislation mandates systematic evaluation of proposed projects before approval. International conventions, such as the Stockholm Convention on Persistent Organic Pollutants, the Convention on Biological Diversity, and the Paris Agreement on climate change, guide transboundary environmental risk management. Regulatory frameworks ensure that industries and governments implement mitigation strategies, conduct monitoring, and adopt best practices to minimize environmental harm. Policies also promote transparency, public participation, and accountability in environmental decision-making.

Challenges in Risk Assessment and Mitigation

Several challenges complicate effective environmental risk assessment and mitigation. Uncertainty in data, especially regarding emerging pollutants, complex ecosystems, or long-term impacts, can hinder accurate risk characterization. Limited resources and technical capacity in developing regions may reduce the ability to monitor and mitigate environmental risks. Conflicting stakeholder interests, regulatory gaps, and enforcement limitations may also compromise mitigation efforts. Additionally, climate change introduces dynamic and unpredictable environmental stressors, requiring adaptive and forward-looking approaches. The integration of multidisciplinary data, evolving scientific knowledge, and socio-economic considerations is essential to overcome these challenges and design effective mitigation strategies.

- Environmental risk assessment identifies, evaluates, and characterizes potential hazards to ecosystems and human health.
- Risk assessment involves hazard identification, dose-response analysis, exposure assessment, and risk characterization.
- Environmental risks include chemical, biological, physical, socioeconomic, and cumulative stressors.

- Mitigation strategies can be preventive, corrective, or adaptive, depending on the nature of the risk.
- Life cycle assessment evaluates the environmental impact of products or processes from production to disposal.
- Geographic information systems and ecological modeling facilitate spatial and predictive risk analysis.
- Regulatory frameworks, environmental impact assessments, and international conventions guide risk management.
- Emerging technologies, such as remote sensing, AI, and big data analytics, enhance risk prediction and mitigation.
- Stakeholder engagement and participatory assessment improve decision-making and accountability.
- Sustainable solutions, including nature-based interventions and green technologies, increase resilience and reduce environmental damage.

CHAPTER 6

Ethical, Legal, and Social Implications (ELSI)

6.1 ETHICAL ISSUES IN SYNTHETIC BIOLOGY

6.1.1 "Playing God": Engineering Life Forms

Engineering life forms represents one of the most profound and transformative capabilities of modern science. It offers enormous potential to address pressing challenges in medicine, agriculture, environment, and industry while simultaneously raising ethical, ecological, and societal questions. The notion of "playing God" captures the tension between human ingenuity and the moral responsibility associated with altering life.

Balancing innovation with ethical reflection, regulatory oversight, public engagement, and environmental stewardship is essential for the responsible development of engineered organisms. Advances in technology, coupled with thoughtful governance and ethical frameworks, will determine whether humanity harnesses the power to engineer life responsibly, maximizing societal benefit while minimizing risk. By navigating these complex dimensions, society can embrace the promise of engineered life forms while respecting the inherent responsibilities that accompany the profound ability to shape life itself.

Historical Context of Engineering Life

Human fascination with modifying life has deep historical roots, from selective breeding in agriculture to early genetic experimentation in the twentieth century. Traditional breeding and domestication allowed humans to shape crops, livestock, and companion animals to meet specific needs, effectively manipulating the traits of living organisms without directly altering their genetic code. The discovery of DNA as the molecular basis of heredity, coupled with the development of recombinant DNA technology in the 1970s, marked a turning point, allowing precise intervention at the genetic level. Techniques such as gene cloning, gene editing, and transgenic organism creation provided unprecedented control over life's building blocks. The advent of synthetic biology in the twenty-first century extended these capabilities further, enabling the construction of entirely novel genetic circuits, synthetic genomes, and engineered microorganisms. Each technological leap has intensified public discourse on the ethical and philosophical dimensions of manipulating life, framing the notion of "playing God" in a modern scientific context.

Synthetic Biology and Life Engineering

Synthetic biology represents the forefront of engineering life, combining principles of biology, engineering, and computational design to create programmable biological systems. Synthetic biologists design genetic circuits, construct synthetic genomes, and rewire cellular behavior to produce desired outputs such as biofuels, pharmaceuticals, or biosensors. By treating cells as programmable entities, researchers can construct novel life forms or substantially modify existing ones to perform specific tasks. For example, engineered bacteria have been developed to degrade environmental pollutants, synthesize therapeutic compounds, or detect disease markers in human bodies. While these innovations hold enormous potential, they also underscore the ethical tension inherent in the power to reconfigure life, highlighting questions about human stewardship, responsibility, and the boundaries of scientific intervention.

Ethical Dimensions of "Playing God"

The ethical implications of engineering life forms are multifaceted, encompassing moral, societal, and ecological considerations. At the core is the question of whether humans possess the right to create or fundamentally alter life, and whether doing so might violate moral or natural laws. Critics argue that manipulating life may disrupt natural processes, compromise ecological balance, or lead to unforeseen consequences, especially when synthetic organisms are released into the environment. Ethical concerns also include issues of consent, particularly in human applications such as gene therapy, germline modification, or designer babies. Conversely, proponents contend that engineering life can be ethically justified when it addresses pressing challenges such as disease eradication, food security, or environmental sustainability. Ethical frameworks in synthetic biology often emphasize principles such as beneficence, non-maleficence, justice, and transparency, encouraging responsible innovation that balances societal benefit with moral accountability.

Religious and Philosophical Perspectives

Religious and philosophical perspectives have significantly influenced public perceptions of "playing God." Many religious traditions caution against human hubris in assuming divine powers, viewing the creation or alteration of life as an ethically sensitive domain. For instance, in some interpretations of Christianity and Islam, life is sacred and its creation is considered a divine prerogative, raising concerns about human intervention in fundamental natural processes. Philosophical debates often center on questions of moral responsibility, the limits of human knowledge, and the precautionary principle. Philosophers argue that engineering life entails profound responsibility for potential consequences, including ecological disruptions, unintended health impacts, or ethical slippery slopes. While religious and philosophical perspectives vary, they converge on the importance of reflection, caution, and societal discourse when engaging in life engineering, reinforcing the notion that scientific capability must be tempered by ethical judgment.

Technological Advances and Implications

Technological advances have dramatically expanded the capacity to engineer life. CRISPR-Cas9 and other genome-editing tools enable precise modifications of DNA sequences, allowing scientists to correct genetic disorders, enhance agricultural traits, or create novel organisms. Synthetic genome construction allows the assembly of entire bacterial genomes in vitro, effectively creating life forms from scratch. These tools accelerate innovation but also intensify concerns about unintended consequences, biosecurity, and ethical responsibility. Applications in human health, such as gene therapy or xenotransplantation, illustrate the tension between transformative potential and moral responsibility. While engineered life forms offer solutions to previously intractable problems, their creation also necessitates rigorous oversight, ethical reflection, and risk management to prevent misuse, ecological harm, or societal inequities.

Environmental and Ecological Considerations

Environmental and Ecosystem Considerations

The engineering of life forms carries significant environmental and ecological considerations. Synthetic organisms released into natural ecosystems may interact with existing species, potentially leading to disruption of food webs, unintended competition, or horizontal gene transfer.

Engineered microorganisms designed for environmental remediation or biofuel production must be carefully monitored to ensure that their introduction does not destabilize local ecosystems. Risk assessment frameworks, containment strategies, and post-release monitoring are essential to mitigate potential ecological impacts. Additionally, synthetic life can contribute positively to environmental sustainability, such as through carbon capture, biodegradable material production, or remediation of contaminated sites. Balancing these benefits with potential ecological risks underscores the need for comprehensive environmental governance and responsible innovation in life engineering.

Social and Public Perception

Public perception of "playing God" plays a critical role in shaping policy, regulation, and scientific practice. Media coverage, ethical debates, and societal values influence how synthetic biology and genetic engineering are perceived by the public. Concerns about safety, morality, and unnatural manipulation of life can lead to skepticism or opposition, while understanding of potential benefits may foster support for research and commercialization. Public engagement initiatives, educational programs, and transparent communication are essential to build trust, address misconceptions, and foster informed societal dialogue. Social acceptance is crucial for the adoption of engineered life forms in healthcare, agriculture, and environmental applications, demonstrating that ethical and societal considerations are inseparable from scientific progress.

Regulatory Frameworks and Oversight

Regulation and oversight are critical in addressing the ethical and safety concerns associated with engineering life forms. Different regions have developed frameworks to govern the creation, manipulation, and release of synthetic organisms. In the United States, agencies such as the FDA, EPA, and USDA provide case-by-case oversight for genetically engineered organisms and synthetic biology applications.

The European Union emphasizes precautionary principles, extensive risk assessment, and harmonized regulatory standards through EFSA and member state authorities. International coordination, guidelines, and standards from organizations like the WHO and OECD support biosafety, biosecurity, and ethical compliance. Regulatory frameworks aim to balance innovation with protection of human health, the environment, and societal values, ensuring that engineered life forms are developed responsibly and ethically.

Potential Benefits and Societal Impact

Despite ethical concerns, engineering life forms offers substantial societal benefits. In medicine, gene therapies and engineered cells provide treatments for genetic disorders, cancer, and infectious diseases. In agriculture, synthetic biology enables the development of crops with enhanced nutritional value, pest resistance, and climate resilience. Environmental applications include bioremediation, sustainable biofuel production, and carbon capture. Industrial biotechnology benefits from engineered microorganisms capable of producing enzymes, chemicals, and materials with high efficiency. These innovations can address critical global challenges such as food security, public health, and environmental sustainability, highlighting the transformative potential of life engineering when conducted responsibly.

Challenges and Controversies

The creation of engineered life forms is accompanied by significant challenges and controversies. Technical challenges include controlling gene expression, preventing unintended mutations, and ensuring stability and predictability of synthetic organisms. Ethical controversies center on the moral legitimacy of altering life, potential inequities in access to technology, and long-term consequences for society and ecosystems. Safety concerns encompass biosecurity risks, accidental release of engineered organisms, and dual-use potential for harmful applications. Governance challenges include harmonizing regulations across regions, adapting frameworks to rapid technological change, and integrating public input into decision-making. These challenges underscore the complexity of balancing innovation, safety, ethics, and societal acceptance in the engineering of life forms.

Table: Considerations in Engineering Life Forms

Aspect	Implications
Ethical	Moral responsibility, societal consent, natural order concerns
Environmental	Ecological balance, risk of unintended consequences, biosafety
Technological	Genome editing, synthetic organisms, predictability challenges
Regulatory	Oversight, biosafety protocols, international coordination
Societal	Public perception, acceptance, education, and communication

6.1.2 Genetic Privacy and Ethical Boundaries

Unique Challenges of Genetic Privacy

Genetic privacy is uniquely vulnerable because the information is both personal and shared among relatives, raising complex questions about consent and disclosure. Unlike other personal data that can be changed or anonymized effectively, genetic information is permanent and potentially re-identifiable through familial connections or genomic databases. Advances in data mining and cross-referencing techniques have shown that even anonymized genomic data can often be traced back to individuals, increasing risks of unauthorized access and misuse. The widespread availability of direct-to-consumer genetic testing and public genetic repositories has democratized access to genomic information but concomitantly increased exposure to privacy breaches and unintended familial disclosures. Consequently, safeguarding genetic privacy demands robust data protection regulations, consent protocols sensitive to familial implications, and technological tools such as encryption, differential privacy, and synthetic data generation to mitigate risks.

Ethical Considerations in Genetic Information Use

Ethical concerns around genetic information revolve around autonomy, confidentiality, equity, and justice. Individuals have a fundamental right to make informed decisions about whether and how their genetic data is collected, stored, and shared. Yet, ethical tensions arise when balancing individual privacy against potential benefits to public health or scientific advancement. Issues include whether relatives ought to be informed of inherited risks discovered incidentally, how to handle unexpected findings, and who owns genetic data once it enters research or commercial realms. Furthermore, concerns about genetic discrimination by employers or insurers undermine trust and willingness to participate in genomic initiatives. Ethical frameworks emphasize confidentiality and the minimization of harm, advocating transparency, equitable access, and protections against misuse to uphold these principles. The justice dimension calls attention to disparities in genetic research participation and access to genomic medicine between different populations, insisting on inclusivity and fairness.

Ethical Boundaries in Synthetic Biology

Synthetic biology amplifies ethical questions by introducing the creation of novel genetic constructs and organisms that transcend traditional boundaries of natural species and biological functions. The ability to design and synthesize synthetic DNA sequences, including potentially entire genomes, prompts reflections on the moral status of synthetic life forms and the appropriate uses of engineered biology. Philosophical debates question whether synthetic organisms should be regarded merely as engineered machines or as entities meriting moral consideration, given their living qualities. The creation of synthetic human genomes, although currently theoretical, evokes concerns related to human dignity, identity, and potential eugenics applications, raising profound societal and ethical implications. Addressing these questions involves defining responsible stewardship, applying precautionary principles, and ensuring that research and applications promote public welfare without compromising ethical standards.

Governance and Legal Frameworks

Governance of genetic privacy and synthetic biology ethics involves multi-layered legal and regulatory frameworks that seek to harmonize innovation with safety, security, and respect for rights. International guidelines, national laws, and institutional review processes regulate genetic data use, synthetic organism development, and associated research practices.

Examples include the Genetic Information Nondiscrimination Act (GINA) addressing workplace and health insurance discrimination, data protection laws like GDPR ensuring privacy rights, and biosafety regulations governing synthetic biology practices. These frameworks emphasize informed consent, data security, risk assessment, and public transparency. Challenges include keeping pace with rapidly evolving technologies, addressing gaps in enforcement, and fostering global coordination to prevent disparities and misuse. Adaptive governance approaches that integrate ethical deliberation, stakeholder engagement, and anticipatory regulation are vital to robust oversight.

Societal Implications and Public Engagement

Public perception and societal values play a pivotal role in shaping the ethical and privacy landscapes of genetics and synthetic biology. Misinformation, cultural differences, and historic abuses in genetic research add layers of complexity to public trust and policy acceptance. Engaging diverse communities through education, dialogue, and participatory decision-making mechanisms enhances legitimacy, mitigates fears, and informs ethical frameworks that resonate with societal norms. Public deliberation frameworks encourage transparency regarding potential risks and benefits, promoting ethical reflexivity among scientists and policymakers. Addressing issues of equity requires ensuring marginalized and indigenous communities have a voice in governance and equitable access to benefits arising from genetic and synthetic biology innovations.

6.1.3 Ethical Concerns in Human Genetic Modifications

Human genetic modification is a double-edged sword, offering unprecedented power to heal and improve while posing significant ethical perils. The decisions we make today will shape the future of humanity in ways that are difficult to fully comprehend. A precautionary approach is essential, prioritizing safety, equity, and justice.

Robust, adaptive, and inclusive governance frameworks are needed at local, national, and international levels to guide research and application. Continuous, transparent, and global public engagement is crucial to ensure that diverse perspectives are heard and that the development of these technologies reflects societal values. The goal should not be to stifle innovation but to direct it wisely, ensuring that genetic modification is used to alleviate suffering, promote human flourishing, and uphold the inherent dignity and equality of all people. As we stand on this precipice, the choices we make will determine whether genetic technology becomes a powerful force for good or a source of new forms of injustice and division.

Human Genetic Modification: From Therapy to Enhancement

Human genetic modification encompasses a spectrum of techniques aimed at altering an individual's DNA. Somatic gene therapy targets non-reproductive cells, aiming to treat or cure diseases in the individual patient without affecting future generations. This is the most widely accepted form of genetic intervention, exemplified by approved therapies for certain inherited immunodeficiencies and retinal diseases. Germline editing, in contrast, involves modifying sperm, eggs, or embryos, resulting in changes that can be passed down to offspring.

This form is far more controversial due to its heritable nature and the ethical implications of altering the human gene pool. Gene enhancement pushes the boundaries further, involving modifications to improve traits like intelligence, athleticism, appearance, or longevity, even in the absence of disease. This distinction between therapy (restoring normal function) and enhancement (exceeding normal function) is central to many ethical debates. The development of these technologies has accelerated dramatically, moving from theoretical concepts to clinical applications in a relatively short time, intensifying the urgency of ethical scrutiny.

Core Ethical Concerns: A Multifaceted Dilemma

The ethical concerns surrounding human genetic modification are deeply interconnected and resist simple resolution. They touch upon fundamental questions about human rights, societal values, and our relationship with nature.

- **Safety and Unintended Consequences:** Perhaps the most immediate concern is safety. Genetic editing, especially using tools like CRISPR, can cause off-target effects—unintended mutations in parts of the genome not targeted for editing. These could lead to cancer, other diseases, or unpredictable health consequences. For somatic therapies, this risk is borne by the consenting patient. However, for germline editing or enhancements, the risks extend to future individuals who cannot consent. Ensuring the long-term safety of genetic modifications is a significant technical challenge, and the potential for unforeseen health problems raises serious ethical red flags. The principle of **non-maleficence** ("do no harm") demands that the benefits must demonstrably outweigh the risks before proceeding.
- **Informed Consent and Autonomy:** Informed consent is a cornerstone of medical ethics. For somatic gene therapy, obtaining valid consent

from patients or their guardians is possible, though the complexity of the science can make truly informed consent difficult. The situation becomes far more problematic with germline editing or enhancements. Future generations affected by heritable changes cannot consent to the modifications made to their DNA. This violates the principle of autonomy for those yet unborn. Furthermore, for children or vulnerable populations, ensuring consent is free from coercion (e.g., parental pressure for enhancement) is critical.

- **Equity, Justice, and Access:** Genetic technologies are likely to be expensive initially, raising serious concerns about **access and equity**. If only the wealthy can afford genetic enhancements or cures for genetic diseases, it could exacerbate existing social inequalities, creating a genetic divide between the "enhanced" and the "unenhanced." This could lead to new forms of discrimination and social stratification. The principle of **justice** requires that the benefits of genetic medicine be distributed fairly. Internationally, the gap between developed and developing nations could widen if these technologies are inaccessible to the global poor, potentially violating principles of global health equity.
- **Germline Editing and Heritable Changes:** Modifying the human germline is arguably the most ethically fraught application. It represents a permanent alteration to the human gene pool, with consequences echoing through countless generations. This raises fears of irrevocable mistakes—errors that could become fixed in the human population. Ethicists also worry about eugenics—the deliberate attempt to improve the human species through selective breeding or genetic control. While modern intentions may be therapeutic, the history of eugenics movements, marked by coercion and human rights abuses, casts a long shadow. Many argue that humans should not "play God" by altering the fundamental blueprint of humanity. The **He Jiankui incident** (see below) starkly highlighted the risks and ethical violations associated with heritable germline editing.
- **Identity, Diversity, and Human Nature:** Genetic modification challenges concepts of human identity and diversity. If we can edit genes to eliminate diseases or enhance traits, what does it mean to be human? Could it lead to a loss of genetic diversity, which is crucial for the health and adaptability of the human population? Some fear that a societal push for "perfection" could devalue people with disabilities or genetic differences, undermining principles of inclusivity and respect for human variation. Enhancement, in particular, might redefine human

nature, prioritizing certain traits over others and potentially eroding empathy for natural human limitations.

- **Societal and Eugenic Concerns:** Beyond individual ethics, genetic modification has broader societal implications. It could lead to social pressure to conform to certain genetic standards, particularly for enhancement. Parents might feel compelled to edit their children's genes for better health, intelligence, or appearance to give them a competitive edge, creating a new form of coercion. This could fundamentally alter family dynamics and societal values. The potential for misuse by authoritarian regimes or malicious actors to create biological weapons or designer humans is also a significant security concern. Furthermore, the line between therapy and enhancement is often blurry and culturally dependent, making global consensus difficult.

Case Studies: Lessons from the Front Lines

Real-world events have dramatically shaped the ethical discourse:

- **The He Jiankui Incident:** In 2018, Chinese biophysicist He Jiankui announced the birth of the world's first CRISPR-edited babies, twin girls whose genomes were altered to confer resistance to HIV. This act was met with near-universal condemnation from the scientific community and the public. Key ethical violations included:
  - Performing germline editing on embryos for reproduction without sufficient safety data or understanding of long-term risks.
  - Lack of proper informed consent (participants were reportedly misled about the nature of the research).
  - Ignoring international consensus (at the time) that germline editing for reproduction was irresponsible and should not be attempted.
  - Exacerbating health risks (the edited gene, CCR5, has been linked to increased susceptibility to other diseases like West Nile virus and potentially altered cognitive function in some studies).
  - This incident served as a wake-up call, highlighting the urgent need for stricter oversight, enforceable moratoria, and global dialogue on germline editing.
- **Gene Therapy Successes and Setbacks:** Approved somatic gene therapies, like Luxturna for inherited blindness or Zolgensma for spinal muscular atrophy, demonstrate the life-changing potential of the technology. These therapies, while expensive, offer cures where none existed before. However, serious setbacks, such as the death of Jesse Gelsinger in a 1999 gene therapy trial due to an immune reaction, and

cases of leukemia linked to early viral vector therapies, underscore the critical importance of rigorous safety testing, phased clinical trials, and transparent risk communication. These events reinforce the ethical principle that patient safety must remain paramount.

Navigating the Ethical Landscape: Governance and Regulation

The complex ethical challenges necessitate robust governance frameworks at multiple levels:

- **International Guidelines and Treaties:** The Oviedo Convention (Council of Europe, 1997) is one of the few binding international treaties addressing human genetic modification, prohibiting germline interventions for reproduction. However, key nations like the US, China, and Russia are not signatories. The UNESCO Declaration on Bioethics and Human Rights (2005) provides ethical principles but is not legally binding. The WHO Expert Advisory Committee on Developing Global Standards for Governance and Oversight of Human Genome Editing is actively working to establish international standards, particularly in the wake of the He Jiankui incident. There are ongoing calls for a global moratorium or ban on heritable germline editing.
- **National Regulatory Frameworks:** Countries have developed varying regulatory approaches:
  - **United States:** The FDA regulates somatic gene therapies strictly, requiring extensive preclinical and clinical data. Germline editing for reproduction is effectively banned, with no federal funding for such research and restrictions on clinical applications. The NIH does not fund research involving heritable germline editing.
  - **United Kingdom:** The Human Fertilisation and Embryology Authority (HFEA) strictly regulates research involving human embryos, including CRISPR editing. While basic research is permitted under license, clinical application for reproduction is currently prohibited.
  - **China:** Following the He Jiankui incident, China tightened regulations, imposing stricter penalties for unethical biomedical research and proposing new laws to ban the clinical application of heritable germline editing. However, basic research continues.
  - **European Union:** The EU has stringent regulations on GMOs, and its framework on advanced therapies (ATMPs) covers gene therapies, requiring rigorous centralized authorization.
- **The Role of Institutional Review Boards (IRBs) and Ethics Committees:** These bodies play a crucial role in reviewing and approving research protocols involving human subjects, including genetic mod-

ification. They assess risks and benefits, ensure informed consent, and evaluate the ethical implications of proposed studies. Their effectiveness depends on their independence, expertise, and adherence to evolving ethical standards.

Slippery Slope: Enhancement vs. Therapy

The line between using genetic modification to treat disease and to enhance human traits is notoriously difficult to define and police. Is correcting a genetic mutation that causes blindness therapy, while enhancing night vision enhancement? What about editing genes to increase muscle mass in a healthy individual? Many ethicists argue that slippery slope concerns are valid: allowing somatic enhancement could normalize the practice and eventually lead to societal pressure for germline enhancement or more extreme modifications. This could divert resources from essential healthcare, exacerbate social inequalities, and fundamentally alter human values. Distinguishing therapy from enhancement is crucial for regulation, but it remains ethically contentious and culturally relative. Some societies may prioritize collective well-being, while others emphasize individual autonomy and choice.

Global Perspectives and Cultural Variations

Ethical norms and public acceptance of human genetic modification vary significantly across cultures:

- **Western Individualism:** In countries like the US, there is often a stronger emphasis on individual autonomy and the right to choose enhancement for oneself or one's children, coupled with significant concerns about safety and corporate control.
- **Communitarian Values:** In many Asian and African cultures, there may be greater emphasis on family and societal benefits, potentially leading to different attitudes towards germline editing for hereditary disease prevention, but also concerns about collective identity and eugenics.
- **Religious Perspectives:** Religious traditions hold diverse views. Some interpretations within Abrahamic faiths (Judaism, Christianity, Islam) may oppose germline editing on the grounds that it usurps God's role or disrupts natural order, while others may support it for alleviating suffering. Dharmic religions (Hinduism, Buddhism) may emphasize non-harm (ahimsa) and the sanctity of life, influencing views on both therapy and enhancement.
- **Global South Perspectives:** There is often heightened concern in the Global South about exploitation, neocolonialism, and being used as testing grounds for unproven technologies. Ensuring equitable access and inclusive decision-making is a major priority.

6.2 LEGAL FRAMEWORK FOR SYNTHETIC BIOLOGY

6.2.1 Global Standards and Regulation of Synthetic Biology

Global standards and regulation play a pivotal role in the development, commercialization, and societal integration of synthetic biology. As a field with profound implications for healthcare, agriculture, energy, and environmental sustainability, synthetic biology requires frameworks that balance innovation with safety, ethics, and equitable access. Standardization enhances reproducibility, interoperability, and data sharing, while regulatory oversight ensures compliance with biosafety, biosecurity, and ethical guidelines. International cooperation, adaptive governance, and harmonized technical practices are essential to address challenges posed by rapid technological advancement and diverse regulatory environments. By fostering responsible innovation, robust oversight, and global collaboration, standards and regulations ensure that synthetic biology contributes positively to societal, environmental, and economic objectives. The continued evolution of global frameworks will support a safe, transparent, and equitable synthetic biology ecosystem capable of addressing pressing global challenges.

Importance of Global Standards in Synthetic Biology

Global standards in synthetic biology provide consistent frameworks for research, development, and commercialization. Standardization ensures that synthetic biological parts, such as DNA sequences, enzymes, and genetic circuits, can be reliably reproduced and shared across laboratories worldwide. Standardized protocols for gene synthesis, assembly, and characterization enhance reproducibility, reduce errors, and accelerate innovation. Moreover, standards facilitate interoperability of biological systems, enabling researchers and companies to integrate modular parts effectively. Beyond technical aspects, global standards also encompass biosafety, biosecurity, ethical practices, and data reporting, creating a comprehensive structure that promotes responsible research. The development of internationally recognized standards enhances collaboration, reduces barriers to technology transfer, and supports regulatory compliance across borders.

Regulatory Challenges in Synthetic Biology

Regulating synthetic biology is inherently complex due to the novelty and diversity of the field. Engineered organisms may not fit neatly into existing regulatory categories, which were primarily designed for conventional biotechnology, pharmaceuticals, or chemical products. The rapid pace of innovation often outstrips the ability of regulators to develop and implement appropriate guidelines. Key challenges include assessing biosafety risks, such as accidental release of genetically modified organisms, and biosecurity threats, including potential misuse for harmful purposes. Ethical considerations, such as gene editing in humans, environmental release of synthetic organisms, and equitable access to technologies, further complicate regulatory decisions. Moreover, global disparities in regulatory infrastructure and capacity create challenges for harmonization and consistent enforcement.

International Regulatory Frameworks

Several international frameworks guide the regulation of synthetic biology, focusing on biosafety, biosecurity, and ethical standards. The Cartagena Protocol on Biosafety under the Convention on Biological Diversity provides guidelines for the safe transfer, handling, and use of living modified organisms that may have adverse effects on biodiversity. The World Health Organization (WHO) provides guidance on the development and regulation of genetically engineered vaccines and therapeutics. The Organisation for Economic Co-operation and Development (OECD) offers standards for environmental safety testing and risk assessment of biotechnology products. International cooperation is crucial to address transboundary issues, harmonize safety standards, and facilitate trade while ensuring that synthetic biology applications meet ethical and environmental requirements.

National Regulatory Approaches

Individual countries have developed regulatory frameworks tailored to their legal, environmental, and social contexts. In the United States, oversight involves multiple agencies, including the Food and Drug Administration (FDA), the Environmental Protection Agency (EPA), and the Department of Agriculture (USDA), each responsible for different aspects of synthetic biology products, such as pharmaceuticals, genetically engineered crops, and industrial microbes. The European Union relies on a precautionary approach, integrating biosafety, risk assessment, and public consultation under directives related to genetically modified organisms. In Asia and Africa, regulatory structures are evolving, with many countries developing biosafety laws, ethical guidelines, and oversight committees to manage synthetic biology research responsibly. Harmonization across national frameworks is vital to facilitate global collaboration and prevent regulatory conflicts.

Biosafety and Biosecurity Standards

Biosafety standards aim to prevent accidental exposure to or release of synthetic organisms that could harm humans, animals, or the environment. Laboratory containment levels (e.g., BSL-1 to BSL-4) provide protocols for handling organisms of varying risk levels. Risk assessment frameworks evaluate the likelihood and severity of adverse events, informing containment and monitoring requirements. Biosecurity standards focus on preventing deliberate misuse of synthetic biology, including bioterrorism or creation of harmful pathogens. Guidelines include access control, personnel screening, and secure storage of genetic material. Both biosafety and biosecurity measures rely on international best practices and continuous updating to address emerging risks in synthetic biology.

Protocols and Standards for Biosafety and Biosecurity

Ethical and Social Considerations

Ethical, social, and legal implications are central to regulating synthetic biology. Gene editing, synthetic organisms, and novel bioproducts raise questions about environmental stewardship, human health, equity, and social acceptance. Regulatory frameworks often incorporate ethical review processes, public consultation, and stakeholder engagement to balance innovation with societal values. Transparency in research, responsible communication of risks, and mechanisms for benefit sharing are critical to maintain public trust. Social considerations also include intellectual property rights, equitable access, and benefit sharing, particularly in low- and middle-income countries where synthetic biology technologies may have high societal relevance.

Standardization Efforts in Technical Practices

Technical standardization enhances reproducibility, interoperability, and efficiency in synthetic biology research. Organizations such as the International Organization for Standardization (ISO) and the BioBricks Foundation have developed standards for biological parts, measurement units, and laboratory practices.

Examples include standardized genetic modules, DNA assembly methods, and performance characterization protocols. Measurement standards, such as relative promoter activity or protein expression levels, allow consistent comparison across laboratories. Standardized databases and registries, including the iGEM Registry of Standard Biological Parts, promote sharing and reuse of validated components. Technical standardization supports regulatory compliance, facilitates international collaboration, and accelerates innovation by reducing experimental variability.

Challenges to Global Harmonization

Global harmonization of synthetic biology standards and regulations faces multiple challenges. Differences in legal systems, risk tolerance, scientific capacity, and public perception create diverse regulatory landscapes. The rapid pace of technological advancement often exceeds the capacity of regulatory bodies to respond, leading to gaps or inconsistencies.

Proprietary technologies, intellectual property rights, and commercial competition can limit data sharing, hindering standardization. Additionally, cultural, ethical, and societal perspectives influence regulatory priorities and public acceptance. Achieving harmonization requires ongoing dialogue, capacity building, international collaboration, and flexible regulatory frameworks capable of adapting to new scientific developments.

- Synthetic biology combines biology, engineering, and computational design to create novel biological systems.
- Global standards ensure reproducibility, interoperability, safety, and ethical research practices.
- Intellectual property and regulatory frameworks incentivize innovation while protecting public interest.
- Biosafety measures prevent accidental release of synthetic organisms, with containment levels and risk assessment frameworks.
- Biosecurity standards mitigate risks of deliberate misuse, including bioterrorism and harmful pathogens.
- International agreements, such as the Cartagena Protocol, guide cross-border biosafety and biotechnology regulations.
- National frameworks vary, with agencies overseeing pharmaceuticals, genetically engineered crops, and industrial microbes.
- Technical standardization, including BioBricks and iGEM registries, facilitates sharing and consistent measurement of biological parts.
- Ethical, social, and equity considerations are integral to regulation and public acceptance.

- Future directions emphasize adaptive governance, harmonization, open-access standards, and stakeholder engagement.

6.2.2 Intellectual Property Rights and Patent Law

Intellectual property rights and patent law are essential instruments for promoting innovation, protecting creative efforts, and fostering economic growth in a knowledge-based society. Patents, as a central component of IPR, provide inventors with exclusive rights in exchange for public disclosure, creating a balance between private incentive and societal benefit. The historical evolution of intellectual property frameworks, international harmonization efforts, and legal enforcement mechanisms have shaped a global system that supports innovation while addressing ethical, social, and economic considerations. Challenges remain, particularly in adapting to emerging technologies, ensuring equitable access, and resolving disputes in a complex, interconnected world. As science and technology continue to advance, intellectual property law must evolve to address novel inventions, ethical concerns, and societal needs. By balancing the rights of inventors with public interest, IPR and patent law provide a legal and ethical foundation that encourages creativity, drives technological progress, and contributes to sustainable economic and social development.

Historical Development of Intellectual Property Rights

The concept of protecting intellectual creations dates back several centuries, with early forms emerging in Europe during the Renaissance. Guilds and monarchs granted privileges to inventors and authors to encourage innovation and dissemination of knowledge. The Statute of Monopolies enacted in England in 1624 is often considered one of the first legislative frameworks formalizing patent protection for inventions, laying the foundation for modern patent law. Over time, these protections evolved to cover a broader spectrum of intellectual creations, including trademarks, copyrights, and industrial designs. International agreements, such as the Paris Convention for the Protection of Industrial Property in 1883 and the Berne Convention for the Protection of Literary and Artistic Works in 1886, established the basis for cross-border recognition of intellectual property rights. In the contemporary era, global trade agreements, such as the Agreement on Trade-Related Aspects of Intellectual Property Rights (TRIPS), have further harmonized patent and intellectual property protections, ensuring consistency in enforcement and recognition across jurisdictions.

Principles of Patent Law

Patent law is founded on several fundamental principles that define the scope, duration, and enforceability of intellectual property protection. The primary criteria for patentability include novelty, inventive step (non-obviousness), and industrial applicability. Novelty requires that an invention must not be publicly disclosed prior to filing a patent application, ensuring that only new innovations receive protection. Inventive step implies that the invention must involve a technical advancement or solution that is not obvious to a person skilled in the relevant field. Industrial applicability ensures that the invention has practical utility and can be produced or applied in an industrial context. Patents grant the inventor exclusive rights to prevent others from making, using, selling, or importing the patented invention without authorization, typically for a period of twenty years from the filing date. This exclusivity creates a legal incentive for innovation while the requirement of disclosure ensures that technical knowledge enters the public domain, facilitating further research and technological progress.

Types of Intellectual Property Rights

Intellectual property rights encompass a range of legal protections beyond patents. Copyright protects original literary, artistic, and musical works, granting authors exclusive rights to reproduce, distribute, and perform their creations. Trademarks safeguard distinctive signs, symbols, or logos that identify goods or services, ensuring brand recognition and consumer trust. Industrial designs protect the aesthetic features of functional objects, while trade secrets provide confidentiality for commercially valuable information. Utility models, plant breeders' rights, and geographical indications are additional categories that offer specialized protection for innovations in specific sectors. Together, these rights create a comprehensive framework that addresses various forms of creativity and innovation, providing legal mechanisms to enforce ownership, prevent unauthorized use, and incentivize continued invention.

Patent Application Process

The patent application process is a structured legal procedure designed to assess the patentability of an invention and grant protection if the criteria are met. The process typically begins with the preparation of a detailed patent specification, which includes a description of the invention, claims defining the scope of protection, and drawings or diagrams if necessary. The application is submitted to a national or regional patent office, such as the United States Patent and Trademark Office (USPTO) or the European Patent Office (EPO). The patent office conducts a formal examination to verify compliance with procedural requirements, followed by a substantive examination to assess novelty, inventive step, and industrial applicability. The examination process may involve correspondence with the applicant to clarify claims or address objections. Once the patent office determines that the invention meets the criteria, a patent is granted, providing the inventor with enforceable rights. Maintenance fees and periodic renewal requirements ensure that patents remain in force while encouraging active use of protected technologies.

Global Framework and International Treaties

Intellectual property rights operate within a complex global framework, reflecting the need for international harmonization in an increasingly interconnected economy. Treaties and agreements facilitate cross-border recognition, enforcement, and cooperation among patent offices. The TRIPS agreement, under the World Trade Organization, sets minimum standards for protection, enforcement, and dispute resolution for intellectual property rights across member countries.

The Patent Cooperation Treaty (PCT) enables inventors to file a single international application that can be recognized in multiple jurisdictions, streamlining the process for global patent protection. Regional frameworks, such as the European Patent Convention (EPC), allow centralized examination and granting of patents within member states. International coordination ensures that inventors can access protection in multiple countries while maintaining consistency in standards, legal interpretation, and enforcement mechanisms.

Challenges in Patent Law

Patent law faces several challenges that affect its effectiveness, accessibility, and relevance in the modern era. One key challenge is the tension between encouraging innovation and ensuring public access to essential technologies, particularly in the pharmaceutical and healthcare sectors. High costs and exclusivity may limit access to life-saving drugs or essential technologies, raising ethical and policy concerns. Patent litigation and infringement disputes can be complex, costly, and time-consuming, often requiring specialized legal expertise.

The emergence of rapidly evolving technologies, such as artificial intelligence, biotechnology, and synthetic biology, challenges traditional definitions of patentability, inventive step, and ownership. Additionally, patent thickets, overlapping claims, and strategic patenting can create barriers for innovators, particularly small enterprises and researchers, by increasing costs and limiting freedom to operate. These challenges necessitate ongoing legal reform, policy adjustment, and international dialogue to balance innovation incentives with public interest.

Patent Enforcement and Infringement

Enforcing patent rights is critical to maintaining the value and integrity of intellectual property protection. Patent holders have the right to prevent unauthorized manufacturing, use, sale, or importation of their inventions. Enforcement mechanisms typically involve civil litigation, where the patent holder may seek injunctions, damages, or royalties from infringers. Some jurisdictions also provide administrative remedies, such as opposition procedures or patent office reexamination, to resolve disputes without resorting to litigation. Effective enforcement requires clear documentation of ownership, well-defined claims, and evidence of infringement. International enforcement is particularly challenging due to differences in legal systems, jurisdictional boundaries, and enforcement mechanisms. Cross-border cooperation, harmonized legal standards, and alternative dispute resolution mechanisms help address these complexities and ensure that patent rights are respected globally.

Impact on Innovation and Economy

Intellectual property rights and patent law have a profound impact on innovation, economic growth, and technological development. Patents incentivize research and development by providing exclusive rights and potential financial returns, encouraging private investment and long-term innovation. Intellectual property protection fosters competition, drives technological diffusion, and facilitates commercialization of new products and services. In sectors such as pharmaceuticals, biotechnology, information technology, and renewable energy, strong patent systems enable firms to recover development costs and secure funding for further research. However, the balance between protection and accessibility is critical; overly restrictive or broad patents may stifle innovation, limit collaboration, and impede the dissemination of knowledge. Policymakers and legal systems must therefore carefully calibrate patent law to promote innovation while safeguarding public interest and access to essential technologies.

Ethical and Social Considerations

Ethical and social considerations are central to debates surrounding patent law and intellectual property rights. Questions arise regarding the moral justification of granting exclusive rights over life-saving medicines, genetically modified organisms, or digital knowledge. The tension between profit incentives and public welfare, particularly in developing countries, highlights the need for equitable access to technology and information. Ethical frameworks, public policy, and international agreements aim to balance the interests of inventors, consumers, and society at large. Transparency, public engagement, and ethical oversight are increasingly recognized as essential components of an effective intellectual property system, ensuring that innovation serves societal needs without compromising moral responsibility or social equity.

Emerging Trends in Intellectual Property

Emerging technologies, globalization, and digital transformation are reshaping the landscape of intellectual property and patent law. Artificial intelligence-generated inventions, synthetic biology, and advanced materials pose novel legal and ethical questions regarding ownership, inventorship, and patentability. Digital rights management, blockchain-based IP tracking, and automated patent analysis are transforming the way intellectual property is monitored, enforced, and licensed. Collaborative innovation, open-source models, and patent pools are emerging as alternative strategies to balance proprietary rights with shared technological advancement. International harmonization and adaptive legal frameworks are essential to address these trends while maintaining incentives for innovation and protecting public interest.

Table: Comparison of Intellectual Property Types

Type	Protection Scope	Duration
Patents	Inventions, processes, products	20 years from filing
Copyright	Literary, artistic, musical works	Life of author + 70 years (varies)
Trademarks	Signs, logos, brand identifiers	Renewable indefinitely
Industrial Designs	Aesthetic and functional features	Typically, 15–25 years
Trade Secrets	Confidential commercial information	Unlimited as long as secrecy is maintained

6.3 PUBLIC PERCEPTION AND SOCIETAL IMPACT

6.3.1 Engaging the Public in Synthetic Biology Dialogue

Engaging the public in synthetic biology dialogue is fundamental to the responsible advancement and societal integration of this potent technological field. Constructive and inclusive dialogue promotes shared understanding, addresses ethical and safety concerns, and enhances democratic governance processes. Multi-faceted engagement strategies, employing both traditional and innovative communication platforms, allow scientists, policymakers, and citizens to collaboratively shape synthetic biology's trajectory. By embracing transparency, respecting diverse perspectives, and institutionalizing engagement practices, the synthetic biology community can foster public trust and harness the technology's transformative potential ethically and sustainably for the benefit of all.

Importance of Public Dialogue in Synthetic Biology

Synthetic biology's transformative potential in health, agriculture, energy, and environment is accompanied by uncertainties and ethical issues that have sometimes sparked public skepticism or controversy. Incorporating public perspectives early and continuously helps preempt misunderstandings, dispel fears, and identify socially acceptable pathways for development. Engaged publics can shed light on diverse values and priorities, enriching risk-benefit analyses and fostering legitimacy for research and regulatory decisions. Moreover, public dialogue promotes scientific literacy by demystifying complex concepts such as genome editing, synthetic constructs, and biological engineering. This educational aspect supports informed citizenry capable of participating in democratic processes linked to biotechnology governance.

Methods and Approaches to Public Engagement

A variety of approaches are employed to engage different publics in synthetic biology discussions. These range from formal public forums, citizen juries, and deliberative workshops to informal science education activities, online platforms, interactive exhibits, and media campaigns.

Hands-on experiences using models, virtual reality, or DIY biology kits enable experiential learning and foster curiosity. Dialogue-centric formats emphasize mutual learning where scientific experts share knowledge and listen actively to public concerns and questions. Inclusive engagement strategies strive to reach underrepresented groups by addressing cultural sensitivities, language barriers, and accessibility. Tailoring communication to specific audiences—be it policymakers, students, or local communities—maximizes relevance and impact.

Challenges in Public Engagement

Engaging publics in synthetic biology presents unique challenges related to complexity, rapidly evolving science, and diverse societal contexts. The novelty and interdisciplinarity of synthetic biology can overwhelm or confuse participants, necessitating carefully designed user-friendly explanations devoid of jargon. Balancing optimism about benefits with frank discussion of uncertainties and risks requires nuanced facilitation to avoid extremes of hype or fear.

Distrust in scientific institutions or previous negative experiences with biotechnology may hinder productive dialogue. Moreover, identifying and reaching representative sections of the public is difficult given varying levels of interest, knowledge, and civic engagement. Sustaining engagement over long development timelines, and translating dialogue outcomes into tangible policy or research changes, represent ongoing logistical and evaluative challenges.

Themes Emerging from Public Feedback

Public conversations about synthetic biology often reveal consistent themes such as concerns about safety, environmental impacts, ethical boundaries, equitable access, and long-term societal consequences. Many recognize the potential benefits of synthetic biology for medicine, environmental remediation, and sustainable manufacturing but seek assurances regarding regulation, oversight, and precautionary measures.

Questions arise about "playing God," unnatural modification of life, and implications for biodiversity and ecosystem integrity. Equity and global justice emerge as important issues, with calls for inclusive benefit-sharing and preventing exacerbation of existing inequalities. Transparency, accountability, and responsible governance feature prominently as conditions for public acceptance.

Institutional and Policy Implications

Insights from public engagement shape institutional strategies and policy frameworks governing synthetic biology. Governments and funding agencies increasingly mandate or encourage public engagement components in grant applications and research projects. Regulatory bodies incorporate public input into developing guidelines, standards, and ethical oversight mechanisms. Multi-stakeholder advisory panels blend scientific, ethical, and public viewpoints to guide synthetic biology governance. Engagement outcomes can influence approval processes, labelling decisions, and intellectual property considerations. International bodies also recognize the importance of public perspectives in harmonizing transnational regulations and addressing shared global concerns. Institutionalizing engagement as a routine aspect of synthetic biology ensures responsiveness and adaptability to evolving societal demands.

Role of Scientists and Educators in Public Engagement

Scientists bear a critical responsibility in engaging publics authentically and effectively. Beyond communicating scientific facts, researchers should foster dialogue that respects divergent values and integrates societal concerns. Training in science communication and ethics equips scientists with skills needed for meaningful engagement.

Collaborations with social scientists, ethicists, and communication specialists enhance design and evaluation of engagement activities. Educators can embed synthetic biology topics into curricula with emphasis on societal dimensions, empowering future generations with knowledge and critical thinking skills. Scientist-led citizen science projects and community partnerships promote inclusive participation and democratization of knowledge generation.

Technologies Enhancing Public Engagement

Digital platforms and emerging technologies expand the reach and interactivity of synthetic biology public engagement. Social media, podcasts, webinars, and online forums provide accessible spaces for dialogue across demographic and geographic boundaries. Virtual reality, augmented reality, and gamification create immersive educational experiences that simulate synthetic biology research and applications. Artificial intelligence tools personalize learning pathways and curate scientifically accurate content. Online collaborative platforms enable crowdsourcing of questions, concerns, and innovation ideas contributing to participatory governance. Harnessing these technologies requires attention to digital divides and misinformation mitigation.

Advancing Public Engagement through Technology

Evaluating and Sustaining Engagement Impact

Systematic evaluation of public engagement initiatives is essential to measure effectiveness, identify areas for improvement, and embed lessons learned into best practices. Metrics include changes in knowledge, attitudes, trust, and willingness to participate in policy processes. Qualitative assessments capture nuanced insights into values, emotions, and concerns expressed by publics. Ongoing stakeholder feedback mechanisms, adaptive management, and transparent reporting promote trust and sustained dialogue. Institutional commitment to iterative engagement fosters continuous learning and responsiveness, ensuring synthetic biology development aligns with public values long-term.

6.3.2 Risk Communication and Managing Public Fears

Effective risk communication and management of public fears stands as an essential competency for navigating the complex challenges facing contemporary society, from emerging diseases to climate change, from technological risks to social upheavals. The evolution of risk communication from a technical exercise in information transmission to a sophisticated practice incorporating psychology, sociology, cultural studies, and communication science reflects growing recognition of its critical importance. Success requires not only scientific accuracy but also emotional intelligence, cultural sensitivity, and strategic thinking about how messages interact with existing beliefs, social dynamics, and information ecosystems.

The lessons learned from recent global crises underscore that risk communication cannot be an afterthought but must be integrated into risk management planning from the beginning. Building resilient societies capable of responding appropriately to risks requires long-term investment in public education, trust-building with communities, and communication infrastructure that can function effectively during crises. As the nature of risks continues to evolve with technological and social change, risk communication strategies must likewise adapt while maintaining core principles of transparency, empathy, and respect for public agency. The ultimate goal of risk communication extends beyond managing immediate fears to empowering individuals and communities with the knowledge and confidence to make informed decisions in an uncertain world.

Psychology of Risk Perception

Individuals and communities perceive risk forms the foundation for effective risk communication, as human risk perception often diverges significantly from statistical probabilities and expert assessments. Psychological research reveals that people evaluate risks through mental shortcuts and emotional responses rather than careful statistical analysis, leading to systematic biases in how different hazards are perceived and prioritized. The availability heuristic causes people to overestimate the likelihood of risks that are easily recalled or vividly imagined, explaining why rare but dramatic events like plane crashes often generate more fear than statistically greater risks like automobile accidents.

Factors such as voluntariness, controllability, familiarity, and catastrophic potential profoundly influence risk perception, with involuntary and uncontrollable risks generating disproportionate fear even when their actual probability is low. The affect heuristic demonstrates that emotional responses to hazards often drive risk judgments more than analytical thinking, making it insufficient to simply present statistical information without addressing emotional concerns. Social amplification of risk occurs when psychological, social, institutional, and cultural processes interact to intensify or attenuate public responses to hazards, sometimes creating widespread fear about minimal risks while dangerous hazards remain ignored. Understanding these psychological dynamics enables communicators to craft messages that resonate with how people actually process risk information rather than how experts think they should.

Principles of Effective Risk Communication

Successful risk communication requires adherence to established principles that respect public concerns while providing accurate, actionable information that enables informed decision-making. Transparency and openness about what is known, unknown, and uncertain builds credibility and prevents the erosion of trust that occurs when organizations appear to withhold information or mislead the public. Early announcement of risks, even when information is incomplete, demonstrates respect for the public's right to know and ability to make decisions, though this must be balanced with avoiding premature or alarmist communications. Message clarity requires translating technical information into language accessible to diverse audiences without oversimplifying complex issues or condescending to public intelligence. Acknowledging uncertainty honestly rather than projecting false confidence helps maintain credibility when situations evolve and initial assessments prove incorrect. Two-way communication that involves listening to public concerns, answering questions, and incorporating feedback creates dialogue rather than monologue, increasing message acceptance and compliance with recommended actions. Consistency across different messengers and channels prevents confusion and contradictory information that can undermine public confidence and compliance. Actionable guidance that provides specific steps people can take to protect themselves empowers individuals and reduces feelings of helplessness that amplify fear.

Challenges in Modern Risk Communication

Contemporary risk communication faces unprecedented challenges arising from technological change, social fragmentation, and the evolving information ecosystem that shapes public discourse. The speed of information spread through digital channels often outpaces the ability of authoritative sources to verify facts and craft appropriate responses, creating windows for misinformation to take hold and spread. Information overload from multiple sources makes it difficult for people to identify credible information and can lead to confusion, paralysis, or disengagement from important risk messages. Political polarization increasingly affects how risk information is received and interpreted, with partisan affiliations influencing acceptance of scientific evidence and expert recommendations. The decline in traditional institutional authority means that official sources no longer command automatic trust, requiring more effort to establish and maintain credibility. Global interconnectedness means that local risks quickly become international concerns, requiring coordination across jurisdictions with different communication norms, languages, and cultural contexts. Scientific uncertainty and evolving knowledge create communication dilemmas when recommendations must change based on new evidence, potentially appearing as incompetence or dishonesty to skeptical publics. The presence of hostile actors deliberately spreading disinformation for political or economic gain adds a adversarial dimension to risk communication that traditional approaches were not designed to address.

Role of Media and Digital Platforms

Media channels and digital platforms play pivotal roles in shaping public understanding of risks, acting as both amplifiers and filters of risk information with profound implications for public fear and behavior. Traditional news media's tendency toward dramatic coverage of unusual events while ignoring ongoing risks can distort public risk perception, creating fear about statistically minor threats while obscuring significant hazards. Social media platforms enable rapid information sharing but also facilitate the spread of rumors, conspiracy theories, and misinformation that can overwhelm accurate risk communication. Algorithm-driven content distribution on digital platforms can create echo chambers where risk perceptions become amplified within like-minded groups while alternative perspectives remain invisible. The democratization of information production means that anyone can position themselves as a risk expert online, making it difficult for the public to distinguish authoritative sources from unreliable ones. Visual media's power to evoke emotional responses through dramatic imagery can overwhelm rational risk assessment, particularly when images are taken out of context or manipulated. Influencers and opinion leaders on social media can significantly shape risk perception among their followers, sometimes contradicting official guidance and creating competing narratives. The economic incentives of media organizations toward sensationalism and engagement can conflict with public health and safety goals of accurate, measured risk communication.

Building Trust and Credibility

Trust serves as the foundation upon which all effective risk communication rests, determining whether messages are accepted, believed, and acted upon by their intended audiences. Credibility must be earned through consistent, honest communication over time and can be quickly destroyed by perceived deception, incompetence, or conflicts of interest. Acknowledging past mistakes and demonstrating learning from them paradoxically builds more trust than claiming infallibility or refusing to admit errors. Local voices and trusted community leaders often carry more weight than distant experts, making partnerships with community organizations essential for reaching diverse populations. Cultural competence in risk communication requires understanding how different communities perceive authority, process information, and make decisions about risks. Demonstrating empathy and acknowledging the legitimacy of public concerns, even when they seem irrational from a technical perspective, helps build rapport and openness to risk messages. Transparency about decision-making processes, including how risks are assessed and why certain recommendations are made, helps publics understand and accept guidance even when they might prefer different outcomes. Pre-event relationship building creates reservoirs of trust that can be drawn upon during crises when there is no time to establish credibility from scratch.

Crisis Communication and Emergency Management

Crisis situations requiring immediate risk communication present unique challenges that demand specialized strategies for managing public fears while motivating appropriate protective actions. The initial hours and days of a crisis often determine the trajectory of public response, making prepared communication protocols and trained spokespersons essential for effective emergency risk communication. Message coordination across multiple agencies and jurisdictions prevents contradictory information that can paralyze public response and erode confidence in emergency management. Regular updates, even when new information is limited, maintain public engagement and prevent information voids that become filled with speculation and rumors. Acknowledging emotions and providing psychological support alongside practical information helps people process fear and anxiety in constructive ways. Clear chains of command and designated spokespersons prevent the cacophony of voices that can emerge during crises and confuse public understanding. Post-crisis communication that explains what happened, what was learned, and what will be done differently maintains relationships and prepares for future events. Recovery phase communication must address long-term fears and help communities rebuild confidence while maintaining appropriate vigilance for ongoing risks.

Misinformation and Conspiracy Theories

The proliferation of misinformation and conspiracy theories about risks poses fundamental challenges to public health and safety, requiring sophisticated strategies that go beyond simple fact-checking. Understanding the psychological and social needs that conspiracy theories fulfill for believers helps explain their persistence and suggests more effective intervention approaches. Prebunking techniques that inoculate audiences against misinformation before exposure prove more effective than debunking after false beliefs have taken hold. Avoiding repetition of false claims, even when refuting them, prevents inadvertent reinforcement of misinformation through increased familiarity. Providing alternative explanations that address the underlying concerns or observations that conspiracy theories purport to explain offers believers a face-saving way to revise their views. Building resilience to misinformation through media literacy education helps populations develop critical thinking skills for evaluating risk information. Engaging with communities where misinformation circulates, rather than dismissing them, creates opportunities for dialogue and trust-building. Legal and platform-based interventions against malicious disinformation must balance free speech concerns with public safety imperatives.

Cultural and Social Dimensions

Risk communication must account for the profound influence of cultural values, social identities, and group dynamics on how different communities understand and respond to risk information. Cultural worldviews shape what risks people pay attention to and how they interpret evidence, with individualistic and communitarian orientations leading to systematically different risk perceptions. Social identity and group membership influence which messengers are considered trustworthy and which risks are seen as relevant or concerning. Historical experiences of discrimination, exploitation, or betrayal by institutions affect how marginalized communities receive official risk communication. Language barriers extend beyond translation to include cultural concepts and metaphors that may not have equivalents across cultural contexts. Religious and spiritual beliefs about causation, fate, and appropriate responses to danger must be respectfully addressed in risk communication. Community participation in developing risk communication strategies ensures messages resonate with local values and address specific concerns. International risk communication must navigate diverse cultural norms about authority, uncertainty, and appropriate public behavior.

6.4 GOVERNANCE OF SYNTHETIC BIOLOGY

6.4.1 International Collaboration and Policy Development

International collaboration and policy development are indispensable tools for addressing complex global challenges that transcend national boundaries. By fostering cooperation, knowledge sharing, and joint action, international collaboration enhances scientific innovation, environmental stewardship, public health, and economic development.

Effective policy development ensures that collaborative efforts are structured, accountable, and sustainable, balancing national interests with global priorities. Core principles, including transparency, inclusivity, flexibility, and evidence-based decision-making, guide the formulation and implementation of policies across diverse sectors. Technological advancements and digital platforms further enhance collaboration, enabling efficient communication, data sharing, and monitoring. While challenges such as political differences, resource disparities, and legal variations persist, structured frameworks, capacity building, and adaptive governance mechanisms provide pathways to overcome these obstacles. As global challenges continue to evolve, the integration of international collaboration and robust policy development will remain essential to achieving sustainable, equitable, and effective solutions that benefit societies worldwide.

Importance of International Collaboration

International collaboration fosters the sharing of expertise, technology, and best practices among nations, accelerating scientific progress and policy innovation. In fields such as public health, global disease surveillance and joint research initiatives enable rapid responses to pandemics and emerging pathogens. In climate science, collaborative research programs allow for comprehensive data collection, modeling, and the development of mitigation strategies across diverse ecosystems. Collaborative approaches in space exploration, renewable energy, and biotechnology enhance innovation by leveraging complementary strengths of multiple countries. Beyond technical benefits, international collaboration promotes diplomacy, mutual understanding, and the equitable distribution of resources and knowledge. By building networks of cooperation, countries can address complex problems more effectively while reducing duplication of effort and resource wastage.

Role of Policy Development

Policy development is essential for guiding international collaboration, ensuring that cooperative efforts are structured, sustainable, and legally enforceable. Policies provide clear objectives, rules of engagement, funding mechanisms, and accountability frameworks for multinational initiatives. International treaties, agreements, and conventions serve as instruments of policy, setting standards for behavior, cooperation, and compliance. For example, the Paris Agreement on climate change establishes binding commitments for greenhouse gas reductions, while the International Health Regulations (IHR) provide frameworks for global disease reporting and response. Policy development also addresses ethical considerations, intellectual property rights, and equity in access to resources and technology. Without coherent policies, international collaboration may be fragmented, inefficient, or prone to conflict, undermining collective goals.

Mechanisms of International Collaboration

Various mechanisms facilitate effective international collaboration across research, governance, and industry. Multilateral organizations such as the United Nations, World Health Organization (WHO), and World Trade Organization (WTO) provide platforms for dialogue, standard-setting, and coordinated action. Bilateral agreements between countries establish partnerships for trade, research, or technological development. International research consortia bring together academic institutions and industries to tackle specific scientific challenges. Public-private partnerships combine governmental oversight with private sector innovation for large-scale infrastructure or health initiatives. Global funding mechanisms, including the Green Climate Fund or international development banks, enable resource mobilization for collaborative projects. These mechanisms provide structured avenues for coordination, monitoring, and accountability in global initiatives.

Policy Frameworks and Standards

Developing effective policies for international collaboration requires harmonized frameworks and standards. Standardization ensures interoperability, compliance, and equitable participation among participating countries. For instance, global health policies rely on standardized disease reporting systems, vaccine protocols, and laboratory procedures to ensure consistency across regions. Trade policies often include standardized regulations on tariffs, safety standards, and environmental compliance to facilitate fair international commerce. In technology and innovation, policies regarding intellectual property, cybersecurity, and ethical research practices are essential for protecting interests while promoting collaboration. Harmonized frameworks reduce uncertainty, enhance trust, and encourage broader participation in multinational initiatives.

Challenges in International Collaboration

Despite its benefits, international collaboration faces numerous challenges. Differences in legal systems, regulatory standards, and governance structures can hinder alignment of policies and procedures. Resource disparities between high-income and low-income countries may create inequalities in access, influence, or benefits derived from collaboration. Political tensions, trade restrictions, and conflicting national interests can undermine cooperative efforts. Cultural differences and language barriers may also complicate communication and consensus-building. In scientific research, data sharing may be restricted due to privacy, security, or intellectual property concerns. Addressing these challenges requires robust negotiation, trust-building, capacity development, and adaptive governance mechanisms.

Case Studies of Effective Collaboration

Several examples illustrate the success of international collaboration in addressing global challenges. The Global Polio Eradication Initiative unites governments, non-governmental organizations, and private donors to coordinate vaccination campaigns, surveillance, and research, significantly reducing polio incidence worldwide. The Intergovernmental Panel on Climate Change (IPCC) combines scientific assessments from experts across more than 120 countries to inform global climate policy. In biotechnology, the International Genetically Engineered Machine (iGEM) competition fosters collaboration among students and researchers globally, advancing synthetic biology innovations through shared standards and protocols. These cases demonstrate how structured collaboration, supported by clear policies, can achieve substantial societal and scientific outcomes.

Principles for Policy Development

Effective international policy development is guided by several core principles. Transparency ensures that objectives, processes, and outcomes are clearly communicated to stakeholders. Inclusivity allows participation from diverse countries, institutions, and communities, promoting equity and shared ownership. Accountability requires mechanisms to monitor compliance and evaluate the effectiveness of policies. Flexibility allows policies to adapt to emerging scientific, technological, or geopolitical developments. Sustainability emphasizes long-term environmental, economic, and social considerations. Evidence-based decision-making ensures that policies are grounded in scientific data and best practices. Risk management addresses uncertainties and potential negative impacts proactively. Collaboration and coordination enhance synergies and reduce duplication. Ethical compliance safeguards human rights, privacy, and equity. Capacity building supports the development of skills, infrastructure, and resources for effective participation.

Role of Technology in Facilitating Collaboration

AV Technology Facilitating Effective Remote Collaboration

Technological advancements play a critical role in enabling international collaboration and policy implementation. Digital communication platforms allow real-time coordination among global teams. Data-sharing repositories facilitate access to research, environmental monitoring, and health information. Artificial intelligence and predictive analytics support decision-making by identifying trends, risks, and resource needs.

Blockchain technology enhances transparency and accountability in financial transactions, supply chains, and international agreements. Remote sensing, satellite monitoring, and GIS enable collaborative environmental assessment and disaster management. By integrating technology, international collaborations become more efficient, transparent, and responsive to emerging challenges.

- International collaboration enables pooling of resources, expertise, and knowledge to address global challenges.
- Policy development provides regulatory and strategic frameworks to guide cooperative efforts.
- Multilateral organizations, bilateral agreements, and public-private partnerships facilitate structured collaboration.
- Harmonized standards and frameworks ensure interoperability, compliance, and equitable participation.
- Transparency, inclusivity, accountability, flexibility, and sustainability are core principles of policy development.
- Challenges include legal disparities, resource inequality, political tensions, cultural differences, and data-sharing restrictions.
- Technology, including AI, digital platforms, blockchain, and GIS, enhances collaboration and decision-making.
- Case studies like the IPCC, GPEI, and iGEM illustrate successful international collaboration.
- Evidence-based and risk-informed policies promote effective global governance and innovation.
- Future directions emphasize adaptive governance, stakeholder engagement, capacity building, and ethical standards.

6.4.2 Balancing Innovation and Safety

Balancing innovation and safety is an enduring and complex challenge that lies at the intersection of technology, ethics, regulation, and societal expectation. Innovation drives progress, economic growth, and solutions to global challenges, but it inherently carries risk and uncertainty. Ensuring safety requires structured frameworks, ethical oversight, risk management strategies, and proactive public engagement. Historical experiences, regulatory practices, and technological safeguards illustrate that the two objectives are not mutually exclusive but can be integrated to achieve responsible progress. Effective balancing involves continuous assessment, adaptive governance, and collaboration among stakeholders to ensure that innovation advances within a framework that protects human health, environmental integrity, and societal trust. By fostering a culture that values both creativity and caution, societies can harness the transformative potential of innovation while minimizing harm, creating a sustainable and ethically responsible pathway for technological advancement.

Nature of Innovation

Innovation encompasses the introduction of novel products, processes, or ideas that significantly improve existing systems or create entirely new possibilities. It can occur in diverse sectors, including technology, medicine, biotechnology, information systems, energy, and transportation. Innovation is often characterized by experimentation, uncertainty, and rapid iteration. While these characteristics enable breakthroughs, they also increase the likelihood of unforeseen consequences, operational failures, or societal disruption. The process of innovation is inherently dynamic, involving not only technical and scientific advancement but also economic, regulatory, and social dimensions. Effective innovation requires a willingness to explore new frontiers, adapt to emerging challenges, and leverage creative problem-solving, all while remaining cognizant of the ethical, legal, and safety implications of novel developments.

Safety as a Guiding Principle

Safety, in the context of innovation, refers to the proactive management of risks associated with new technologies, processes, or systems. It encompasses physical, biological, environmental, and societal dimensions, ensuring that innovation does not result in harm or unintended consequences. Safety is guided by principles such as precaution, redundancy, containment, and monitoring. In engineering, safety standards regulate the design, construction, and operation of systems to prevent failure or accidents. In medicine and biotechnology, rigorous clinical trials, preclinical testing, and regulatory oversight ensure that new therapies or interventions are safe for human use. In environmental applications, impact assessments, sustainability evaluations, and containment strategies prevent ecological disruption. Prioritizing safety does not necessarily inhibit innovation; instead, it provides a framework within which novel ideas can be responsibly tested, refined, and deployed.

Tension between Innovation and Safety

The pursuit of innovation often conflicts with the imperatives of safety, creating a delicate tension for regulators, developers, and society at large. Rapid technological advancement may outpace the ability of existing regulatory frameworks to anticipate and manage risks. For example, emerging fields such as synthetic biology, artificial intelligence, and nanotechnology present complex safety challenges that were unforeseen when earlier regulations were developed. Innovators may face pressure to accelerate development for competitive, economic, or societal reasons, potentially compromising safety assessments. Conversely, overly stringent safety regulations can slow innovation, restrict experimentation, and limit the commercialization of beneficial technologies. Balancing this tension requires adaptive governance, evidence-based risk assessment, and a flexible approach that integrates both precaution and opportunity.

Risk Assessment and Management

Risk assessment is a central tool in balancing innovation with safety. It involves identifying potential hazards, evaluating their likelihood and impact, and determining appropriate measures to mitigate risks. Risk management extends these assessments into actionable strategies, including containment measures, safety protocols, monitoring systems, and contingency planning. In biotechnology, for instance, risk assessment may evaluate the potential ecological or health impacts of genetically modified organisms, while in engineering, it may assess the likelihood of structural failure or operational hazards. Scenario analysis, modeling, and simulations are often employed to anticipate complex interactions and outcomes. Effective risk assessment and management enable innovators to make informed decisions, reducing the probability of adverse events while allowing experimentation and progress.

Regulatory Frameworks and Oversight

Regulatory frameworks play a pivotal role in harmonizing innovation and safety by establishing standards, guidelines, and enforcement mechanisms. Agencies such as the Food and Drug Administration (FDA), the Environmental Protection Agency (EPA), and international bodies like the World Health Organization (WHO) provide oversight, approval processes, and monitoring systems for novel technologies. Regulations often require rigorous testing, documentation, and review prior to commercialization. These frameworks aim to prevent harm, ensure compliance with ethical standards, and provide legal recourse in the event of safety breaches. Adaptive regulatory approaches, including conditional approvals, phased deployment, and post-market surveillance, allow innovation to proceed while maintaining safety oversight. Such frameworks also facilitate public trust, as stakeholders can be confident that novel developments are subject to careful evaluation and accountability.

Ethical Considerations in Balancing Innovation and Safety

Ethics is central to navigating the trade-offs between innovation and safety. Ethical principles such as beneficence, non-maleficence, justice, and autonomy guide decision-making in situations where risks and benefits must be carefully weighed. In medical research, for instance, clinical trials must balance potential therapeutic benefits against possible harm to participants. In environmental innovation, ethical considerations involve weighing societal benefits against ecological risks. Transparency, informed consent, and stakeholder engagement are essential to ethical decision-making. Ethical frameworks complement regulatory oversight by providing normative guidance that shapes policies, informs public debate, and reinforces societal trust in innovation. By integrating ethics into risk management and regulatory practices, innovators can pursue progress responsibly while minimizing harm.

Technological Solutions for Safety

Advances in technology offer tools to enhance safety without stifling innovation. Automation, sensor networks, and real-time monitoring enable rapid detection of anomalies and hazards in industrial and research environments. Computational modeling and simulations allow virtual testing of new processes, reducing the need for high-risk experimentation. Fail-safe mechanisms, redundancy in design, and modular system architectures increase resilience against failure. In biotechnology, containment strategies, genetic safeguards, and controlled expression systems minimize unintended ecological or human impacts. These technological solutions exemplify how innovation itself can be leveraged to enhance safety, demonstrating that the two objectives are complementary rather than mutually exclusive.

Public Engagement and Societal Trust

Public perception and engagement are critical components of balancing innovation and safety. Societal trust influences the acceptance, adoption, and regulation of new technologies. Open communication, stakeholder consultation, and participatory decision-making foster transparency and legitimacy. Addressing public concerns, misconceptions, and ethical questions proactively can prevent backlash, improve compliance, and enhance the societal benefits of innovation. Public engagement is particularly important in high-impact fields such as synthetic biology, artificial intelligence, and nuclear technology, where potential risks are complex, long-term, or uncertain. By involving communities, experts, and policymakers in the decision-making process, innovators can align technological development with societal values, creating a responsible pathway for progress.

Case Studies in Innovation and Safety

Historical and contemporary case studies illustrate the importance of balancing innovation and safety. The development of genetically modified crops demonstrates the interplay between technological advancement, regulatory oversight, and public acceptance. Rigorous safety assessments, field trials, and post-release monitoring have enabled widespread adoption in some regions, while public concern and regulatory caution have limited adoption in others. In the pharmaceutical industry, the introduction of new drugs involves extensive preclinical and clinical testing to ensure safety, demonstrating a structured approach to managing risk while enabling life-saving innovation. Similarly, nuclear energy development has highlighted the consequences of inadequate safety measures, reinforcing the need for robust risk management and regulatory frameworks. These examples emphasize that innovation without appropriate safety measures can have far-reaching societal and environmental consequences, while well-managed innovation can yield transformative benefits.

Economic Considerations

Economic factors also influence the balance between innovation and safety. Investments in research and development are often significant, and regulatory compliance, safety testing, and risk management represent additional costs. While stringent safety requirements can increase development timelines and expenses, inadequate safety measures can result in costly accidents, liability claims, and loss of public trust. Companies, governments, and investors must therefore consider both the immediate and long-term economic implications of safety in innovation. Cost-benefit analysis, strategic risk investment, and insurance mechanisms are tools that can help integrate economic considerations into decision-making, ensuring that innovation is both financially viable and socially responsible.

Global Perspectives and Harmonization

Balancing innovation and safety is a global concern, particularly as technologies increasingly transcend national boundaries. International standards, treaties, and cooperative frameworks provide guidance on risk assessment, safety protocols, and regulatory harmonization.

Table: Strategies for Balancing Innovation and Safety

Strategy	Description
Risk Assessment	Identifying hazards, evaluating probability and impact, and planning mitigation
Regulatory Oversight	Implementing standards, approval processes, and monitoring mechanisms
Ethical Frameworks	Applying moral principles to guide responsible innovation
Technological Safeguards	Using automation, containment, and fail-safe mechanisms to reduce risk
Public Engagement	Involving stakeholders to build trust and ensure societal alignment

Organizations such as the World Health Organization, the Organisation for Economic Co-operation and Development, and the International Organization for Standardization develop guidelines that promote consistency and best practices worldwide. Cross-border collaboration in research and regulatory alignment ensures that innovations are deployed safely while facilitating international trade, technological transfer, and global problem-solving. Harmonization reduces duplicative regulatory burdens, improves safety standards, and fosters innovation in a coordinated and responsible manner.

CHAPTER 7

Future Directions of Synthetic Biology

7.1 ADVANCEMENTS IN SYNTHETIC GENOME DESIGN

7.1.1 Synthetic Genomes and Xenobiology

Synthetic genomes and xenobiology represent paradigm-shifting dimensions of synthetic biology, enabling the creation of life forms and genetic systems beyond natural boundaries. By designing and constructing synthetic chromosomes and incorporating unnatural genetic components, scientists gain unprecedented control over biology's fundamental language, unlocking new potentials for innovation and discovery.

These advancements offer profound opportunities across sectors including medicine, energy, environment, and materials science. Yet, their deployment demands careful consideration of ethical, safety, and regulatory challenges unique to designing life itself. Through continued technological innovation coupled with thoughtful governance and public engagement, synthetic genomes and xenobiology hold promise to reshape bioengineering and biotechnology, opening new realms of possibility for the benefit of global society.

Defining Synthetic Genomes and their Characteristics

Synthetic genomes involve the design and assembly of DNA sequences that represent either exact replicas of the genomes of existing organisms or extensively modified constructs containing novel elements. Unlike conventional genetic modification, which edits specific genes within a natural genome, synthetic genomics strives to synthesize entire chromosomes or genomes from chemically-made oligonucleotides, stitching them together using advanced molecular techniques.

These genomes can then be transplanted into host cells to create living organisms that operate according to the synthetic instructions encoded within. A hallmark achievement was the synthesis of a minimal bacterial genome, Mycoplasma mycoides JCVI-syn3.0, featuring only the essential genes required for life. More recently, efforts have produced redesigned Escherichia coli genomes with reduced codon usage, demonstrating the ability to recode natural genomes entirely.

Synthetic genomes facilitate both fundamental biological research—shedding light on minimal requirements for life—and applications such as custom microbial chassis optimized for industrial or medical tasks.

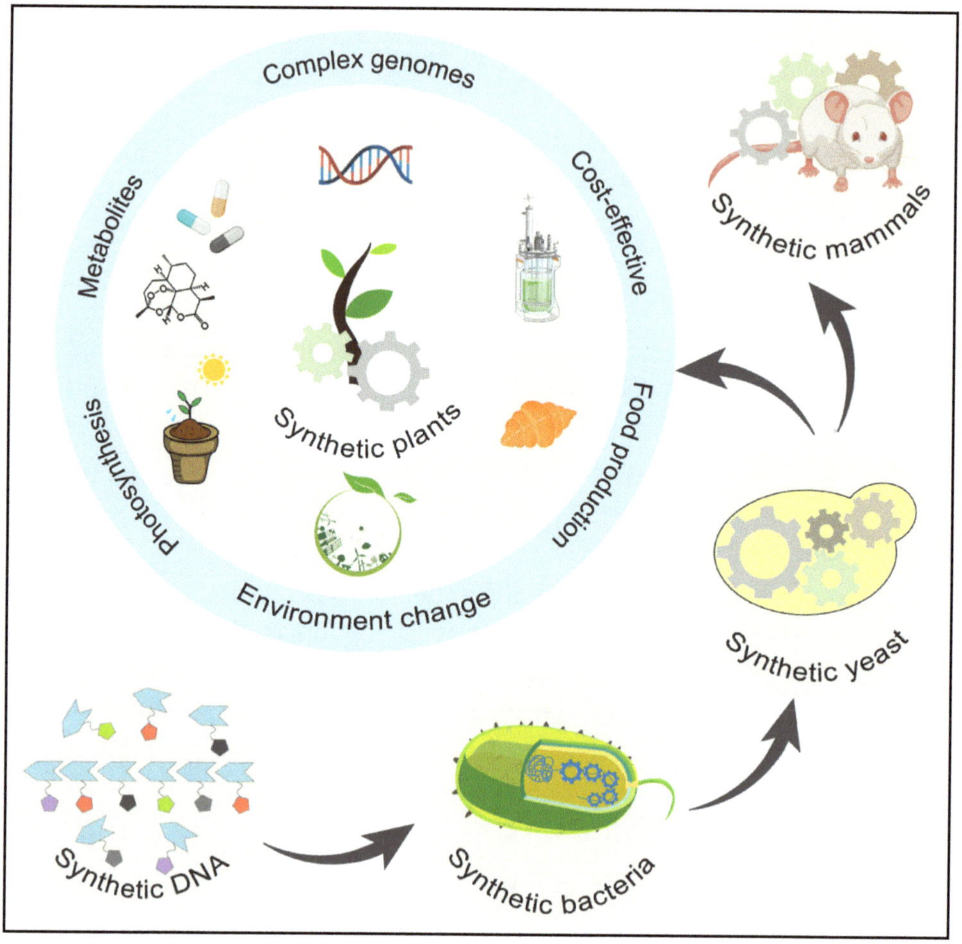

Engineered Genetic Systems and their Characteristics

7.1.2 Synthetic Organisms with New Genetic Codes

The development of synthetic organisms with new genetic codes marks a watershed moment in humanity's relationship with the living world, demonstrating our growing ability to transcend the constraints that have governed life for billions of years. This technology represents far more than an incremental advance in genetic engineering; it constitutes a fundamental reimagining of the molecular basis of life itself. The practical applications emerging from this field promise to address critical challenges in medicine, industry, and environmental management while providing unprecedented biocontainment capabilities that address safety concerns about synthetic biology. Yet the implications extend beyond immediate applications to fundamental questions about the nature of life, the possibilities for directed evolution, and humanity's role in shaping the future of biology on Earth and potentially beyond. As these technologies continue to advance, society must grapple with the profound responsibilities that come with the power to create truly novel forms of life. The successful development and deployment of organisms with synthetic genetic codes will require not only continued technological innovation but also thoughtful consideration of ethical implications, safety protocols, and governance frameworks. Ultimately, synthetic genetic codes may represent one of humanity's most powerful tools for addressing global challenges while simultaneously opening entirely new frontiers in our understanding and manipulation of living systems.

7.2 THE CONVERGENCE OF SYNTHETIC BIOLOGY WITH OTHER TECHNOLOGIES

7.2.1 Integration with Artificial Intelligence and Machine Learning

The integration of artificial intelligence and machine learning into various sectors represents a profound technological evolution that is reshaping how humans interact with data, systems, and decision-making processes. By enhancing predictive capabilities, automating repetitive tasks, and generating insights from vast datasets, AI and ML empower organizations and researchers to tackle complex challenges more efficiently and accurately.

Across healthcare, industry, finance, agriculture, environmental management, and scientific research, integration drives innovation, optimizes resource use, and improves outcomes. However, challenges related to data quality, transparency, ethics, and regulation require careful consideration and proactive management. Future developments will focus on explainable AI, real-time data integration, secure collaboration, and adaptive governance, ensuring that AI and ML integration continues to advance human capability while adhering to ethical, societal, and environmental standards. By fostering responsible adoption, transparent practices, and inclusive policies, AI and ML can sustainably enhance innovation and problem-solving across the globe.

Integration in Agriculture and Environmental Management

AI and ML are increasingly integrated into agriculture and environmental management to enhance productivity, sustainability, and resource efficiency. Precision agriculture employs AI-driven sensors and drones to monitor soil quality, crop health, and water usage. ML algorithms analyze data to optimize irrigation, fertilization, and pest control strategies. Environmental monitoring systems leverage AI to track climate patterns, detect pollution levels, and predict natural disasters. In conservation, AI-powered image recognition identifies species in ecosystems, supporting biodiversity studies and habitat protection. This integration ensures that agricultural and environmental practices are efficient, data-informed, and environmentally responsible.

Agriculture with Environmental Management Strategies

Integration in Scientific Research

Scientific research benefits significantly from AI and ML integration by enabling faster hypothesis generation, data analysis, and predictive modeling. In genomics, ML algorithms process vast sequences of genetic information to identify patterns associated with diseases or traits. In physics and chemistry, AI models predict molecular interactions, optimize experiments, and accelerate material discovery. Climate modeling and epidemiology rely on AI to simulate complex systems and forecast future scenarios. AI integration reduces the time and cost of research, enhances accuracy, and allows scientists to tackle problems that were previously computationally or analytically intractable.

Challenges in Integration

Despite its benefits, integrating AI and ML into various domains poses several challenges. Data quality and availability are critical, as biased, incomplete, or noisy datasets can lead to inaccurate predictions or unethical outcomes. Algorithm transparency and interpretability are often limited, creating difficulties in understanding decision-making processes. Integration may require significant infrastructure investments, technical expertise, and changes in organizational workflows. Ethical concerns arise around privacy, security, accountability, and potential job displacement. Additionally, regulatory frameworks often lag behind technological advancements, creating uncertainty in compliance and governance. Addressing these challenges requires careful planning, robust oversight, and ongoing evaluation of AI and ML systems.

- AI refers to computer systems capable of simulating human intelligence, while ML allows systems to learn from data.
- Integration of AI and ML enhances data analysis, prediction, and decision-making across multiple sectors.
- In healthcare, AI supports diagnostics, predictive analytics, drug discovery, and personalized treatment.
- In industry, AI enables automation, predictive maintenance, quality control, and supply chain optimization.
- Financial institutions use AI for fraud detection, risk assessment, algorithmic trading, and customer service.
- Agriculture and environmental management benefit from AI-driven precision farming, climate modeling, and conservation monitoring.
- Scientific research leverages AI to accelerate hypothesis generation, genomics analysis, molecular discovery, and climate simulations.
- Challenges include data quality, algorithm transparency, infrastructure needs, ethical concerns, and regulatory gaps.
- Future integration focuses on explainable AI, IoT connectivity, federated learning, and multi-modal data analysis.
- Policy, governance, and ethical frameworks are essential for responsible, equitable, and sustainable AI integration.

7.2.2 Quantum Computing for Synthetic Biology

Quantum computing offers a transformative opportunity for synthetic biology by providing unprecedented computational power to tackle complex molecular, genetic, and cellular problems. By accelerating molecular modeling, optimizing genetic circuits, enhancing predictive modeling, and streamlining drug discovery, quantum computing addresses critical computational bottlenecks in synthetic biology. The integration of these technologies requires careful consideration of ethical, safety, and societal implications, ensuring responsible innovation and minimizing risks. Interdisciplinary collaboration, regulatory oversight, and technological development will be key factors in realizing the potential of quantum-enhanced synthetic biology. As quantum computing matures, its application promises to redefine the scope and efficiency of synthetic biology, enabling more precise, rapid, and innovative solutions to global challenges in health, industry, and the environment. The synergy between these cutting-edge fields represents a frontier of scientific exploration where computational power meets biological creativity, offering the possibility of groundbreaking advancements in our understanding and engineering of life.

7.3 SYNBIO AND THE FUTURE OF PERSONALIZED MEDICINE

Synthetic biology profoundly influences the future of personalized medicine by transcending traditional therapeutic approaches with engineered biological systems capable of targeted sensing, precise control, and adaptive treatment of diseases. Through programmable synthetic cells, closed-loop therapeutic devices, refined gene and cell therapies, and smart biomaterials, SynBio delivers customized healthcare solutions optimized to individual genetic and physiological profiles.

These technologies promise to increase treatment efficacy, reduce adverse effects, and empower real-time disease management. Challenges related to safety, ethics, regulation, and societal acceptance remain, demanding collaborative efforts across disciplines. As these hurdles are surmounted, synthetic biology will play a central role in shaping personalized medicine, transforming patient care from reactive to proactive, and enabling a new paradigm of precision health innovation.

7.3.1 Synthetic Cells as Programmable Therapeutic Platforms

At the core of SynBio-enabled personalized medicine are synthetic cells engineered to function as autonomous therapeutic platforms. These cells can be reprogrammed with sophisticated genetic circuits that sense specific biomarkers indicative of disease states—such as cancer-specific antigens, inflammatory molecules, or metabolic imbalances—and trigger tailored therapeutic responses accordingly. For example, engineered T cells expressing synthetic gene switches can identify tumor cells with high specificity, releasing cytotoxic agents only in tumor microenvironments, thereby sparing healthy tissues and reducing adverse effects typically seen in chemotherapy or conventional immunotherapy. The ability to construct complex gene networks incorporating logic gates, feedback loops, and memory functions allows these living therapies to adapt to temporal and spatial variations in disease biomarkers, improving treatment precision and reducing off-target toxicity.

7.3.2 Synthetic Biology in Diagnostic Innovation

Synthetic biology revolutionizes diagnostics by enabling the creation of biosensors and synthetic circuits that provide real-time, multiplexed, and highly sensitive detection of disease biomarkers. These innovations include cell-based biosensors engineered to produce detectable outputs such as fluorescence or electrical signals upon recognition of specific proteins, nucleic acids, or small molecules associated with pathological conditions. Paper-based or cell-free synthetic gene circuits further facilitate point-of-care testing that is rapid, cost-effective, and deployable outside traditional clinical settings. Such diagnostic platforms allow earlier detection of diseases like cancer, infectious diseases, and metabolic disorders, enabling timely and personalized medical interventions. The integration of biosensing with therapeutic modules in "smart" synthetic cells represents an emerging frontier—diagnostics that not only detect but dynamically respond to disease states in real time.

7.3.3 Pharmacogenetics and Synthetic Biology

Synthetic biology enriches pharmacogenetics and pharmacogenomics, which study how genetic variability among individuals influences drug metabolism, efficacy, and toxicity. Through modular design of biosynthetic pathways and synthetic gene circuits, synthetic biology allows the creation of tailor-made drugs optimized for specific genetic backgrounds, minimizing adverse reactions and maximizing therapeutic effects.

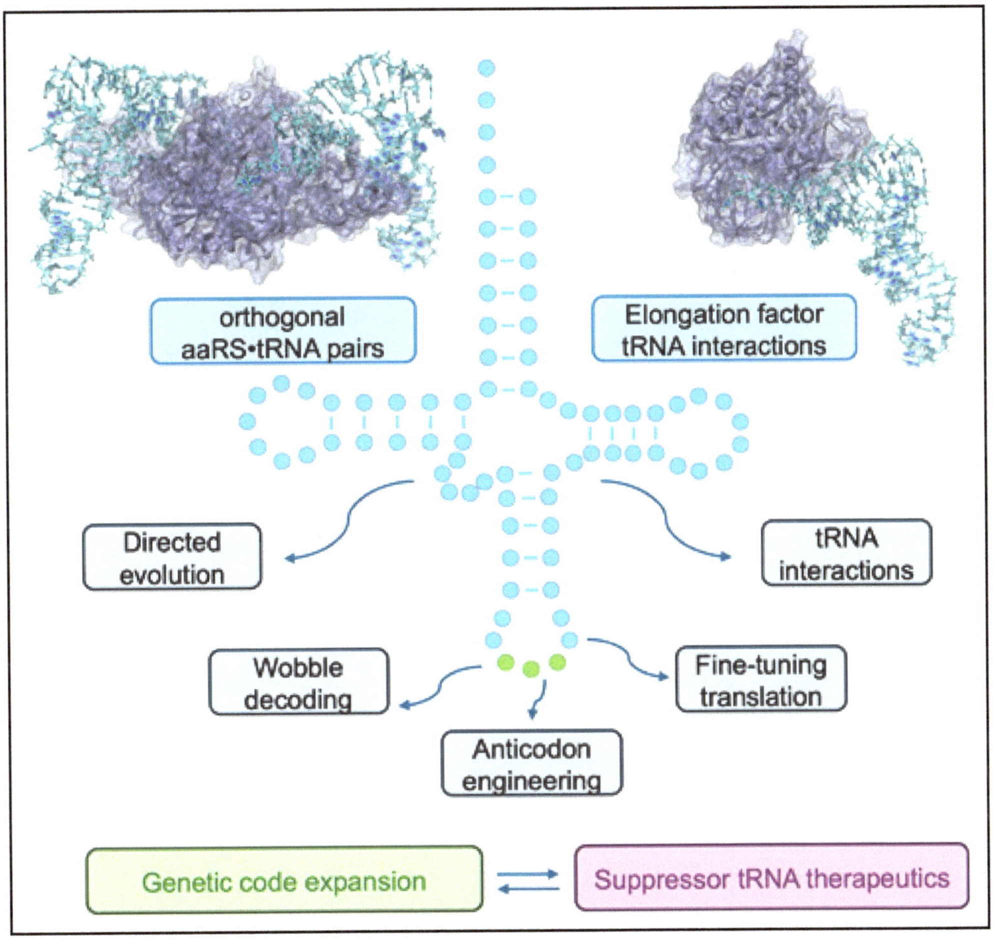

Synthetic Biology Applications in Personalized Medicine

Pathway modeling and genome-scale metabolic engineering enable the production of personalized drug molecules or prodrugs activated specifically in diseased tissues. Synthetic gene and cell therapies further refine drug delivery, allowing controlled and localized release triggered by patient-specific biomarkers. Such approaches exemplify the fusion of synthetic biology tools with personalized medicine principles to increase treatment precision and patients' quality of life.

7.3.4 Closed-Loop Therapeutic Systems and Feedback Control

A hallmark of SynBio's impact on personalized medicine is the development of closed-loop therapeutic systems—autonomous treatment devices that continuously monitor physiological or biochemical parameters and adjust therapeutic output accordingly. Inspired by natural homeostatic mechanisms, these engineered cellular or biomolecular systems incorporate sensor modules that detect disease severity or biomarker fluctuations, computational modules that integrate signals using synthetic gene circuits, and effector modules that release or modulate therapeutic agents. For chronic diseases such as diabetes or autoimmune disorders, these closed-loop devices maintain drug concentrations within optimal therapeutic windows, reducing side effects and eliminating the need for frequent clinical monitoring. Their programmability enables fine-tuning of treatment regimens tailored to individual responses and disease dynamics, epitomizing the personalization of medicine at the molecular level.

7.3.5 Advancements in Cell and Gene Therapies

Synthetic biology accelerates innovations in cell and gene therapies, which are pivotal components of personalized medicine. Advances in constructing synthetic gene circuits enable precise spatial and temporal control over gene expression, improving safety, efficacy, and durability of therapeutic interventions. For instance, CAR-T cells engineered with synthetic switches can be programmed to recognize multiple tumor markers and activate only under defined microenvironmental conditions, reducing risks of cytokine release syndrome and off-target effects. Genome editing techniques combined with synthetic biology allow correction of genetic defects at unprecedented precision tailored to patients' unique mutations. The integration of synthetic DNA constructs enables the programming of patient-derived cells with new functionalities, such as enhanced migration, immune modulation, or resistance to inhibitory signals, fostering personalized and adaptive cellular therapies.

7.3.6 Synthetic Biology and Biomaterials for Personalized Solutions

Beyond genetic engineering, synthetic biology contributes to the design of novel biomaterials and drug delivery systems personalized to patient needs. Synthetic biomaterials embedded with responsive elements can modulate drug release kinetics, degrade in specific tissue environments, or interact dynamically with cellular receptors. By incorporating synthetic gene circuits into biomaterial scaffolds, it becomes feasible to create smart implants and tissue engineering constructs that sense local physiological conditions and adjust therapeutic outputs accordingly. These personalized biomaterials promise superior integration with host tissues, improved regeneration, and reduced complications. Biofabrication technologies leveraging synthetic biology enable the design of patient-specific implants and therapeutic formulations with enhanced efficacy and biocompatibility.

7.3.7 Challenges and Ethical Considerations

Despite the promising advances, the application of synthetic biology in personalized medicine raises numerous challenges. Ensuring the safety, robustness, and controllability of synthetic biological systems in complex human environments remains paramount. Potential immunogenicity, off-target effects, and long-term stability of engineered cells or gene circuits require comprehensive preclinical and clinical evaluations. Ethical concerns include equitable access to these advanced therapies, consent regarding genetic data use, and implications of modifying human cells with synthetic systems. Regulatory frameworks struggle to keep pace with rapid technological innovation, necessitating adaptive governance models that balance patient safety with innovation incentives. Public acceptance and trust are integral to clinical translation, emphasizing the need for transparent communication, engagement, and ethical stewardship.

7.3.8 Prospects and Impact on Healthcare

The convergence of synthetic biology and personalized medicine heralds a transformative era for healthcare. Future developments will likely encompass fully programmable therapeutic cells capable of multi-modal sensing and complex decision-making, integrated with wearable devices for continuous health monitoring. Advances in multi-omics, machine learning, and systems biology will refine patient stratification, enabling the design of individualized synthetic biology interventions with unprecedented precision. Cost reductions through scalable biomanufacturing and automation will enhance accessibility, moving personalized synthetic biology therapies beyond specialized centers. Ultimately, SynBio empowers a new age of precision healthcare, where therapies are dynamically tailored and optimized throughout the course of an individual's lifetime, improving outcomes and redefining medical practice.

7.4 THE ROLE OF SYNTHETIC BIOLOGY IN SPACE EXPLORATION

Synthetic biology stands poised to transform space exploration from a resource-intensive endeavor dependent on Earth support to a sustainable expansion of human presence throughout the solar system. The biological technologies under development today will enable future astronauts to produce food, materials, and medicines wherever they travel, fundamentally changing the economics and feasibility of space colonization. The integration of living systems into spacecraft and habitats represents a philosophical shift toward viewing space settlements not as isolated mechanical outposts but as living ecosystems that include humans as integral components. As synthetic biology capabilities continue to advance, the boundary between science fiction and achievable reality steadily shifts, bringing closer the day when humans can truly live off the land on other worlds.

The challenges that remain are substantial but not insurmountable, requiring continued innovation, international cooperation, and careful consideration of safety and ethical implications. The synergy between synthetic biology and space exploration drives advances in both fields, with technologies developed for space applications offering benefits for terrestrial challenges and vice versa. Ultimately, synthetic biology may prove to be the key technology that enables humanity to become a multi-planetary species, carrying life beyond Earth to flourish throughout the cosmos.

7.4.1 Life Support and Resource Recycling

The development of biological life support systems using synthetic biology offers revolutionary improvements over traditional physicochemical approaches for maintaining habitable environments during space missions. Engineered microorganisms can perform multiple functions within regenerative life support systems, including carbon dioxide removal, oxygen production, water purification, and waste processing, all within compact bioreactors that require minimal energy input.

Synthetic biology enables the optimization of photosynthetic organisms such as cyanobacteria and algae to maximize oxygen production while minimizing resource requirements, potentially achieving higher efficiency than mechanical systems. Modified bacteria can break down human waste and other organic materials, converting them into useful products such as nutrients for food production or precursors for manufacturing, creating closed-loop systems that dramatically reduce the need for resupply missions. The engineering of extremophile organisms adapted to survive radiation, temperature fluctuations, and other space conditions ensures system reliability even in harsh extraterrestrial environments. Biofilms engineered to coat surfaces within spacecraft could continuously purify air and water while preventing the growth of harmful microorganisms that might threaten crew health. The integration of multiple engineered organisms into symbiotic communities creates robust ecosystems that can maintain stability through mutual support and redundancy, mimicking Earth's natural biogeochemical cycles in miniature.

7.4.2 Food Production and Nutrition

Synthetic biology revolutionizes space food systems by enabling the production of nutritious, palatable food from minimal resources, addressing one of the most significant challenges for long-duration missions. Engineered microorganisms can convert simple inputs such as carbon dioxide, water, and minerals into complete proteins, essential fatty acids, vitamins, and other nutrients required for human health, eliminating the need to transport large quantities of food from Earth. Modified plants optimized for growth in controlled environments can produce higher yields with reduced water and nutrient requirements, while engineered traits such as enhanced nutritional content and extended shelf life improve their suitability for space agriculture.

Synthetic biology enables the production of cultured meat and other animal products without the resource-intensive requirements of raising livestock, providing dietary variety and psychological comfort to crew members on extended missions. The development of organisms capable of utilizing in-situ resources such as Martian regolith or lunar soil as growth substrates could enable food production using local materials, crucial for establishing self-sufficient colonies. Personalized nutrition becomes possible through engineered organisms that can adjust their production of specific nutrients based on individual crew member needs and health status. The creation of novel food organisms that combine optimal nutrition with appealing taste and texture addresses both physiological and psychological aspects of crew wellbeing during isolation in space.

7.4.3 Biomanufacturing in Space

The application of synthetic biology to manufacturing in space enables the production of essential materials, tools, and equipment using biological processes that would be impossible or impractical with traditional manufacturing methods. Engineered microorganisms can produce a vast array of materials including bioplastics, adhesives, textiles, and even electronic components, reducing the need to launch manufactured goods from Earth. The ability to program organisms to produce specific materials on demand provides flexibility to respond to unexpected needs or equipment failures without waiting for resupply missions.

Biomineralization processes engineered into organisms could generate construction materials from regolith and other local resources, enabling the building of habitats and infrastructure on planetary surfaces. Self-assembling biological systems could construct complex structures through programmed growth patterns, potentially creating living habitats that can self-repair and adapt to environmental conditions.

The production of pharmaceuticals and medical supplies through engineered organisms ensures continuous availability of essential medicines without concerns about shelf life or storage conditions. Synthetic biology enables the recycling and upcycling of waste materials into valuable products, supporting circular economy principles essential for sustainable space settlements.

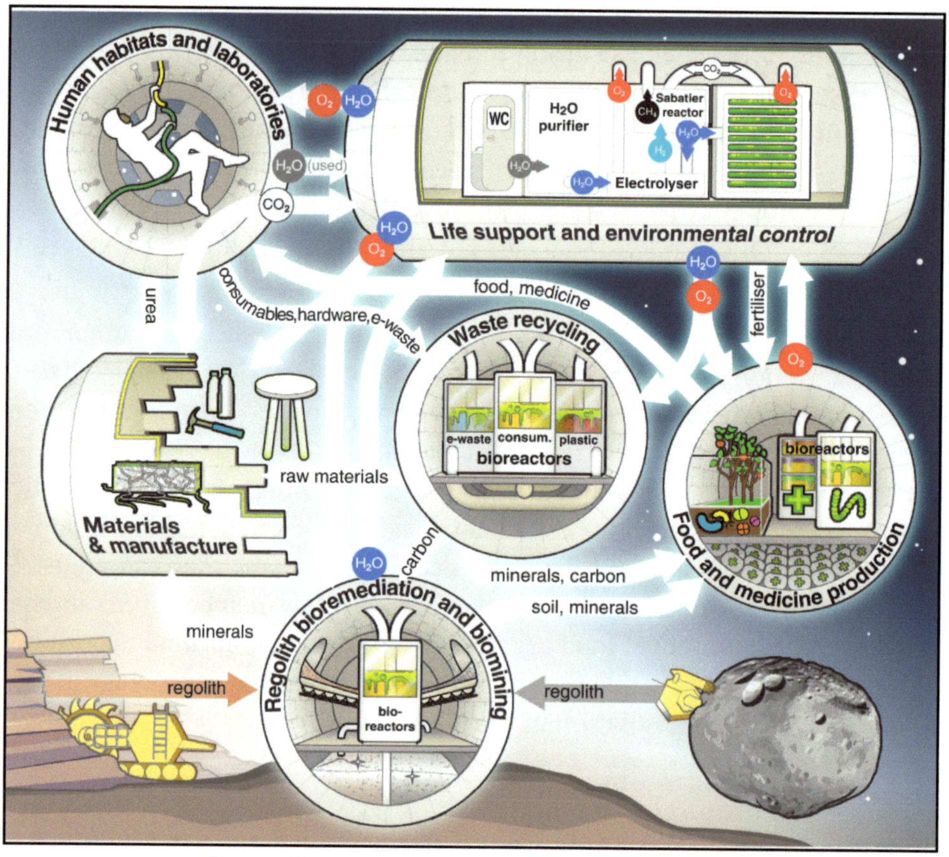

Advanced Biomanufacturing for Space Missions

7.4.4 Medicine and Healthcare Applications

Synthetic biology provides innovative solutions for maintaining crew health and treating medical conditions during space missions where traditional medical infrastructure is unavailable. Engineered probiotics can monitor and maintain astronaut health by detecting biomarkers of disease, producing therapeutic compounds, and modulating the immune system to counteract the effects of microgravity and radiation.

Biosensors based on synthetic biological circuits can provide real-time health monitoring and early disease detection, crucial for preventing medical emergencies when evacuation to Earth is impossible. The ability to produce personalized medicines on-demand using engineered organisms eliminates the need to stock extensive pharmaceutical supplies and ensures treatments remain effective despite long storage periods. Synthetic biology approaches to wound healing include engineered skin substitutes and biological sutures that promote tissue regeneration in the altered healing environment of space. Radiation resistance genes incorporated into therapeutic organisms or even potentially into human cells could provide protection against cosmic radiation, one of the most significant health risks for space travelers. The development of biological countermeasures for bone loss, muscle atrophy, and other physiological changes associated with spaceflight could maintain crew health during extended missions.

7.4.5 Habitat Construction and Materials

The role of synthetic biology in constructing and maintaining space habitats extends beyond simple material production to encompass living architecture that provides multiple functions simultaneously. Engineered organisms could grow structural materials in predetermined shapes and configurations, creating self-assembling habitats that require minimal human intervention to construct. Biocement produced by modified bacteria could bind regolith particles into solid construction materials, enabling the use of local resources for building protective structures against radiation and micrometeorite impacts.

Living walls incorporating engineered organisms could provide air purification, temperature regulation, and even lighting through bioluminescence, creating more Earth-like environments for psychological wellbeing. Self-healing materials incorporating biological components could automatically repair micro-fractures and damage, crucial for maintaining pressure integrity in the harsh space environment. The integration of synthetic biology with architectural design enables adaptive structures that can respond to changing environmental conditions or mission requirements. Biomaterials with properties tailored for specific extraterrestrial environments, such as extreme cold resistance for Europa or acid resistance for Venus, expand the possibilities for human presence throughout the solar system.

7.4.6 Terraforming and Planetary Engineering

While remaining largely theoretical, synthetic biology offers the most plausible pathway toward terraforming other planets to support human life without life support systems. Engineered extremophile organisms could begin the process of atmospheric transformation on Mars, producing oxygen while sequestering carbon dioxide and releasing greenhouse gases to warm the planet.

Synthetic organisms designed to thrive in Martian conditions could establish the foundation of a biosphere, creating soil from regolith and establishing nutrient cycles necessary for more complex life. The engineering of organisms capable of surviving in the clouds of Venus or the subsurface oceans of icy moons expands the potential for biological transformation of diverse celestial bodies. Synthetic biology could accelerate geological processes that would naturally take millions of years, such as the weathering of rocks to release minerals and the formation of organic-rich soils. The careful design of artificial ecosystems ensures stability and prevents unintended consequences that could derail terraforming efforts. While the ethical and practical challenges of planetary-scale engineering remain enormous, synthetic biology provides the tools that make such ambitious projects theoretically achievable.

7.4.7 Challenges and Technical Hurdles

Despite its enormous potential, the application of synthetic biology to space exploration faces significant technical, safety, and regulatory challenges that must be overcome. The effects of microgravity, radiation, and other space conditions on engineered organisms remain poorly understood, potentially affecting their performance and stability in unpredictable ways. Containment and biosafety become critically important in the confined environment of spacecraft, where contamination events could prove catastrophic with no possibility of evacuation.

The evolution and adaptation of engineered organisms over time could lead to loss of desired functions or emergence of unwanted traits, requiring robust genetic safeguards and monitoring systems. Resource limitations in space constrain the complexity of biological systems that can be maintained, necessitating extremely efficient and reliable organisms. International space law and planetary protection protocols restrict the introduction of Earth organisms to other celestial bodies, requiring careful consideration of synthetic biology applications. The integration of biological and traditional spacecraft systems presents engineering challenges related to incompatible operating conditions and potential interference.

7.4.8 Current Research and Mission Planning

Space agencies and research institutions worldwide are actively developing synthetic biology applications for upcoming space missions, with several proof-of-concept experiments already conducted or planned. The International Space Station serves as a testbed for synthetic biology experiments, investigating how engineered organisms behave in microgravity and developing protocols for their safe use in space. NASA's Space Synthetic Biology program funds research into biological solutions for in-situ resource utilization, life support, and manufacturing for future Mars missions.

Private space companies increasingly recognize synthetic biology as essential for their commercial space ventures, investing in research and development of biological technologies. International collaborations combine expertise in synthetic biology with space engineering to accelerate the development of practical applications. Ground-based analog missions in extreme environments test synthetic biology systems under conditions that simulate aspects of space exploration. The establishment of dedicated facilities for developing and testing space-compatible biological systems indicates growing institutional commitment to this approach.

7.5 SYNTHETIC BIOLOGY IN GLOBAL HEALTH AND PANDEMIC PREPAREDNESS

Synthetic biology has emerged as a critical tool in global health and pandemic preparedness, offering unprecedented capabilities to detect, prevent, and respond to infectious disease outbreaks. Its applications in diagnostics, vaccine development, therapeutics, and pathogen surveillance enable rapid, precise, and scalable interventions, enhancing public health resilience. By integrating synthetic biology with AI, predictive modeling, and advanced manufacturing, global health systems can respond more effectively to emerging pathogens and evolving disease threats.

However, challenges related to equity, infrastructure, regulatory oversight, and biosecurity must be addressed to ensure safe, ethical, and inclusive application. International collaboration, standardized platforms, and transparent data-sharing networks are essential to maximize the societal impact of synthetic biology in pandemic preparedness. Moving forward, continued innovation, ethical governance, and capacity building will strengthen global health systems, reduce vulnerability to infectious diseases, and improve preparedness for future pandemics, ensuring that the benefits of synthetic biology reach populations worldwide.

7.5.1 Role of Synthetic Biology in Diagnostics

Rapid and accurate diagnostics are essential for identifying infectious diseases and controlling outbreaks. Synthetic biology enables the development of novel diagnostic tools that are highly sensitive, specific, and adaptable to emerging pathogens. For example, CRISPR-based diagnostics can detect viral RNA in patient samples within hours, offering a faster and more portable alternative to traditional PCR-based methods. Biosensors and engineered genetic circuits can identify pathogen-specific biomarkers in environmental or clinical samples, allowing for real-time monitoring. Additionally, synthetic biology supports the development of modular diagnostic platforms that can be quickly reconfigured for new diseases, enhancing global readiness for emerging infectious threats.

7.5.2 Vaccine Development and Production

Synthetic biology has revolutionized vaccine development by enabling rapid design, testing, and manufacturing of novel vaccines. mRNA vaccines, which were instrumental in controlling COVID-19, rely on synthetic biology techniques to design and produce sequences encoding viral antigens. Synthetic biology also facilitates the engineering of viral vectors, nanoparticles, and protein-based vaccines, allowing for improved efficacy and stability. Modular vaccine platforms enable quick adaptation to new viral strains, reducing the time from pathogen identification to vaccine deployment. Furthermore, synthetic biology supports scalable production processes, ensuring that vaccines can be manufactured in large quantities and distributed globally, particularly during health emergencies.

7.5.3 Therapeutics and Antimicrobial Strategies

Beyond vaccines, synthetic biology contributes to the development of targeted therapeutics and antimicrobial agents. Engineered microbes and synthetic enzymes can produce antiviral compounds, monoclonal antibodies, or small-molecule drugs with high efficiency. Gene-editing technologies, such as CRISPR-Cas systems, are being explored as potential therapies for viral infections, allowing precise targeting and disruption of pathogen genomes. Synthetic biology also enables the creation of synthetic antibodies, peptides, and immunomodulatory molecules to treat infections or prevent immune evasion by pathogens. These approaches enhance the speed, specificity, and adaptability of therapeutic interventions during pandemics.

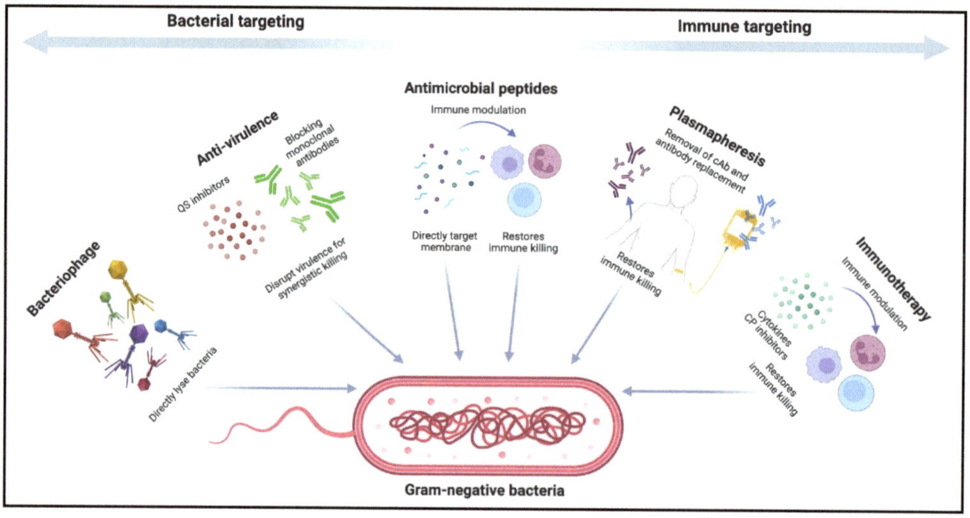

Therapeutic Strategies in Combating Microbial Infections

7.5.4 Pathogen Surveillance and Early Warning Systems

Effective pandemic preparedness relies on timely detection and monitoring of emerging infectious diseases. Synthetic biology tools enable advanced pathogen surveillance by integrating molecular diagnostics, biosensors, and environmental monitoring. Engineered organisms can serve as sentinels that detect specific pathogens in water, air, or soil, providing early warning signals of outbreaks. High-throughput sequencing and synthetic biology-based assays allow for rapid characterization of viral evolution and the identification of mutations that may affect transmissibility or vaccine efficacy. Predictive models generated from synthetic biology data can inform public health interventions, enabling proactive measures to contain outbreaks and prevent global spread.

7.5.5 Scaling and Manufacturing Capabilities

Pandemic preparedness requires the capacity to rapidly produce and distribute vaccines, diagnostics, and therapeutics. Synthetic biology enhances manufacturing scalability by enabling modular, automated, and highly efficient production processes. Cell-free systems, engineered microbes, and synthetic bioreactors can produce biologics at scale while maintaining quality and consistency. These technologies reduce dependence on traditional production methods that may be slow or resource-intensive. By combining synthetic biology with advanced manufacturing platforms, public health agencies and pharmaceutical companies can respond quickly to pandemics, ensuring that essential medical countermeasures reach populations in need.

7.5.6 Global Collaboration and Data Sharing

Synthetic biology facilitates international collaboration by providing standardized tools, platforms, and data-sharing mechanisms. Global health networks can share genetic sequences, diagnostic protocols, and vaccine designs in real-time, enabling coordinated responses to emerging threats. Initiatives such as open-source bioinformatics databases and international registries of synthetic biological parts support collaboration across countries, academic institutions, and industry partners. Transparent sharing of synthetic biology innovations accelerates research, reduces duplication of effort, and enhances global preparedness. Collaborative frameworks also help address disparities in access to technologies, ensuring equitable deployment of diagnostic, therapeutic, and preventive measures.

7.5.7 Ethical, Regulatory, and Safety Considerations

The integration of synthetic biology into global health raises important ethical, regulatory, and safety considerations. Engineering pathogens or biological systems carries inherent biosafety and biosecurity risks, including accidental release or misuse. Regulatory frameworks must balance innovation with risk management, ensuring that research and deployment adhere to ethical standards, environmental safeguards, and international guidelines. Intellectual property policies, equitable access, and benefit-sharing mechanisms are critical to prevent disparities in the availability of synthetic biology-based solutions. Public engagement and transparency are essential for building trust and promoting the responsible application of these technologies in pandemic preparedness.

7.5.8 Challenges in Implementation

Despite its promise, the integration of synthetic biology in global health faces several challenges. Limited infrastructure and technical expertise in low- and middle-income countries may restrict the adoption of synthetic biology tools. High costs and intellectual property restrictions can hinder equitable access to diagnostics, vaccines, and therapeutics. Data privacy concerns and cybersecurity risks arise when sharing genetic and health-related information across borders. Rapidly evolving pathogens may outpace the ability of synthetic biology platforms to adapt, requiring continuous innovation and surveillance. Addressing these challenges requires global coordination, capacity-building initiatives, and investment in robust infrastructure and workforce development.

- Synthetic biology enables rapid development of diagnostics, vaccines, therapeutics, and surveillance systems.
- CRISPR-based diagnostics and biosensors improve pathogen detection and outbreak monitoring.
- mRNA and modular vaccine platforms accelerate vaccine design and scalable production.
- Engineered microbes and gene-editing technologies facilitate targeted therapeutics and antimicrobial strategies.
- Synthetic biology enhances pathogen surveillance through environmental monitoring and high-throughput assays.
- Scalable production platforms, including cell-free systems and bioreactors, ensure rapid deployment of medical countermeasures.
- Global collaboration and data-sharing networks accelerate research, innovation, and equitable access.
- Ethical, regulatory, and safety frameworks are essential to mitigate biosafety, biosecurity, and equity risks.
- Challenges include infrastructure limitations, high costs, intellectual property barriers, and rapidly evolving pathogens.
- Future directions emphasize automation, AI integration, decentralized manufacturing, and adaptive policy frameworks.

CHAPTER 8

Case Studies in Synthetic Biology

8.1 ENGINEERING E. COLI FOR PHARMACEUTICAL PRODUCTION

Engineering E. coli for pharmaceutical production exemplifies the intersection of microbial biology, genetic engineering, and industrial biotechnology. Through sophisticated genetic, metabolic, and bioprocess engineering strategies, E. coli can be transformed into a versatile production platform capable of synthesizing a wide array of therapeutic molecules. While challenges such as protein folding, metabolic burden, and regulatory compliance remain, advances in synthetic biology, computational modeling, and fermentation technology continue to enhance the efficiency and reliability of this approach. E. coli remains a foundational organism in biopharmaceutical manufacturing, offering scalable, cost-effective, and safe solutions to meet global healthcare needs. The continued integration of innovation, safety, and regulatory oversight will ensure that engineered E. coli strains contribute to the development of novel therapies, improved patient outcomes, and a more sustainable biopharmaceutical industry.

8.1.1 Historical Background and Development

The use of E. coli for recombinant protein production began in the late 1970s with the advent of recombinant DNA technology. Early successes included the expression of human insulin and growth hormone, which demonstrated that E. coli could produce biologically active human proteins. These breakthroughs transformed pharmaceutical manufacturing, offering scalable and cost-effective alternatives to traditional extraction methods from animal or plant sources. Over time, advances in molecular cloning, promoter design, plasmid stability, and fermentation technology further enhanced E. coli's utility as a production host. Regulatory approval of recombinant therapeutics derived from E. coli validated its safety and efficacy, paving the way for broader applications in biopharmaceuticals. Today, E. coli continues to serve as a versatile and highly adaptable platform for producing both simple and complex pharmaceutical molecules.

8.1.2 Genetic Engineering Strategies

Engineering E. coli for pharmaceutical production begins with precise genetic modifications to optimize expression and functionality of target molecules. Plasmid-based expression systems are commonly used to introduce recombinant genes under the control of strong promoters, enabling high-level protein production. Selection of codon usage, optimization of ribosome binding sites, and elimination of rare codons improve translation efficiency. Chromosomal integration of target genes offers an alternative approach to enhance stability and reduce plasmid-associated metabolic burden. Advanced techniques, such as CRISPR-Cas genome editing and recombineering, allow targeted modification of metabolic pathways, regulatory networks, and host genome architecture. These strategies enable the fine-tuning of transcriptional and translational machinery, reduce byproduct formation, and enhance overall yield, ensuring that E. coli can efficiently synthesize high-quality pharmaceuticals.

8.1.3 Metabolic Engineering for Enhanced Production

Metabolic engineering is critical to optimizing E. coli for pharmaceutical production. By manipulating native metabolic pathways, researchers can redirect cellular resources toward the biosynthesis of desired molecules. For example, overexpression of precursor biosynthesis enzymes or deletion of competing pathways increases the flux toward target compounds.

Cofactor availability, energy balance, and redox state are also critical considerations, as they influence the efficiency of enzymatic reactions. Systems biology approaches, including genome-scale metabolic modeling and flux balance analysis, allow researchers to predict the effects of genetic modifications on cellular metabolism, identify bottlenecks, and design optimized production strains. By integrating metabolic engineering with genetic engineering, E. coli can be transformed into a highly productive microbial factory capable of synthesizing complex pharmaceuticals at industrial scale.

8.1.4 Protein Expression and Folding Challenges

Producing recombinant proteins in E. coli presents challenges related to expression, folding, and post-translational modifications. High-level expression of heterologous proteins can lead to the formation of inclusion bodies, aggregates of misfolded proteins that are biologically inactive.

Strategies to overcome these issues include co-expression of molecular chaperones, fusion protein tags to enhance solubility, and fine-tuning of expression conditions such as temperature, inducer concentration, and growth medium composition. While E. coli lacks the ability to perform complex eukaryotic post-translational modifications such as glycosylation, advances in glycoengineering and co-expression of modifying enzymes have enabled partial reconstruction of these pathways. Correct protein folding and functionality are essential for therapeutic efficacy, particularly for enzymes, antibodies, and biologics, making optimization of the expression system a key focus in pharmaceutical strain engineering.

8.1.5 Fermentation and Bioprocess Optimization

Efficient pharmaceutical production requires not only optimized genetic constructs but also carefully controlled fermentation processes. Bioreactor design, growth medium composition, oxygen supply, pH, and temperature all influence E. coli productivity. Fed-batch and continuous culture strategies help maintain optimal growth and reduce metabolic stress. Monitoring systems using sensors and real-time data analytics allow precise control over nutrient availability, metabolite accumulation, and cellular health. Scale-up from laboratory to industrial production introduces additional challenges, such as maintaining homogeneity and preventing contamination. Advances in bioprocess engineering, including automated control systems, high-cell-density cultivation, and process modeling, have enabled reliable and reproducible production of pharmaceuticals in E. coli at commercial scales.

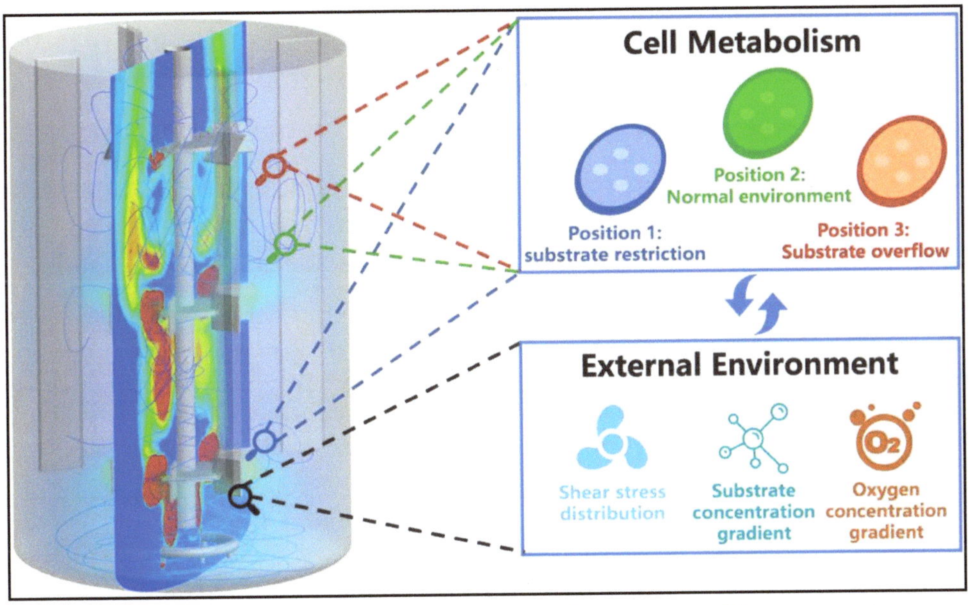

Fermentation Process Optimization and Bioprocess Management

8.1.6 Applications in Therapeutics

E. coli has been successfully engineered to produce a wide range of therapeutic molecules. Recombinant insulin, produced in E. coli since the early 1980s, remains one of the most widely used examples of microbial pharmaceutical production. Other protein therapeutics, including growth hormones, interferons, interleukins, and vaccine antigens, have also been produced using engineered E. coli strains. Small-molecule drugs and precursors, such as antibiotics and amino acid derivatives, can be synthesized through tailored metabolic pathways. Additionally, advances in synthetic biology have enabled the production of more complex biologics, including antibody fragments, therapeutic enzymes, and peptide hormones. These applications demonstrate the versatility of E. coli as a production host and its central role in the biopharmaceutical industry.

8.1.7 Regulatory and Safety Considerations

The production of pharmaceuticals in E. coli is subject to stringent regulatory oversight to ensure safety, efficacy, and quality. Good Manufacturing Practices (GMP) provide guidelines for facility design, process control, quality assurance, and documentation. Regulatory agencies, including the FDA, EMA, and WHO, evaluate production strains, genetic modifications, fermentation processes, and final product quality. Containment measures prevent accidental release of genetically modified organisms, and thorough purification processes remove endotoxins, host proteins, and other contaminants. Safety assessments also include evaluation of potential immunogenicity and allergenicity of the recombinant product. Compliance with regulatory standards is essential not only for market approval but also to maintain public trust in microbial biotechnology for pharmaceutical applications.

8.1.8 Challenges and Limitations

Despite its advantages, engineering E. coli for pharmaceutical production faces several challenges. The lack of complex post-translational modifications limits its utility for certain eukaryotic proteins. Metabolic burden and stress responses can reduce growth rates and productivity.

Table: Strategies for Engineering E. coli in Pharmaceutical Production

Strategy	Purpose
Genetic Engineering	Optimize gene expression, codon usage, and regulatory elements
Metabolic Engineering	Redirect metabolic flux toward target compounds and increase precursor availability
Protein Folding Optimization	Improve solubility, stability, and biological activity of recombinant proteins
Fermentation Control	Maximize yield through optimized growth conditions and bioprocess strategies
Safety and Regulatory Compliance	Ensure GMP adherence, purification of products, and containment of genetically modified strains

Genetic instability of recombinant constructs, plasmid loss, and mutations can compromise long-term production. Additionally, endotoxin contamination from the E. coli cell wall requires rigorous purification processes, adding cost and complexity. Addressing these challenges involves continuous strain improvement, process optimization, and integration of alternative microbial hosts or cell-free systems when necessary. Advanced computational modeling, machine learning, and high-throughput screening provide tools to overcome these limitations and enhance the efficiency and reliability of pharmaceutical production in E. coli.

8.2 SYNTHETIC BIOLOGY FOR BIODEGRADABLE PLASTICS

Synthetic biology offers revolutionary approaches for producing biodegradable plastics that address the pressing environmental impacts of conventional plastic pollution. By engineering microbial cell factories and optimizing biosynthetic pathways, it enables efficient conversion of renewable feedstocks into tunable biodegradable polymers with diverse applications. Combined with advances in feedstock valorization, biodegradation technologies, and process optimization, synthetic biology paves the way toward a circular bioeconomy with reduced ecological footprints. While commercialization challenges persist, continuous technological innovation and supportive policies forecast a promising future where synthetic biology-produced biodegradable plastics play a pivotal role in sustainable material production and environmental stewardship.

8.2.1 Microbial Cell Factories for Bioplastic Synthesis

Microorganisms constitute the foundation of synthetic biology-enabled bioplastic production. Species such as Escherichia coli, Pseudomonas putida, and various yeasts have been engineered to convert lignocellulosic biomass, agricultural residues, or industrial waste streams into monomers and polymers of biodegradable plastics. Notably, synthetic biology tools have enabled the introduction and refinement of biosynthetic pathways to produce polyhydroxyalkanoates (PHAs), polylactic acid (PLA), polybutylene succinate (PBS), and polybutylene adipate-co-terephthalate (PBAT). By modifying gene expression and enzyme activity levels within these microbes, researchers improve yield, productivity, and polymer quality. Recent advances involve metabolic pathway balancing, protein engineering to enhance enzyme efficiency, and optimizing cofactor recycling to maintain energetic and redox homeostasis, all crucial for industrial feasibility.

8.2.2 Synthetic Biology Strategies in Yeast Cell Factories

Yeasts such as Saccharomyces cerevisiae and Yarrowia lipolytica are increasingly explored as robust platforms for biodegradable plastic production. Their inherent tolerance to stressors such as low pH and high temperature complements industrial scale fermentation requirements. Synthetic biology advances assert the integration of heterologous pathways into yeast genomes with precise regulatory controls to maximize production of plastic precursors like lactic acid and polyhydroxyalkanoates. Adaptive laboratory evolution complements rational design by selecting strains with improved growth and product titers under relevant conditions. Furthermore, computational tools assist in the design-build-test-learn cycles to predict and optimize genetic circuit performance in these eukaryotic hosts, which differ mechanistically from bacterial systems.

8.2.3 Feedstock Flexibility and Circular Production Models

Utilization of renewable and waste feedstocks underpins the sustainability of synthetic biology-derived biodegradable plastics. Microbial factories engineered through synthetic biology convert lignocellulosic biomass—derived from non-edible plant materials—into fermentable sugars, which then feed biopolymer biosynthesis. Protein- and energy-rich agro-industrial waste streams also serve as substrates, contributing to resource recovery and waste valorization in circular economy frameworks. By coupling synthetic biology with bioconversion processes, the entire production system retains carbon and energy more efficiently, substantially minimizing environmental impact. Recycling of residual biomass and carbon dioxide fixation by engineered microbes further augment sustainability credentials, aligning with global efforts to mitigate climate change and plastic pollution.

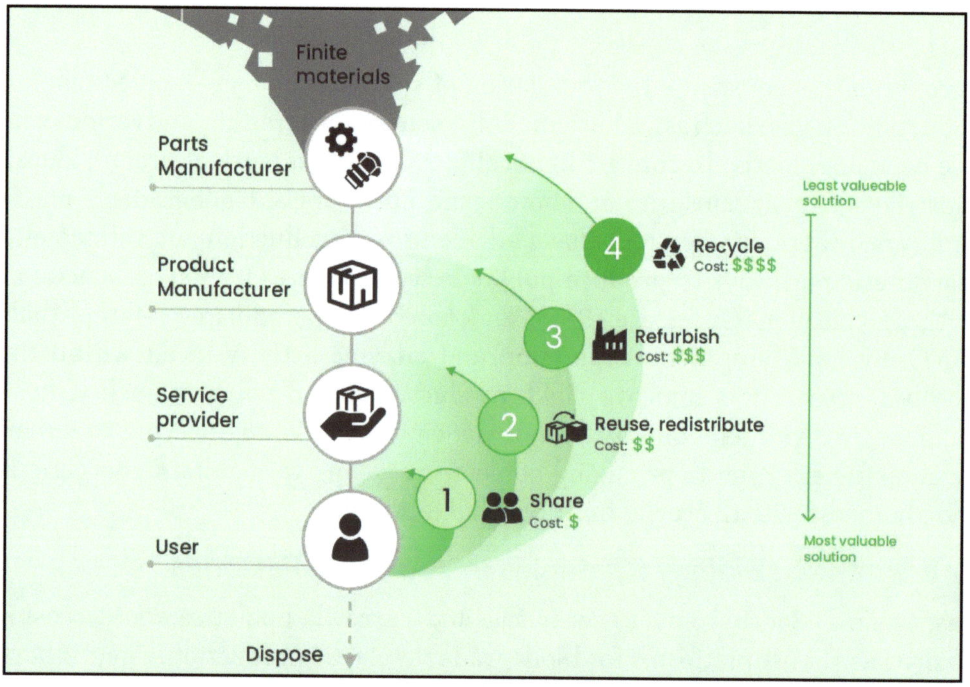

Circular Economy Production Models with Flexibility

8.2.4 Engineering Polymers with Tunable Properties

An exciting aspect of synthetic biology is the capacity to engineer polymers with customizable physicochemical and mechanical properties by controlling monomer composition, chain length, and polymer architecture. Biosynthetic pathways are engineered to produce polyesters with varying side chains, enabling modulation of biodegradability, flexibility, and strength to suit diverse applications from packaging to medical implants. Protein engineering and synthetic metabolic pathways allow fine-tuning of enzyme specificity and activity, fostering precise control over polymer assembly. This tailored approach overcomes limitations of naturally derived bioplastics, broadening their usability and market potential.

8.2.5 Environmental Impact and Biodegradation Pathways

Biodegradable plastics synthesized via synthetic biology demonstrate enhanced degradation profiles compared with conventional plastics, breaking down more rapidly through microbial enzymatic action in soil, marine, or composting environments. Engineered microbial consortia expressing tailored depolymerases degrade bioplastics into harmless metabolites like carbon dioxide and water, closing the material lifecycle loop. Synthetic biology facilitates development of bioplastics compatible with existing recycling infrastructures while enabling new degradation pathways suited for diverse environments. In this way, synthetic biology contributes not only to bioplastic production but also to end-of-life management, maximizing environmental benefits.

8.2.6 Economic and Industrial Considerations

Commercial viability of bioplastics produced via synthetic biology hinges on overcoming production cost, scalability, and process consistency challenges. While the cost of synthetic DNA and metabolic engineering tools has decreased, downstream processing, product purification, and large-scale fermentation remain resource-intensive. Synthetic biology enables development of microbial strains with improved robustness, productivity, and substrate utilization rates to reduce costs. Process optimization through high-throughput screening and machine learning accelerates strain development and industrial transfer. Public policies supporting biobased products, carbon pricing, and plastic waste reduction further drive market adoption, fostering growth in green materials sectors.

8.2.7 Challenges and Opportunities Ahead

Despite its promise, synthetic biology for biodegradable plastics faces challenges including genetic stability of engineered strains, competition with petrochemical plastics on cost and performance, and regulatory hurdles. Maintaining consistent biopolymer quality while scaling to industrial volumes requires meticulous control of genetic constructs and process parameters. Intellectual property complexities and market acceptance remain critical considerations. Nonetheless, ongoing research into novel microbial chassis, synthetic metabolic pathways, and bio-based production platforms continues to unlock new opportunities. Integration of synthetic biology with material science, nanotechnology, and bioprocess engineering will further enhance properties and application breadth of sustainable bioplastics.

8.3 CRISPR-DRIVEN ADVANCES IN HUMAN GENETIC DISEASES

CRISPR technology has transformed the landscape of genetic disease treatment, converting previously incurable conditions into treatable disorders and offering hope to millions of patients worldwide. The rapid progression from laboratory discovery to clinical application demonstrates the transformative power of this technology and the dedication of the scientific and medical communities to translating these advances into patient benefits. Success stories in treating blood disorders have validated the therapeutic potential of gene editing and paved the way for applications in increasingly complex diseases affecting multiple organ systems.

While significant challenges remain, particularly in delivery to certain tissues and ensuring long-term safety, the pace of innovation suggests that many current limitations will be overcome in the coming years. The democratization of CRISPR technology and decreasing costs may eventually make these therapies accessible to patients globally, though addressing equity and affordability remains crucial. As we stand on the cusp of a new era in genetic medicine, CRISPR-driven advances promise not just to treat genetic diseases but to fundamentally change our understanding of what is medically possible, offering the potential for cures where once there was only management of symptoms. The continued responsible development and application of these technologies will require ongoing collaboration between scientists, clinicians, regulators, and society to ensure that the benefits of CRISPR therapeutics are realized while maintaining the highest standards of safety and ethics.

8.3.1 Fundamentals of CRISPR Therapeutics

The application of CRISPR technology to treat human genetic diseases relies on the system's ability to make precise changes to DNA in living cells, either correcting disease-causing mutations or introducing compensatory genetic modifications. The CRISPR-Cas9 system, derived from bacterial immune mechanisms, uses a guide RNA to direct the Cas9 enzyme to specific genomic locations where it creates double-strand breaks that cell repair through various pathways.

These repair mechanisms can be harnessed to delete harmful mutations, insert corrective sequences, or disrupt disease-causing genes. The development of base editors and prime editors has expanded therapeutic possibilities by enabling precise single-letter changes without creating double-strand breaks, reducing the risk of unwanted insertions or deletions. The choice of CRISPR variant depends on the specific genetic defect and therapeutic goal, with some diseases requiring precise correction of point mutations while others benefit from gene disruption or insertion. Delivery methods represent a critical consideration in therapeutic applications, with approaches ranging from extracting cells for ex vivo editing to direct in vivo delivery using viral vectors or lipid nanoparticles. The immunogenicity of CRISPR components and the potential for off-target effects require careful optimization and monitoring to ensure safety and efficacy in clinical applications.

8.3.2 Blood Disorders: Pioneering Success Stories

Blood disorders have emerged as the leading frontier for CRISPR therapeutics, with remarkable clinical successes in treating sickle cell disease and beta-thalassemia marking watershed moments in genetic medicine. These conditions, caused by mutations in the beta-globin gene, affect millions worldwide and have historically required lifelong treatment with limited effectiveness. CRISPR approaches for these diseases employ a clever strategy of reactivating fetal hemoglobin production by disrupting the BCL11A gene, which normally suppresses fetal hemoglobin in adults.

Clinical trials have shown that patients receiving CRISPR-edited hematopoietic stem cells experience dramatic improvements, with many achieving transfusion independence and elimination of pain crises associated with sickle cell disease. The success rates observed in these trials, with some reporting functional cures in over ninety percent of patients, have exceeded expectations and validated the therapeutic potential of gene editing. The ex vivo nature of these treatments, where cells are edited outside the body before reinfusion, provides additional safety by allowing thorough characterization of edited cells before they are returned to patients. These achievements have paved the way for regulatory approvals, with the first CRISPR therapy for sickle cell disease receiving approval in multiple countries, representing a historic milestone in genetic medicine.

8.3.3 Inherited Blindness and Sensory Disorders

CRISPR technology offers new hope for treating inherited forms of blindness and other sensory disorders that result from specific genetic mutations affecting sensory organ function. Leber congenital amaurosis 10, caused by mutations in the CEP290 gene, has become a prime target for CRISPR therapy, with ongoing clinical trials delivering the gene editing components directly to photoreceptor cells in the eye.

This approach represents the first in vivo CRISPR trial for an inherited disease, where the editing occurs within the patient's body rather than in extracted cells. The eye presents unique advantages for gene therapy, including immune privilege that reduces the risk of immune responses and the ability to treat one eye while using the other as a control. Inherited hearing loss, affecting millions globally, is also being targeted with CRISPR approaches aimed at correcting mutations in genes such as TMC1 and OTOF that are essential for auditory function. Preclinical studies have demonstrated restoration of hearing in animal models through local delivery of CRISPR components to the inner ear. The development of more sophisticated delivery methods and improved understanding of sensory organ biology continues to expand the range of sensory disorders amenable to CRISPR intervention.

8.3.4 Neuromuscular and Neurological Applications

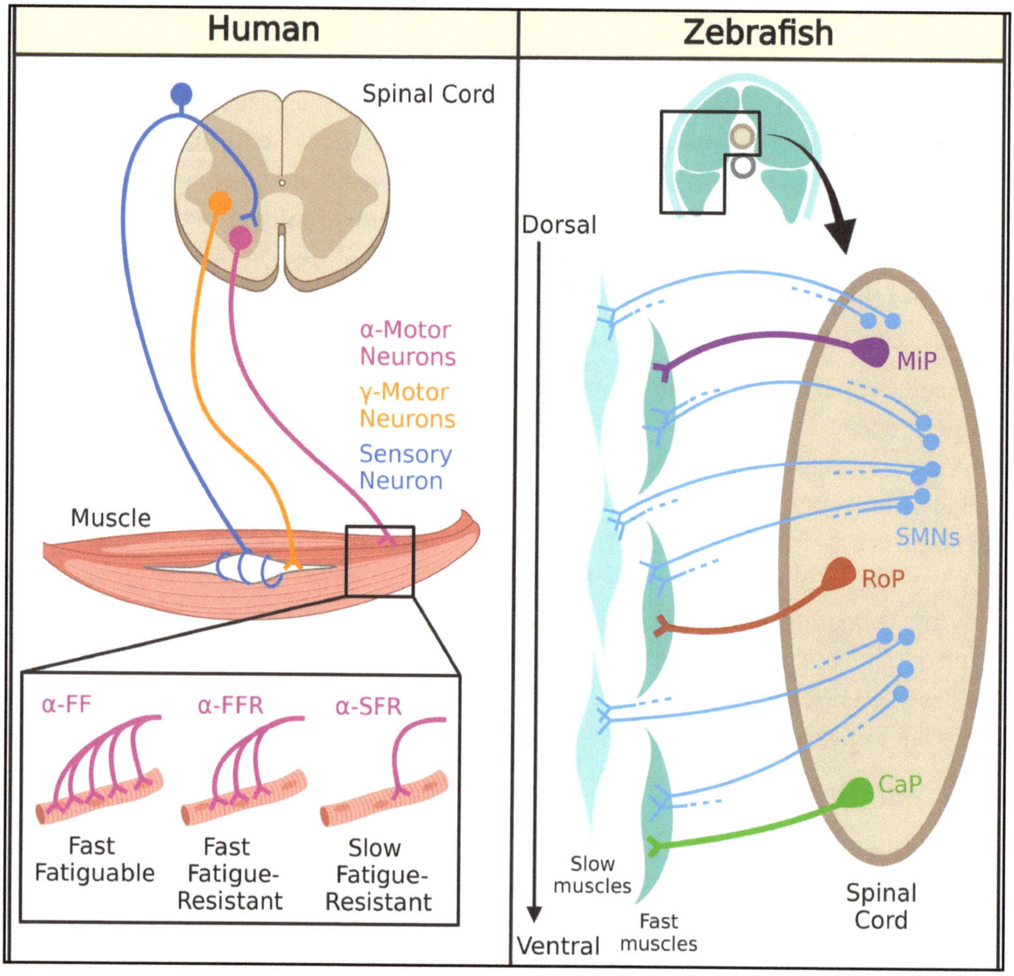

Neurological and Muscular System Applications

The application of CRISPR to neuromuscular and neurological diseases presents both tremendous opportunities and significant challenges due to the complexity of the nervous system and difficulties in delivery to affected tissues. Duchenne muscular dystrophy, caused by mutations in the dystrophin gene, has shown promising results in preclinical studies where CRISPR is used to delete mutated exons and restore the reading frame, producing a shortened but functional dystrophin protein. Clinical trials are underway testing systemic delivery of CRISPR components using adeno-associated virus vectors to reach muscle tissue throughout the body. Neurological conditions such as Huntington's disease and amyotrophic lateral sclerosis are being targeted with CRISPR strategies aimed at reducing the expression of toxic proteins or correcting disease-causing mutations. The blood-brain barrier presents a formidable obstacle for delivering CRISPR components to the central nervous system, spurring development of novel delivery methods including engineered viral vectors with enhanced brain penetration and direct injection into cerebrospinal fluid. The post-mitotic nature of neurons requires strategies that don't rely on cellular division for gene correction, leading to innovations in homology-directed repair enhancement and base editing approaches.

8.3.5 Cancer Immunotherapy and Oncology

CRISPR technology is revolutionizing cancer treatment through the engineering of immune cells to better recognize and eliminate tumor cells, as well as direct targeting of cancer-causing mutations. CAR-T cell therapy, already a breakthrough treatment for certain blood cancers, is being enhanced through CRISPR editing to create more effective and safer therapeutic cells. Multiple edits can be made simultaneously to T cells, including insertion of chimeric antigen receptors, deletion of checkpoint inhibitors like PD-1, and removal of endogenous T cell receptors to prevent graft-versus-host disease.

Clinical trials have shown that CRISPR-edited CAR-T cells demonstrate improved persistence and efficacy compared to conventional CAR-T therapies. Beyond blood cancers, CRISPR is being used to develop therapies for solid tumors by engineering T cells to overcome the immunosuppressive tumor microenvironment. Direct in vivo editing of cancer cells to correct driver mutations or induce synthetic lethality represents another therapeutic avenue under investigation. The heterogeneity of tumors and their ability to evolve resistance mechanisms necessitate combination approaches that target multiple pathways simultaneously.

8.3.6 Delivery Methods and Technical Innovations

The success of CRISPR therapies critically depends on efficient and safe delivery of gene editing components to target cells, driving continuous innovation in delivery technologies. Adeno-associated virus vectors remain the primary choice for in vivo delivery due to their safety profile and ability to target specific tissues, though their limited packaging capacity restricts the size of CRISPR components that can be delivered. Lipid nanoparticle formulations, similar to those used in COVID-19 mRNA vaccines, show promise for delivering CRISPR components to the liver and other organs, with advantages including reduced immunogenicity and repeat dosing capability. Ex vivo editing approaches using electroporation or nucleofection achieve high editing efficiency in extracted cells but are limited to accessible cell types such as blood cells and stem cells. Novel delivery methods under development include engineered extracellular vesicles, cell-penetrating peptides, and targeted antibody conjugates that could enable precise delivery to specific cell types. The development of smaller CRISPR systems, such as those based on Cas12 and Cas13 enzymes, expands delivery options by fitting within the constraints of various delivery vehicles.

8.3.7 Clinical Trial Landscape and Regulatory Progress

The clinical translation of CRISPR therapies has accelerated dramatically, with dozens of trials underway worldwide targeting various genetic diseases and cancers. Early-phase trials have generally demonstrated good safety profiles, with adverse events primarily related to the conditioning regimens required for cell therapies rather than the gene editing itself.

Regulatory agencies have developed frameworks for evaluating CRISPR therapies, considering unique aspects such as off-target effects, genomic integrity, and long-term consequences of genetic modifications. The approval of the first CRISPR therapy for sickle cell disease has established regulatory precedents that will facilitate future approvals while maintaining rigorous safety standards. International harmonization of regulatory requirements remains a work in progress, with different countries taking varied approaches to oversight of gene editing therapies. Patient advocacy groups play increasingly important roles in clinical trial design and regulatory discussions, ensuring that patient perspectives are considered in risk-benefit assessments. The high cost of current CRISPR therapies raises important questions about accessibility and healthcare equity that must be addressed as these treatments become more widely available.

8.4 AGRICULTURAL BIOTECHNOLOGY: CREATING DROUGHT-RESISTANT CROPS

Agricultural biotechnology offers powerful tools for addressing the global challenge of water scarcity through the creation of drought-resistant crops. By leveraging genetic engineering, molecular biology, and advanced genomics, scientists have developed crop varieties capable of maintaining productivity under water-limited conditions. These innovations improve food security, conserve water resources, enhance environmental sustainability, and stabilize farmer incomes in drought-prone regions.

The integration of physiological, agronomic, and molecular strategies ensures that drought-resistant crops are robust and adaptable across diverse environments. However, successful deployment requires careful attention to regulatory compliance, ethical considerations, and equitable access to technology. As climate change continues to exacerbate drought conditions worldwide, the continued advancement of agricultural biotechnology, combined with collaborative research and policy support, will be critical to sustaining global food systems and supporting resilient, sustainable agriculture for future generations.

8.4.1 Importance of Drought-Resistant Crops

Drought-resistant crops are vital for ensuring food security in regions prone to water scarcity. These crops can maintain yields under limited water availability, reducing the vulnerability of farmers and communities to climate-related agricultural shocks. Improved drought tolerance also conserves water resources, supports sustainable farming, and reduces the reliance on irrigation, which is often costly and environmentally unsustainable. Furthermore, drought-resistant crops can mitigate the socio-economic impacts of drought, including food price volatility, loss of livelihoods, and increased rural poverty. As climate models predict more frequent and severe droughts, biotechnological approaches to enhancing crop resilience have become increasingly important for global agricultural sustainability.

8.4.2 Molecular Basis of Drought Tolerance

The creation of drought-resistant crops relies on understanding the molecular mechanisms that enable plants to survive water stress. Key processes include osmotic adjustment, regulation of stomatal closure, accumulation of compatible solutes, activation of stress-responsive genes, and production of protective proteins such as dehydrins and heat-shock proteins. Biotechnology allows the identification and manipulation of specific genes responsible for these adaptive traits. For example, transcription factors like DREB (Dehydration-Responsive Element Binding) and NAC regulate the expression of multiple drought-responsive genes. By enhancing or introducing these genes, scientists can create crops with improved water retention, root architecture, and photosynthetic efficiency under drought conditions.

8.4.3 Genetic Engineering Approaches

Genetic engineering is a cornerstone of creating drought-tolerant crops. Through techniques such as transgenic technology, gene editing, and cisgenesis, desired traits can be introduced, modified, or regulated within plant genomes. Transgenic approaches involve inserting foreign genes that confer drought resistance, such as those encoding osmoprotectants, antioxidants, or stress-responsive transcription factors. Gene-editing tools like CRISPR-Cas9 allow precise modifications of endogenous genes to enhance drought tolerance without introducing foreign DNA. Cisgenic methods utilize genes from the same species or closely related species, reducing regulatory and public concerns associated with transgenic crops. These approaches enable targeted improvements in drought resilience while maintaining crop quality and yield.

8.4.4 Marker-Assisted Selection and Genomics

In addition to direct genetic engineering, marker-assisted selection (MAS) and genomics play essential roles in developing drought-resistant crops. MAS uses molecular markers linked to drought-tolerance traits to accelerate breeding programs, enabling the identification of elite lines without extensive phenotypic evaluation. Genomic selection leverages high-throughput sequencing and bioinformatics to predict plant performance under drought stress, facilitating the selection of superior genotypes. By integrating genomics with traditional breeding, researchers can combine multiple drought-tolerance traits into a single variety, improving robustness and adaptability. These approaches reduce the time and cost of developing drought-resistant crops while ensuring precision and efficiency.

8.4.5 Physiological and Agronomic Enhancements

Beyond genetic improvements, drought-resistant crops benefit from physiological and agronomic enhancements. Modifications that improve root depth, water uptake efficiency, and leaf morphology help plants maintain hydration under limited water conditions. Biotechnological interventions can also enhance photosynthetic efficiency and reduce transpiration rates, allowing crops to sustain growth during drought. Agronomic practices, such as optimized planting density, mulching, and soil moisture management, complement genetic improvements, creating integrated strategies that maximize drought resilience. The combination of molecular, physiological, and agronomic innovations ensures that drought-resistant crops perform effectively in diverse environmental conditions.

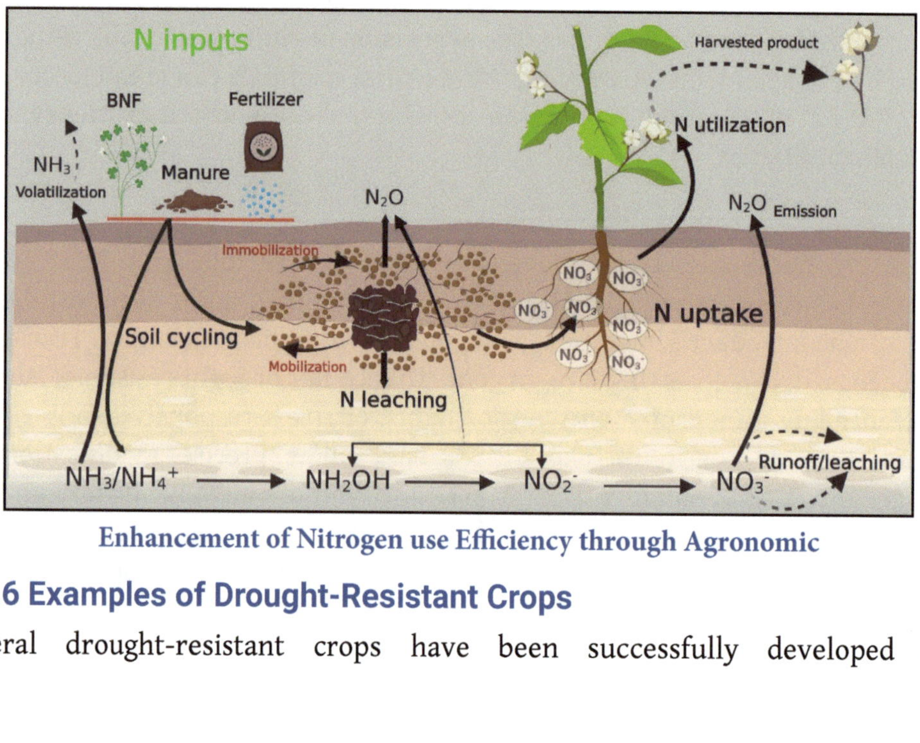

Enhancement of Nitrogen use Efficiency through Agronomic

8.4.6 Examples of Drought-Resistant Crops

Several drought-resistant crops have been successfully developed and commercialized using biotechnology. For instance, drought-tolerant maize varieties have been engineered to maintain yields under water-limited conditions in sub-Saharan Africa. Rice varieties expressing stress-responsive genes such as DREB1A or transcription factors regulating osmoprotectants exhibit improved survival during drought. In wheat, CRISPR-mediated modifications of genes controlling root architecture and stomatal regulation have enhanced drought tolerance. Legumes, including chickpea and soybean, have also been engineered to improve water-use efficiency and stress resilience. These examples demonstrate the potential of biotechnology to create crop varieties that address regional water scarcity and climate challenges.

8.4.7 Environmental and Economic Benefits

The adoption of drought-resistant crops provides both environmental and economic benefits. Reduced reliance on irrigation conserves water resources and lowers energy consumption associated with pumping and irrigation infrastructure. Drought-tolerant crops minimize soil degradation by maintaining root structure and vegetative cover, promoting soil health. Economically, these crops stabilize yields, protect farmer incomes, and contribute to food security in drought-prone regions. Additionally, biotechnology reduces the need for repeated crop cycles due to drought failure, lowering production costs and reducing greenhouse gas emissions associated with agriculture. By combining environmental stewardship with economic resilience, drought-resistant crops offer a sustainable approach to modern agriculture.

8.4.8 Regulatory and Ethical Considerations

The deployment of genetically engineered drought-resistant crops requires careful regulatory and ethical consideration. Biosafety regulations ensure that genetically modified crops do not pose risks to human health, non-target organisms, or ecosystems. Regulatory agencies evaluate the environmental impact, potential gene flow, and food safety of transgenic or gene-edited crops. Ethical considerations include equitable access to technology, farmer rights, and public acceptance. Engagement with stakeholders, transparent communication, and adherence to international standards are essential for responsible development and adoption. Regulatory frameworks balance innovation with safety and societal concerns, ensuring that drought-resistant crops contribute positively to global food systems.

- Drought-resistant crops are essential for food security, water conservation, and climate adaptation.
- Molecular mechanisms such as osmotic adjustment, stress-responsive genes, and protective proteins underpin drought tolerance.
- Genetic engineering introduces or modifies genes to enhance water-use efficiency and stress resilience.

- CRISPR and cisgenic approaches enable precise, targeted modifications in plant genomes.
- Marker-assisted selection and genomic selection accelerate breeding of drought-tolerant varieties.
- Physiological enhancements, including improved root systems and leaf morphology, complement genetic improvements.
- Biotechnology has produced drought-resistant maize, rice, wheat, and legumes with improved survival under water stress.
- Environmental benefits include reduced irrigation, soil preservation, and lower energy consumption.
- Regulatory and ethical frameworks ensure biosafety, equitable access, and public trust.
- Future directions include multi-gene engineering, AI integration, precision agriculture, and global collaborative research.

8.5 THE ROLE OF SYNTHETIC BIOLOGY IN THE CIRCULAR ECONOMY

Synthetic biology offers transformative potential for advancing the circular economy by enabling the design of biological systems capable of producing renewable materials, converting waste into valuable products, generating bio-based energy, and remediating environmental contaminants. By integrating metabolic engineering, genetic design, and computational modeling, researchers can create microorganisms and enzymes tailored to sustainable industrial processes. These innovations support closed-loop production systems, reduce reliance on finite resources, and enhance environmental regeneration. While technical, regulatory, and economic challenges remain, the synergy between synthetic biology and circular economy principles provides a pathway toward sustainable, efficient, and resilient industrial and societal systems. By harnessing the capabilities of engineered biological systems, the circular economy can move beyond conceptual frameworks to practical implementation, creating a future where economic growth and environmental stewardship coexist in a regenerative and sustainable manner.

8.5.1 Foundations of Synthetic Biology

Synthetic biology builds upon molecular biology, genetic engineering, and systems biology to create programmable biological systems. Key components include synthetic DNA, genetic circuits, and standardized biological parts that can be combined to perform novel functions. This field enables the engineering of microorganisms to produce high-value compounds, degrade pollutants, and sense environmental conditions. Techniques such as CRISPR-Cas genome editing, metabolic pathway optimization, and computational modeling allow precise control over biological behavior. By harnessing these tools, synthetic biology provides a platform for creating organisms and enzymes capable of converting waste streams into usable products, producing biofuels, and manufacturing biodegradable materials, all of which are essential components of a circular economy.

8.5.2 Concepts of the Circular Economy

The circular economy emphasizes resource efficiency, waste reduction, and the maintenance of material value within closed-loop systems. Unlike traditional industrial processes that rely heavily on finite resources, circular strategies focus on reusing, recycling, refurbishing, and regenerating products and materials. Biological processes are particularly relevant to the circular economy because they can transform organic waste into valuable compounds, replace petrochemical-derived materials with bio-based alternatives, and create renewable energy sources. Circular economy principles extend beyond material flows to encompass energy efficiency, sustainable consumption, and system-wide optimization, all of which can be enhanced through the integration of synthetic biology.

8.5.3 Synthetic Biology for Waste Valorization

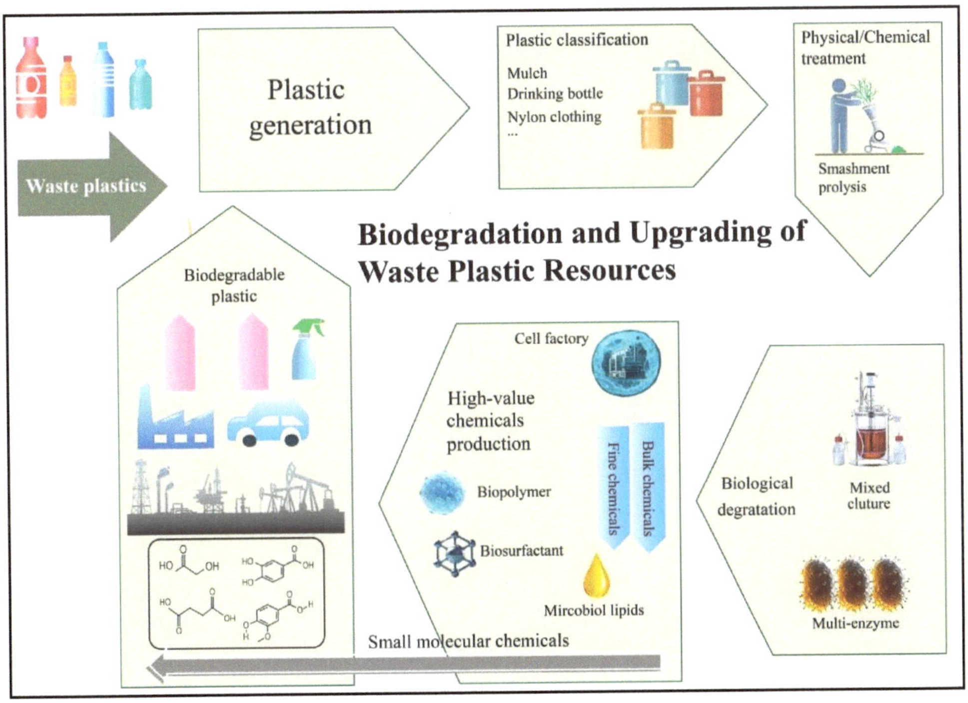

Harnessing Synthetic Biology for Waste Management

One of the most direct applications of synthetic biology in the circular economy is waste valorization—the conversion of waste materials into valuable products.

Engineered microorganisms can metabolize agricultural residues, industrial byproducts, and food waste to produce biofuels, bioplastics, and chemicals. For example, certain genetically modified bacteria and yeast strains have been optimized to convert lignocellulosic biomass into ethanol, organic acids, or lipid-based biofuels.

These processes not only reduce waste accumulation but also provide renewable sources of energy and materials, closing loops that would otherwise be lost in traditional disposal systems. Additionally, synthetic biology enables the creation of enzyme systems that degrade complex polymers, such as plastics or synthetic textiles, into reusable monomers, further supporting the recycling and regeneration of materials within the circular economy.

8.5.4 Bio-Based Material Production

Synthetic biology enables the production of bio-based materials that serve as sustainable alternatives to conventional, non-renewable materials. For instance, microbial fermentation systems can produce biodegradable plastics, such as polyhydroxyalkanoates (PHAs), which can replace petroleum-derived plastics in packaging, textiles, and consumer goods. Engineered microbes can also synthesize fibers, adhesives, and coatings from renewable feedstocks, reducing reliance on fossil fuels and minimizing environmental impact.

The precision offered by synthetic biology allows for tailoring the chemical and mechanical properties of these materials, ensuring that they meet performance requirements while remaining biodegradable or recyclable. By replacing conventional materials with bio-based alternatives, synthetic biology supports the design of products that are compatible with circular economy principles, facilitating the recovery and reuse of resources.

8.5.5 Energy and Biofuel Production

Energy production is another critical area where synthetic biology contributes to circular economy goals. Engineered microorganisms can produce biofuels from organic waste, algae, or carbon dioxide, providing renewable energy sources while simultaneously reducing greenhouse gas emissions. Metabolic engineering techniques optimize the production of ethanol, butanol, biodiesel, and hydrogen, enhancing yield and process efficiency. In addition to liquid fuels, synthetic biology enables the design of microbial systems capable of generating electricity in microbial fuel cells by exploiting the metabolic activity of bacteria to convert chemical energy into electrical energy. Integrating these bio-based energy solutions into industrial and urban systems can reduce dependence on fossil fuels, close energy loops, and contribute to sustainable and self-sufficient circular infrastructures.

8.5.6 Bioremediation and Environmental Restoration

Synthetic biology offers powerful tools for bioremediation, which is essential for maintaining environmental integrity within a circular economy. Engineered microorganisms can degrade pollutants, detoxify heavy metals, and metabolize hazardous chemicals present in soil, water, and industrial effluents. Genetic circuits and biosensors can be designed to detect specific contaminants and activate metabolic pathways that neutralize them. This approach not only mitigates environmental damage but also allows for the recovery of valuable compounds from waste streams, converting pollutants into resources that can re-enter the production cycle. Bioremediation using synthetic biology thus transforms environmental liabilities into economic opportunities, reinforcing the regenerative principles of the circular economy.

8.5.7 Integration with Industrial Processes

Incorporating synthetic biology into industrial processes enables the creation of biorefineries, where multiple waste streams are converted into a variety of products, including biofuels, chemicals, and materials. By designing modular microbial systems, industries can tailor production pathways to specific feedstocks and desired outputs. Closed-loop fermentation systems can recycle nutrients, water, and energy within the production process, minimizing waste and maximizing efficiency. The integration of synthetic biology into industrial workflows supports sustainable production at scale, allowing industries to adopt circular economy practices without compromising productivity or profitability. Furthermore, process optimization through computational modeling and synthetic biology design accelerates innovation, ensuring that industrial systems remain adaptable and resilient.

8.5.8 Economic and Social Implications

The application of synthetic biology to the circular economy has significant economic and social implications. Economically, it enables the valorization of waste streams, reduces dependence on non-renewable resources, and creates new markets for bio-based products. By improving resource efficiency, synthetic biology reduces production costs and environmental liabilities, contributing to sustainable business models. Socially, it promotes environmental stewardship, public health, and sustainable employment in biotechnology sectors. Education and workforce development are critical to ensure that communities can participate in and benefit from these emerging technologies. Furthermore, equitable access to bio-based innovations is essential to avoid exacerbating existing social and economic disparities. By integrating ethical and inclusive considerations, synthetic biology can contribute to a circular economy that is both environmentally and socially sustainable.

8.5.9 Challenges and Limitations

Despite its promise, the application of synthetic biology in the circular economy faces several challenges. Technical limitations include the need for robust and efficient microbial strains, optimization of metabolic pathways, and scalability of bioprocesses. Unintended ecological consequences, such as the accidental release of genetically modified organisms, require careful containment and monitoring strategies. Regulatory frameworks for synthetic biology and bio-based products are still evolving, creating uncertainties for industrial adoption. Economic barriers, including high initial investment costs and market competition with established petroleum-based products, may hinder widespread implementation. Addressing these challenges necessitates interdisciplinary research, policy development, public engagement, and collaboration between industry, academia, and regulatory agencies.

Table: Applications of Synthetic Biology in the Circular Economy

Application	Impact
Waste Valorization	Converts organic and industrial waste into biofuels and chemicals
Bio-Based Materials	Produces biodegradable plastics, fibers, and coatings
Renewable Energy	Generates biofuels and microbial electricity from organic sources
Bioremediation	Detoxifies pollutants and recovers valuable compounds
Industrial Integration	Enables closed-loop bioprocessing and modular production systems

CHAPTER 9

Synthetic Biology in Industry and Commercialization

9.1 COMMERCIALIZATION OF SYNTHETIC BIOLOGY PRODUCTS

9.1.1 Startups and Major Industry Players in Synthetic Biology

Startups and major industry players constitute the dynamic ecosystem propelling synthetic biology from scientific exploration to industrial reality. Early innovators such as Twist Bioscience and Ginkgo Bioworks demonstrate how advanced DNA synthesis and organism engineering platforms underpin wide-ranging applications including healthcare, agriculture, and sustainable materials. Emerging companies focusing on enzyme design, carbon-efficient biomanufacturing, and gene editing amplify this innovation landscape.

Strategic partnerships foster commercialization and regulatory navigation, supported by robust investment fueling sector growth. Navigating complex regulatory, ethical, and social dimensions will be critical for sustained impact. Together, these companies shape the synthetic biology industry's future, driving a new bio-based economy with profound implications for sustainability, health, and innovation.

9.1.2 Investment and Economic Impact of Synthetic Biology

The investment and economic impact of synthetic biology represents one of the most significant economic transformations of the early twenty-first century, comparable to previous technological revolutions but with potentially broader implications given biology's fundamental role in human life and economic activity. The massive capital flows into the sector reflect recognition that synthetic biology offers solutions to critical global challenges while generating substantial economic returns for investors, companies, and regions that successfully develop and deploy these technologies.

The economic impacts extend far beyond direct financial returns to encompass job creation, industrial transformation, environmental benefits, and the development of entirely new economic sectors that didn't exist a decade ago. While risks and challenges remain significant, the trajectory of investment and economic value creation suggests that synthetic biology will become increasingly central to global economic competitiveness and sustainable development. The distribution of economic benefits from synthetic biology will depend on policy choices, investment strategies, and societal decisions about how to develop and deploy these powerful technologies.

As the field matures from promising technology to economic reality, the companies, regions, and nations that successfully navigate the opportunities and challenges of synthetic biology will likely emerge as economic leaders of the bio-based economy. The continued evolution of synthetic biology from scientific curiosity to economic engine demonstrates the power of convergent technologies to create value while addressing humanity's most pressing challenges, suggesting that current economic impacts may be just the beginning of a much larger transformation.

9.2 REGULATORY AND MARKET CHALLENGES FOR SYNTHETIC BIOLOGY COMPANIES

Synthetic biology companies operate in a complex and rapidly evolving environment characterized by both tremendous opportunities and significant challenges. Regulatory hurdles, including fragmented frameworks, biosafety requirements, and international variability, demand careful planning and compliance. Market challenges, including public perception, ethical considerations, high development costs, and competitive dynamics, further complicate commercialization. Companies must strategically navigate intellectual property landscapes, scale production efficiently, and engage transparently with stakeholders to build trust and ensure societal acceptance.

Collaborative partnerships, harmonization of standards, and proactive engagement with regulators and the public are essential for mitigating risks and facilitating sustainable growth. Despite these challenges, synthetic biology holds transformative potential across healthcare, agriculture, industrial manufacturing, and environmental applications. Companies that effectively address regulatory and market barriers while demonstrating social responsibility and technological innovation will be well-positioned to lead in this rapidly advancing field, delivering solutions that benefit both society and the global economy.

9.2.1 Regulatory Complexity in Synthetic Biology

Regulatory oversight is critical for ensuring the safety, ethical deployment, and environmental compatibility of synthetic biology products. However, existing regulations are often fragmented, inconsistent, or insufficiently adapted to novel synthetic constructs. In many countries, regulatory frameworks were initially designed for conventional genetically modified organisms (GMOs) or pharmaceutical products, which may not fully encompass engineered organisms or gene-editing technologies. Companies must navigate multiple agencies with overlapping responsibilities, such as environmental protection, food safety, public health, and industrial production. Compliance requires extensive documentation, risk assessments, and adherence to biosafety and biosecurity guidelines, which can significantly increase development timelines and costs.

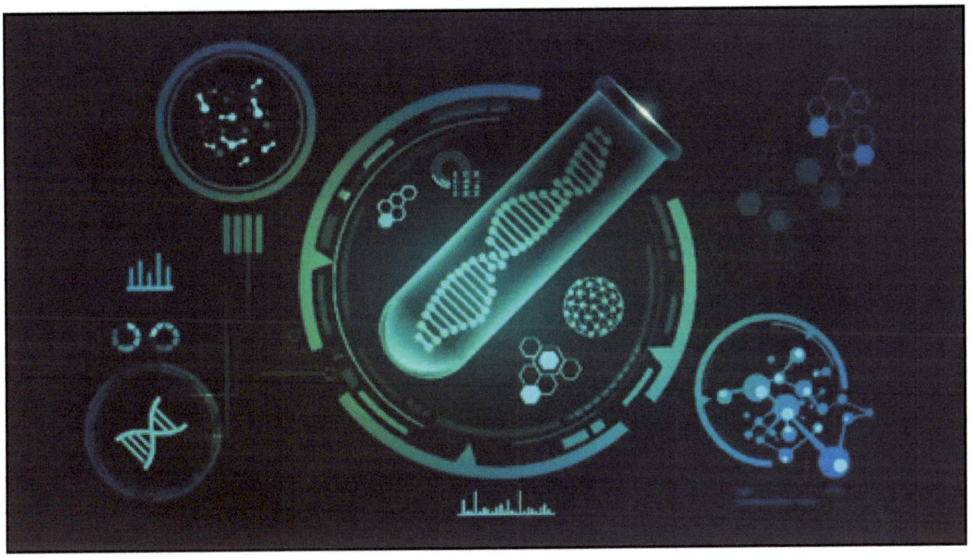

Regulatory Frameworks and Challenges in Synthetic Biology

9.2.2 Intellectual Property and Innovation Constraints

Intellectual property (IP) protection is vital for synthetic biology companies to secure investments, recoup research and development costs, and maintain a competitive advantage. However, IP in synthetic biology is often challenging to define due to the modular, combinatorial nature of biological parts. Patents may cover genes, pathways, synthetic constructs, or computational designs, leading to complex licensing arrangements and potential infringement disputes. Additionally, broad or overlapping patents can stifle innovation by restricting access to foundational tools and components needed for research and product development. Companies must carefully navigate IP landscapes, balancing proprietary protection with collaboration and open-source standards to foster innovation.

9.2.3 Biosafety and Biosecurity Considerations

Synthetic biology involves engineering organisms that may pose potential risks to human health, ecosystems, or biosecurity. Regulatory agencies require companies to conduct rigorous biosafety assessments, implement containment protocols, and adhere to national and international guidelines. Ensuring biosecurity involves preventing misuse of synthetic organisms, such as the creation of harmful pathogens, which raises concerns about dual-use research. Compliance with biosafety and biosecurity regulations can be costly and time-consuming, and failures or accidents can lead to reputational damage, legal liabilities, and loss of public trust. Companies must invest in robust risk management, employee training, and monitoring systems to address these challenges effectively.

9.2.4 Market Adoption and Public Perception

Market challenges for synthetic biology companies extend beyond regulatory compliance. Public perception and societal acceptance play a crucial role in determining market adoption. Synthetic biology products, especially in food, healthcare, or environmental applications, may face skepticism or resistance due to ethical concerns, fear of genetic modification, or lack of understanding. Public engagement, transparent communication, and education are essential to build trust and acceptance. Companies must demonstrate the safety, efficacy, and societal value of their products to gain consumer confidence and achieve market penetration. Failure to address public concerns can lead to regulatory delays, market resistance, or reputational risks.

9.2.5 High Development Costs and Funding Constraints

Developing synthetic biology products requires significant investment in research, laboratory infrastructure, specialized equipment, and skilled personnel. Regulatory compliance, biosafety measures, and long development timelines further increase costs. While venture capital and government grants provide funding, competition for resources is intense, and investors may be cautious due to uncertainties in regulatory approval and market adoption.

Small and medium-sized companies may struggle to sustain operations during prolonged development cycles. Strategic partnerships, public-private collaborations, and phased commercialization approaches can help mitigate financial risks, but high costs remain a major barrier to scaling synthetic biology innovations.

9.2.6 International Regulatory Variability

Synthetic biology companies often operate in a global context, seeking international markets for their products. Regulatory frameworks, however, vary widely across countries, creating challenges for harmonization, product approval, and cross-border trade. Some countries adopt a precautionary approach with stringent biosafety regulations, while others may have less-developed oversight structures.

Navigating these differences requires extensive knowledge of local laws, risk assessments, and compliance procedures. International harmonization efforts, such as ISO standards and guidance from organizations like the WHO or OECD, provide frameworks for consistency, but full alignment remains a challenge. Companies must manage regulatory risks while pursuing global market opportunities.

9.2.7 Supply Chain and Production Challenges

Scaling synthetic biology products from the laboratory to commercial production involves complex supply chain and manufacturing considerations. Engineered organisms, cell-free systems, or bioreactors may require specialized raw materials, controlled environments, and quality management systems. Regulatory oversight often extends to production facilities, requiring certifications and inspections. Any disruption in the supply chain or failure to meet regulatory standards can delay product launch, increase costs, or compromise product quality. Companies must develop robust, flexible, and compliant manufacturing strategies to ensure consistent supply and scalability while adhering to regulatory expectations.

9.2.8 Ethical and Social Responsibility

Ethical considerations are integral to the regulatory and market challenges of synthetic biology. Products must respect societal norms, human rights, environmental stewardship, and equitable access. Companies face scrutiny over dual-use research, potential ecological impacts, and ethical implications of human or animal modifications. Demonstrating social responsibility through transparency, stakeholder engagement, and adherence to ethical guidelines is crucial for maintaining public trust and regulatory support. Companies that fail to address these concerns may face backlash, regulatory hurdles, or loss of market credibility.

9.2.9 Competitive and Market Dynamics

The synthetic biology market is highly competitive, with numerous startups, established biotech firms, and academic spin-offs vying for technological leadership. Rapid innovation cycles, IP disputes, and market uncertainty create challenges for new entrants. Companies must differentiate their products, protect their intellectual property, and navigate regulatory approvals to remain competitive. Strategic alliances, partnerships, and collaborations can provide access to complementary technologies, distribution networks, and regulatory expertise. Understanding market dynamics, regulatory landscapes, and consumer expectations is essential for sustainable growth and commercialization.

- Regulatory frameworks for synthetic biology are complex, fragmented, and evolving, creating compliance challenges.
- Intellectual property protection is difficult due to the modular and combinatorial nature of synthetic biology components.
- Biosafety and biosecurity regulations require rigorous risk assessments, containment measures, and monitoring systems.
- Public perception and ethical concerns influence market adoption and societal acceptance of synthetic biology products.

- High research and development costs, along with regulatory compliance expenses, create financial barriers for companies.
- International regulatory variability complicates cross-border commercialization and requires adaptation to local standards.
- Scaling production involves managing specialized supply chains, quality control, and facility certifications.
- Ethical and social responsibility considerations are critical for maintaining trust, legitimacy, and regulatory support.
- Competitive market dynamics, including innovation cycles and IP disputes, challenge new and existing companies.
- Strategic partnerships, collaborations, and harmonized standards are essential to overcome regulatory and market obstacles.

9.3 CASE STUDIES OF SUCCESSFUL COMMERCIAL PRODUCTS IN SYNTHETIC BIOLOGY

Case studies of successful commercial products in synthetic biology illustrate the transformative potential of this field across healthcare, industrial, environmental, and consumer sectors. Recombinant pharmaceuticals, bio-based materials, biofuels, therapeutic enzymes, and vaccines demonstrate how genetic and metabolic engineering, bioprocess optimization, and computational design converge to create market-ready products. These successes highlight the importance of regulatory compliance, scalability, public acceptance, and ethical considerations in commercialization. While challenges remain, including technical, economic, and societal barriers, the continued development of synthetic biology promises to expand its commercial footprint, driving innovation, sustainability, and societal benefit. These case studies, it becomes clear that synthetic biology is not merely a scientific endeavor but a practical driver of global innovation and a catalyst for sustainable industrial and healthcare solutions.

9.3.1 Recombinant Insulin Production

Recombinant human insulin is among the earliest and most transformative commercial products derived from synthetic biology. Prior to the 1980s, insulin for diabetic patients was extracted from animal pancreases, a process limited by supply, cost, and immunogenicity. The advent of recombinant DNA technology allowed the insertion of human insulin genes into Escherichia coli and later yeast systems. Engineered microorganisms were optimized to produce large quantities of functional insulin, followed by purification processes to ensure safety and efficacy. Recombinant insulin not only provided a reliable, scalable source of this critical hormone but also reduced allergic reactions associated with animal-derived insulin. The commercialization of insulin marked a turning point, demonstrating that synthetic biology could deliver life-saving pharmaceuticals at industrial scale, combining genetic engineering, bioprocess optimization, and regulatory compliance to create a market-ready product.

9.3.2 Artemisinin Production Using Engineered Yeast

The production of the antimalarial compound artemisinin represents a notable example of synthetic biology addressing global health challenges. Traditionally, artemisinin was extracted from the sweet wormwood plant, a method constrained by plant growth cycles, variable yields, and high costs. Researchers employed metabolic engineering to optimize the biosynthetic pathway of artemisinin in Saccharomyces cerevisiae, creating yeast strains capable of producing artemisinic acid, a precursor that could be chemically converted into artemisinin. Commercialization involved scale-up in bioreactors, optimization of fermentation conditions, and integration with chemical synthesis processes. The semi-synthetic production of artemisinin significantly reduced production costs, stabilized supply chains, and enabled broader access to this essential antimalarial drug, illustrating the potential of synthetic biology to provide sustainable solutions for global healthcare.

9.3.3 Synthetic Spider Silk

Spider silk, renowned for its exceptional strength, elasticity, and lightweight properties, has long been sought after for industrial and biomedical applications. Natural spider silk production is limited due to territorial and cannibalistic behavior of spiders, making large-scale harvesting impractical. Synthetic biology enabled the production of recombinant spider silk proteins in microbial hosts such as Escherichia coli, yeast, and even goat milk. Genetic circuits encoding spider silk proteins were optimized for expression, solubility, and assembly into fibers. Commercial ventures have successfully scaled the production of synthetic spider silk for use in textiles, medical sutures, and high-performance materials. This case exemplifies the translation of a naturally occurring biomaterial into a commercially viable product through synthetic biology, highlighting the integration of genetic design, protein engineering, and process optimization.

9.3.4 Synthetic Vanilla and Flavor Compounds

The demand for natural flavor compounds such as vanillin has historically relied on plant extraction, which is limited by environmental factors, cost, and sustainability concerns. Synthetic biology offers an alternative by engineering microbial systems to produce these compounds from renewable feedstocks. For instance, Saccharomyces cerevisiae and Escherichia coli have been modified to express enzymatic pathways that convert glucose or other precursors into vanillin and related flavor compounds. Commercial production of synthetic vanillin provides a consistent, scalable, and environmentally friendly alternative to plant extraction. Similarly, synthetic biology has enabled the production of other high-value flavor and fragrance molecules, such as nootkatone, used in citrus flavoring and insect repellents. These products demonstrate how synthetic biology can deliver bio-based alternatives to traditional chemical or plant-derived compounds, aligning with both sustainability and commercial objectives.

9.3.5 Biofuels and Renewable Chemicals

Synthetic biology has made significant contributions to the production of biofuels and renewable chemicals. Engineered microorganisms, including bacteria, yeast, and algae, have been optimized to convert sugars, lignocellulosic biomass, and waste feedstocks into ethanol, butanol, biodiesel, and other renewable fuels. Companies have commercialized strains capable of high-yield production, enabling competitive alternatives to petroleum-derived fuels. Beyond energy, synthetic biology facilitates the production of renewable chemicals such as succinic acid, lactic acid, and 1,3-propanediol, which serve as building blocks for bioplastics, solvents, and industrial chemicals. These products highlight the potential of synthetic biology to transform traditional chemical manufacturing, reduce environmental impact, and support the development of a circular bioeconomy.

9.3.6 Vaccines and Immunotherapies

Synthetic biology has transformed vaccine development and immunotherapy by enabling the rapid design and production of recombinant antigens, viral vectors, and mRNA constructs. For example, the development of mRNA vaccines for COVID-19 relied on synthetic biology techniques to design stabilized spike protein sequences, optimize codon usage, and engineer lipid nanoparticle delivery systems. Recombinant protein vaccines, produced in engineered yeast or insect cells, have also benefited from synthetic biology approaches to enhance yield, immunogenicity, and scalability. Beyond vaccines, synthetic biology enables the design of engineered T-cells, such as chimeric antigen receptor (CAR) T-cells, for cancer immunotherapy. These applications highlight the capacity of synthetic biology to accelerate therapeutic development, provide targeted treatment options, and respond to emerging public health challenges.

9.3.7 Therapeutic Enzymes and Industrial Proteins

Synthetic biology has enabled the production of enzymes and proteins for both therapeutic and industrial applications. Therapeutic enzymes, such as tissue plasminogen activator for clot dissolution or asparaginase for leukemia treatment, are now produced using engineered microbial systems to ensure consistency, purity, and activity. Industrial enzymes, including cellulases, amylases, and proteases, have been optimized for performance under extreme conditions, such as high temperature or pH, and are widely used in laundry detergents, food processing, and paper production. Genetic modifications improve enzyme stability, catalytic efficiency, and substrate specificity, enabling large-scale production with reduced environmental footprint. The commercial success of these enzymes demonstrates how synthetic biology can tailor biological catalysts to meet specific industrial needs, enhancing process efficiency and sustainability.

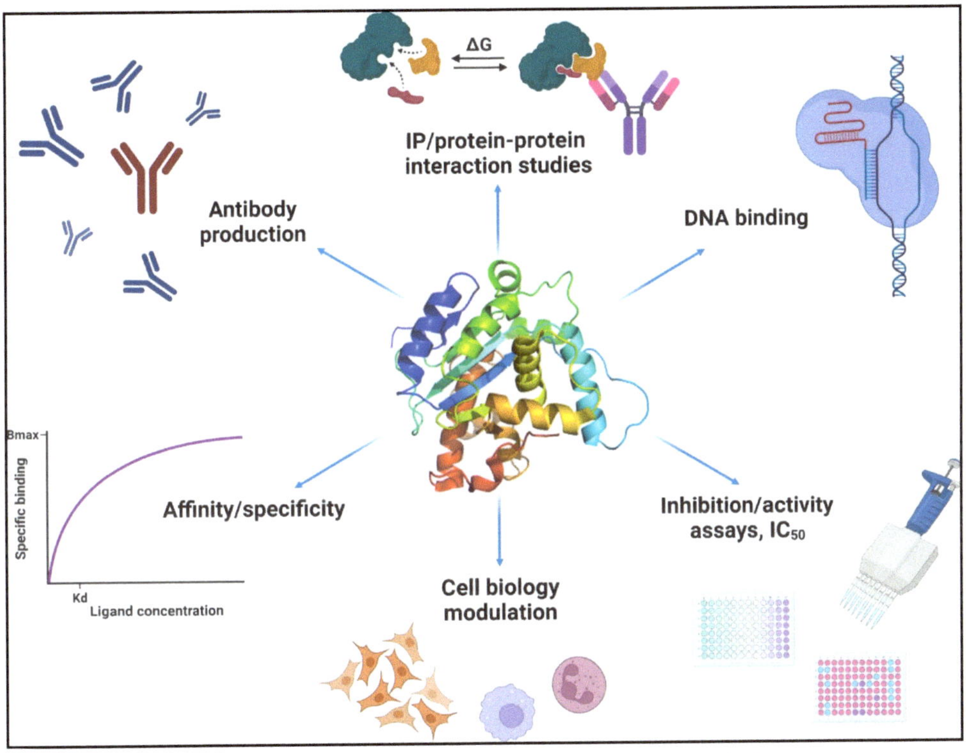

Natural Proteins are Engineered for Research and Therapy

9.3.8 Regulatory and Market Considerations

Commercialization of synthetic biology products requires compliance with regulatory frameworks, quality assurance standards, and safety protocols. Regulatory agencies, including the FDA, EMA, and WHO, evaluate genetically engineered strains, production processes, and final products for safety, efficacy, and environmental impact. Public acceptance, ethical considerations, and market demand also influence the commercial success of synthetic biology-derived products. Companies must navigate intellectual property rights, supply chain logistics, and production scalability to achieve market adoption. Case studies indicate that successful commercialization depends not only on technological innovation but also on rigorous regulatory compliance, strategic partnerships, and the ability to demonstrate safety and consistent quality to consumers and stakeholders.

9.3.9 Economic and Societal Impact

The commercial products of synthetic biology have generated significant economic value by creating new industries, jobs, and markets. Recombinant pharmaceuticals, bio-based chemicals, and sustainable materials provide cost-effective alternatives to traditional production methods, reduce dependence on natural resources, and decrease environmental impact.

Societally, these products improve public health, enhance environmental sustainability, and contribute to global efforts to mitigate climate change. Access to affordable therapeutics, renewable fuels, and biodegradable materials demonstrates how synthetic biology can align commercial objectives with broader societal goals. By integrating economic, environmental, and social considerations, synthetic biology products exemplify the potential of science-driven innovation to address pressing global challenges.

9.3.10 Challenges and Limitations

Despite these successes, several challenges continue to shape the commercialization of synthetic biology products. Technical limitations include achieving high-yield production, stability of engineered strains, and scalability of bioprocesses. Regulatory complexity and public perception present additional hurdles, particularly in areas involving genetically modified organisms or novel therapeutics. Market competition from established chemical and pharmaceutical industries can affect adoption rates and investment. Intellectual property considerations, including patents and licensing agreements, influence the accessibility and development of new products. Addressing these challenges requires continuous technological innovation, strategic collaboration, transparent communication with the public, and adaptive regulatory frameworks that balance safety with innovation.

9.4 FUTURE BUSINESS MODELS AND OPPORTUNITIES IN SYNTHETIC BIOLOGY

Future business models and opportunities in synthetic biology reflect a rapidly shifting landscape fueled by technological innovation, sustainability imperatives, and expanding application domains. Platform-centric models, sustainability-oriented biomanufacturing, personalized medicine, AI integration, and decentralized production exemplify the direction of industry transformation. Collaborative ecosystems and robust investment frameworks accelerate growth, overcoming challenges related to regulation and complexity. Synthetic biology stands poised to underpin a new bioeconomy, delivering environmentally friendly materials, life-saving therapies, and sustainable chemicals that address global needs. As innovation scales and matures, the synthesis of biology, engineering, and data-driven methodologies will generate resilient, dynamic business ecosystems, propelling synthetic biology from groundbreaking science to foundational industry with profound societal impact.

9.4.1 Platform and Ecosystem Business Models

At the heart of emerging business paradigms lies the concept of synthetic biology as a platform technology. Companies develop modular, automated bioengineering platforms that enable diverse applications across multiple verticals, including therapeutics, industrial enzymes, biofuels, and novel materials. This "technology as a service" model abstracts complex biological engineering into standardized processes accessible to various end-users through cloud-based platforms and biofoundries. Customers can design custom organisms or molecules with accelerated turnaround times and reduced risk. Such platform businesses derive value by fostering broad ecosystems of partners—research labs, startups, manufacturers, and distributors—collaborating through shared data, standardized components, and interoperable tools. The scalability and extensibility of platforms underpin sustainable growth, transforming synthetic biology into an enabling infrastructure for innovation much like computing's cloud ecosystem.

9.4.2 Sustainability and Circular Bioeconomy Opportunities

Synthetic biology's capacity to engineer microbial factories and biosynthetic pathways supports emergent business models focused on sustainability and circular economy principles. By developing bio-based alternatives to petroleum-derived materials, chemicals, and fuels, companies reduce carbon footprints and dependence on finite resources. Biomanufacturing approaches integrated with waste valorization and carbon capture create closed-loop systems where biological production cycles align with environmental stewardship. Startups and corporations now collaborate to commercialize biodegradable plastics, renewable specialty chemicals, and sustainable food ingredients produced via synthetic biology. This creates opportunities in emerging markets driven by consumer demand for greener products, regulatory incentives for carbon reduction, and corporate commitments to environmental responsibility. Business models centered on localized, modular biomanufacturing enhance supply chain resilience while contributing to regional economic development and resource optimization.

9.4.3 Personalized Medicine and Therapeutics

The intersection of synthetic biology with healthcare fosters innovative business opportunities tailored to precision medicine and living therapeutics. Companies specializing in engineered cell therapies, synthetic gene circuits, and personalized vaccines employ patient-specific biological design to improve treatment efficacy and safety. Advances in single-cell analysis, CRISPR gene editing, and synthetic circuit design enable highly customized interventions addressing individual genetic and molecular profiles. Business models here emphasize platform-based modularity, enabling rapid adaptation of therapeutic constructs for diverse patient cohorts and indications. Strategic collaborations between biopharmaceutical firms and synthetic biology companies accelerate clinical translation while reducing development timelines and costs. Precision diagnostics coupled with responsive synthetic therapeutics represent integrative healthcare solutions unlocking new markets in chronic disease management, immuno-oncology, and rare genetic disorders.

Medical Practice and Therapeutic Interventions

9.4.4 Data-Driven and AI-Integrated Models

Synthetic biology's complexity and scalability challenges have precipitated business models that tightly integrate machine learning, big data analytics, and automation. AI-enabled design algorithms optimize genetic circuits, metabolic pathways, and genome editing strategies, substantially reducing experimental iterations and accelerating innovation cycles. Data-driven biofoundries employ robotics and lab automation for high-throughput construction, screening, and characterization of synthetic biological constructs. Companies monetize proprietary data, predictive models, and design software embedded within subscription or usage-based services. This shift towards computationally enhanced synthetic biology fosters agility, cost-efficiency, and precision, reshaping business strategies around continuous learning systems. As synthetic biology converges with digital health and bioinformatics, hybrid models blending biological and digital assets emerge, expanding opportunities in biosimulation, digital twin creation, and in silico trial platforms.

9.4.5 Decentralized and On-Demand Biomanufacturing

Recent advances promote decentralization and modularization of biomanufacturing, enabling on-demand, localized production of biomolecules, materials, and therapeutics. Small-scale, containerized biofactories programmed through synthetic biology and automated workflows support agile supply chains resilient to global disruptions.

This approach democratizes biological manufacturing, allowing industries ranging from pharmaceuticals to agriculture to source bio-products closer to end-users, reducing logistical complexity and carbon emissions. New business models emerge involving "manufacturing as a service," where users remotely program biofactories for specific product needs. This paradigm reduces capital investment and facilitates rapid response to dynamic market demands. Integration of real-time monitoring and control systems via IoT and blockchain enhances transparency and traceability, supporting regulatory compliance and consumer trust.

9.4.6 Collaborative and Open Innovation Ecosystems

Synthetic biology's interdisciplinary nature incentivizes collaborative business models that transcend traditional company boundaries. Startups, academic institutions, large corporations, and public-sector organizations increasingly participate in shared innovation ecosystems emphasizing openness, data sharing, and co-development.

Open-source platforms, shared biofoundries, and community tools accelerate technology diffusion and reduce entry barriers. Partnerships leverage complementary competencies—combining fundamental research, engineering expertise, and commercialization capabilities—to foster ecosystem robustness. Business models prioritize network effects, where expanding collaborative nodes enhance innovation velocity and market reach. Moreover, intellectual property frameworks evolve to balance open innovation with commercialization rights, shaping sustainable growth and equitable value capture.

9.4.7 Investment Landscape and Market Growth

The synthetic biology sector has witnessed growing investor interest, driven by its cross-sector impact and technological maturity. Venture capital, corporate investments, and government funding increasingly support platform development, therapeutic innovation, sustainable materials, and synthetic biomanufacturing startups. Market analyses project compounded annual growth rates exceeding 20%, fueled by expanding applications in healthcare, agriculture, chemicals, and environmental technologies. Entrepreneurs and investors recognize synthetic biology's potential to disrupt established industries, create new markets, and deliver scalable bio-solutions addressing global sustainability challenges. Despite technical and regulatory uncertainties, improved platform technologies and successful commercial deployments enhance investor confidence. Strategic prioritization of IP portfolios, market positioning, and regulatory navigation becomes pivotal for startups and incumbents competing in this competitive landscape.

9.4.8 Challenges and Regulatory Considerations

Developing future business models in synthetic biology entails navigating complex technical, ethical, and regulatory challenges. Biological system unpredictability, scale-up difficulties, and manufacturing reproducibility require robust quality control frameworks and flexible business operations. Regulatory agencies worldwide are developing frameworks specifically adapted to synthetic biology products, balancing safety evaluation with innovation facilitation. Compliance with biosafety, biosecurity, data privacy, and ethical standards necessitates investments in governance and transparency. Intellectual property rights management with regard to biological parts and data remains intricate and sometimes contentious. Business models integrating proactive stakeholder engagement, risk assessment, and ethical responsibility are increasingly favored to secure public trust and regulatory approval.

9.4.9 Long-Term Outlook and Strategic Implications

Looking forward, synthetic biology's business models will continue evolving toward integrated, platform-centered, and sustainability-driven paradigms. The synthesis of biology, digitization, and automation will drive next-generation bio-industries capable of producing complex biomolecules and living medicines on demand. Companies that successfully leverage multidisciplinary capabilities, collaborative networks, and data-driven innovation will dominate emerging markets. Synthetic biology's ability to localize production and customize products aligns with broader societal shifts toward resilient, circular economies.

CHAPTER 10

Best Practices in Synthetic Biology Research and Development

10.1 DESIGNING EFFICIENT AND RELIABLE SYNTHETIC BIOLOGICAL SYSTEMS

The design of efficient and reliable synthetic biological systems stands as both a fundamental challenge and enabling capability for the broader synthetic biology field, determining whether engineered biology can deliver on its promise to address global challenges. The integration of engineering principles with biological understanding has produced remarkable advances in our ability to create predictable, robust biological systems, yet significant challenges remain in achieving the level of reliability expected in traditional engineering disciplines. Success requires continued development of design tools, component libraries, and theoretical frameworks that account for the unique properties of biological systems while leveraging their inherent capabilities for self-organization, adaptation, and production of complex molecules.

The evolution from ad hoc genetic engineering to systematic design approaches marks a maturation of the field that will enable increasingly sophisticated applications in medicine, manufacturing, and environmental management. As computational power increases and our understanding of biological systems deepens, the ability to design efficient and reliable synthetic biological systems will improve dramatically, potentially reaching levels of predictability approaching those of electronic circuits. The interdisciplinary nature of this challenge requires continued collaboration between biologists, engineers, computer scientists, and mathematicians to develop integrated solutions that bridge different domains of expertise. Ultimately, mastering the design of efficient and reliable synthetic biological systems will determine the extent to which synthetic biology can transform from a promising technology to a foundational platform for sustainable development and human prosperity.

10.1.1 Fundamental Design Principles

The foundation of efficient and reliable synthetic biological systems rests on core design principles that guide the construction of genetic circuits and metabolic pathways to achieve desired functions with minimal resource consumption and maximum predictability. Orthogonality represents a crucial principle whereby synthetic components operate independently of native cellular processes, preventing unwanted crosstalk that can compromise system performance or host viability. The principle of modularity enables the construction of complex systems from well-characterized, interchangeable parts that can be combined in predictable ways, similar to electronic components in circuit design.

Abstraction hierarchies allow designers to work at different levels of complexity, from individual genetic parts to complete systems, without needing to consider all underlying details simultaneously. Resource allocation must be carefully managed to prevent synthetic systems from overwhelming host metabolism or competing with essential cellular functions for limited resources such as ribosomes, RNA polymerase, and ATP. Fail-safe mechanisms and genetic safeguards ensure that system failures do not result in harmful outcomes, incorporating features such as toxin-antitoxin systems or metabolic dependencies that prevent uncontrolled proliferation. The matching of component strengths prevents bottlenecks where weak promoters, poor ribosome binding sites, or inefficient enzymes limit overall system performance. These principles must be applied holistically, recognizing that optimizing individual components does not necessarily lead to optimal system performance due to emergent properties and non-linear interactions.

10.1.2 Modularity and Standardization Approaches

The implementation of modularity and standardization in synthetic biological systems has revolutionized the design process by enabling the rapid assembly and iteration of complex genetic constructs from characterized components. Standardized biological parts, such as those in the Registry of Standard Biological Parts, provide designers with libraries of promoters, ribosome binding sites, coding sequences, and terminators with known performance characteristics. The development of standard assembly methods including BioBrick, Golden Gate, and Gibson Assembly has eliminated much of the technical complexity of DNA construction, allowing designers to focus on system architecture rather than molecular cloning challenges.

Modular design enables the separation of sensing, processing, and actuation functions into discrete units that can be independently optimized and tested before integration into larger systems. Signal matching between modules requires careful tuning of input and output levels to ensure proper signal propagation without saturation or insufficient activation. Insulation strategies including ribozyme insulators and transcriptional terminators prevent unintended interactions between modules that could compromise system function. The creation of orthogonal resource pools through the engineering of separate transcription and translation machinery for synthetic circuits reduces competition with host processes and improves modularity. Despite significant progress, achieving true modularity remains challenging due to retroactivity, where downstream modules affect upstream behavior, and context effects that cause parts to behave differently in various genetic contexts.

10.1.3 Computational Design and Modeling

The complexity of biological systems necessitates sophisticated computational approaches for designing efficient and reliable synthetic biological systems that can predict behavior before expensive and time-consuming experimental implementation. Mathematical modeling using ordinary differential equations, stochastic simulations, and Boolean logic frameworks enables designers to explore parameter spaces and identify optimal configurations for desired system behavior. Computer-aided design tools specifically developed for synthetic biology, such as Cello, SBOL Designer, and iBioSim, automate aspects of the design process from high-level specifications to DNA sequences.

Machine learning approaches increasingly contribute to design optimization by learning from large datasets of characterized parts and circuits to predict performance and identify non-obvious design solutions. Metabolic flux analysis and genome-scale models help identify metabolic bottlenecks and predict the impact of synthetic pathways on host metabolism, enabling rational optimization of production systems. Design of experiments approaches systematically explore multidimensional design spaces to identify optimal combinations of components and conditions with minimal experimental effort. Digital twins of biological systems enable in silico testing and optimization before physical implementation, reducing development time and cost. The integration of multi-scale models that span from molecular interactions to population dynamics provides comprehensive understanding of system behavior across different biological scales.

10.1.4 Optimization Strategies for Performance

Achieving optimal performance in synthetic biological systems requires systematic approaches to enhance efficiency while maintaining reliability across varying conditions and applications. Pathway balancing through careful tuning of enzyme expression levels prevents accumulation of toxic intermediates and maximizes flux toward desired products without overwhelming cellular resources. Dynamic regulation systems that adjust gene expression in response to metabolic state or environmental conditions optimize resource allocation between growth and production phases. Protein engineering and directed evolution generate improved enzymes with enhanced catalytic efficiency, stability, or substrate specificity that increase overall system performance. Codon optimization and mRNA stability engineering ensure efficient translation and appropriate protein levels while minimizing metabolic burden on the host. Cofactor regeneration systems and metabolic rewiring eliminate bottlenecks in central metabolism that limit the availability of precursors and energy for synthetic pathways.

Compartmentalization strategies using protein scaffolds, metabolosomes, or synthetic organelles concentrate enzymes and substrates to improve reaction efficiency and prevent toxic intermediate diffusion. Load balancing across multiple cells or strains through division of labor approaches distributes metabolic burden and enables more complex synthetic systems than possible in single cells.

10.1.5 Engineering Robustness and Reliability

The development of robust and reliable synthetic biological systems requires explicit consideration of failure modes and implementation of design strategies that maintain function despite perturbations and uncertainties. Redundancy through multiple parallel pathways or backup systems ensures continued function if individual components fail, though this must be balanced against increased resource costs. Feedback control mechanisms including negative feedback for homeostasis and positive feedback for switch-like behavior provide robust performance across varying conditions.

Genetic stability engineering through removal of repetitive sequences, elimination of recombination sites, and use of integration rather than plasmid-based systems prevents evolutionary loss of synthetic functions. Noise management strategies exploit or suppress biological noise as appropriate, using noise-canceling circuit architectures or noise-induced effects for beneficial purposes. Adaptive systems that can sense and respond to performance degradation through self-diagnostics and automatic adjustment maintain long-term reliability. Stress response engineering ensures synthetic systems remain functional under industrial conditions including temperature variations, pH changes, and presence of toxic compounds. Population-level robustness through bet-hedging strategies or distributed functions across heterogeneous populations provides system-level reliability despite individual cell failures.

10.1.6 Validation and Testing Methodologies

Rigorous validation and testing of synthetic biological systems is essential for ensuring that designs perform as intended and maintain reliability across different contexts and scales. Standardized characterization protocols enable comparison of components and systems across different laboratories and provide quantitative performance metrics for design improvement. Multi-scale testing from isolated parts through integrated systems to production conditions identifies context-dependent effects and scaling challenges early in development. Accelerated life testing and stress testing reveal failure modes and degradation patterns that might not appear under normal conditions but could affect long-term reliability.

Statistical analysis of biological variability distinguishes between acceptable performance variation and true system failures requiring design modification. Real-time monitoring using fluorescent reporters, biosensors, and omics technologies provides continuous assessment of system state and early warning of impending failures. Evolutionary stability testing over multiple generations identifies genetic changes that affect system performance and validates strategies for maintaining genetic integrity. Design verification through multiple orthogonal methods ensures that systems meet specifications and perform reliably across the intended operating envelope.

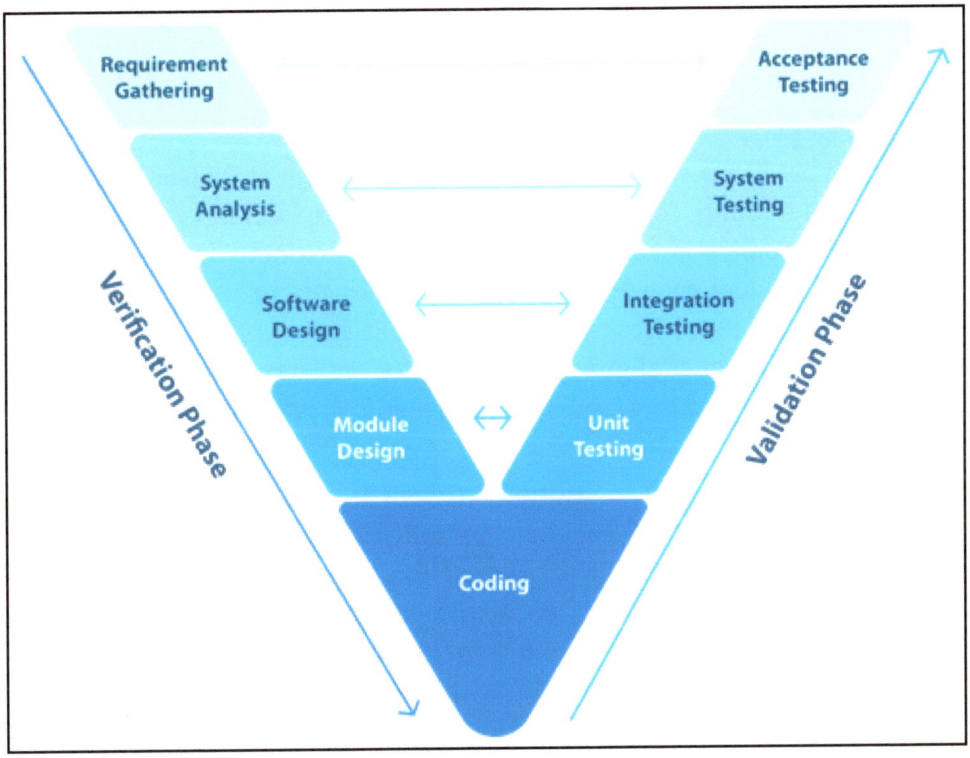

Methodological Approaches for Testing and Validation

10.1.7 Current Limitations and Challenges

Despite significant advances, the design of efficient and reliable synthetic biological systems faces persistent challenges that limit the complexity and applications of engineered biological systems. Biological complexity and incomplete understanding of cellular physiology make it difficult to predict all interactions between synthetic systems and their hosts, leading to unexpected behaviors and failures. Limited part libraries with fully characterized, orthogonal components constrain design options and force compromise between ideal and available solutions. Evolutionary pressure against synthetic functions that impose metabolic burden creates inherent instability that must be actively managed through clever design or selective pressure. Context dependence of biological parts means that characterized components may behave differently in new genetic contexts or host organisms, requiring extensive recharacterization.

Scaling challenges from laboratory to industrial conditions often reveal unexpected problems in efficiency and reliability that were not apparent at smaller scales. The lack of universal standards for characterization, documentation, and data sharing impedes the development of truly modular and reusable biological components. Trade-offs between efficiency, reliability, and other desirable properties such as safety and containment require careful balancing that may limit achievable performance.

10.2 COLLABORATION BETWEEN ACADEMIA, INDUSTRY, AND GOVERNMENT

Collaboration between academia, industry, and government is essential for driving innovation, addressing complex societal challenges, and translating research into practical applications. Each sector brings distinct strengths: academia provides foundational knowledge and skilled human capital, industry delivers commercialization expertise and resources, and government offers regulatory oversight and strategic alignment. Successful collaborations leverage these strengths through structured models such as public-private partnerships, research consortia, and innovation hubs.

While challenges related to culture, timelines, intellectual property, and regulatory compliance exist, they can be mitigated through clear agreements, transparent communication, and shared objectives. Effective collaboration accelerates innovation, enhances economic growth, and ensures that scientific advancements generate tangible benefits for society. By fostering trust, integrating diverse expertise, and maintaining adaptive strategies, academia, industry, and government can work together to address emerging challenges and create sustainable, impactful solutions for the future.

10.2.1 Role of Academia

Academic institutions serve as the foundation for collaborative innovation by generating new knowledge, conducting fundamental research, and training future scientists and professionals. Universities and research institutes drive exploratory studies and theoretical research that lay the groundwork for practical applications. They also provide state-of-the-art laboratories and infrastructure that support experimentation and discovery. In collaborative initiatives, academia often leads the early-stage research, identifying potential technologies or scientific breakthroughs with commercial or policy relevance. Additionally, academic involvement ensures the continuous development of human capital through education and professional training, equipping individuals to contribute effectively to industry and government projects.

10.2.2 Role of Industry

Industry plays a crucial role in transforming academic discoveries into market-ready products, technologies, and services. Companies provide commercialization expertise, production capabilities, and funding for applied research. Industry partners help ensure that research outcomes are scalable, economically viable, and aligned with consumer or societal needs.

Corporate involvement can accelerate technology transfer, bridging the gap between laboratory research and real-world implementation. Companies may also provide mentorship, technical resources, and market insights that guide the development of innovative solutions. By integrating research findings with business strategies, industry ensures that collaborative projects deliver tangible, impactful outcomes.

10.2.3 Role of Government

Government agencies support collaboration by creating regulatory frameworks, funding research, and promoting alignment with public priorities. Policies and grants encourage cooperation between academia and industry while ensuring ethical, safety, and environmental standards.

Governments provide infrastructure such as research parks, incubators, and shared laboratories that facilitate joint projects. Policy initiatives, public-private partnerships, and incentive programs promote translational research and the practical application of scientific discoveries. Governments also oversee regulatory compliance and ensure that collaborative projects advance societal goals while maintaining accountability and transparency.

10.2.4 Models of Collaboration

Collaboration can take multiple forms, from formal partnerships to informal networks. Public-private partnerships (PPPs) allow sharing of resources, expertise, and risk across sectors. Research consortia bring together universities, companies, and government agencies to address large-scale scientific challenges. Technology transfer offices in universities facilitate licensing agreements that move discoveries from lab to market. Innovation hubs, co-working spaces, and incubators provide platforms for regular interaction, knowledge sharing, and joint development. Each model requires clearly defined roles, shared objectives, and governance mechanisms to be successful.

10.2.5 Benefits of Collaboration

The collaboration between academia, industry, and government offers numerous benefits. Joint projects accelerate innovation by combining resources and expertise, reducing duplication of effort. They enhance technology transfer and commercialization, allowing discoveries to reach the market faster. Collaboration supports workforce development, giving students and early-career researchers exposure to industry practices and policy considerations. Cross-sector partnerships also address societal challenges such as public health, energy transition, and climate change more effectively. By leveraging the strengths of all three sectors, collaborative projects maximize the impact of scientific research and investment.

10.2.6 Challenges in Collaboration

Despite its advantages, collaboration faces several challenges. Differences in organizational culture, priorities, and timelines can create friction. Academic research may focus on long-term exploration, whereas industry prioritizes short-term market outcomes. Government agencies emphasize regulatory compliance and societal accountability. Intellectual property disputes, data-sharing restrictions, and publication policies can further complicate partnerships. Financial constraints, bureaucratic procedures, and varying performance metrics also pose obstacles. Effective collaboration requires clear agreements, transparent communication, and mechanisms for conflict resolution to overcome these challenges.

- Academia provides fundamental research, knowledge creation, and workforce development.
- Industry contributes commercialization expertise, investment, and market insights.
- Government ensures regulatory oversight, funding, and alignment with societal priorities.
- Public-private partnerships enable sharing of resources, expertise, and risk.
- Research consortia facilitate multi-institutional collaboration on complex challenges.
- Technology transfer offices bridge the gap between academic research and commercialization.
- Collaboration accelerates innovation, reduces duplication, and enhances societal impact.
- Challenges include differences in culture, IP disputes, regulatory requirements, and funding constraints.

- Successful case studies, such as the Human Genome Project and renewable energy initiatives, highlight effective collaboration.
- Strategies for success include shared goals, transparent communication, clear agreements, joint training, and adaptive project management.

10.3 ETHICAL GUIDELINES FOR SYNTHETIC BIOLOGY RESEARCH

Ethical guidelines for synthetic biology research provide a comprehensive framework to navigate the complex moral, social, and environmental challenges associated with engineering living systems. Principles such as beneficence, non-maleficence, transparency, dual-use awareness, equity, environmental responsibility, informed consent, and accountability guide researchers in conducting work that maximizes societal benefit while minimizing harm.

Implementation of these principles requires robust regulatory oversight, interdisciplinary collaboration, stakeholder engagement, and ongoing ethical training. While challenges remain in balancing innovation with precaution, ethical frameworks are essential to ensure that synthetic biology advances in a manner that is responsible, equitable, and sustainable. By embedding ethics into every stage of research and commercialization, synthetic biology can achieve its transformative potential while maintaining public trust and contributing to the welfare of humanity and the environment.

10.3.1 Principle of Beneficence

One of the fundamental ethical principles guiding synthetic biology research is beneficence, which emphasizes the obligation to maximize benefits while minimizing harm. Research in synthetic biology should aim to provide tangible benefits to society, such as medical therapies, sustainable energy solutions, or environmental restoration. When designing experiments, researchers must evaluate potential outcomes, considering both short-term and long-term impacts on human health, ecosystems, and global communities. Ethical guidelines rooted in beneficence encourage careful assessment of risks and benefits, ensuring that innovation does not compromise public welfare. Projects that fail to demonstrate clear societal value or that introduce disproportionate risks may be reconsidered or modified to enhance positive outcomes.

10.3.2 Principle of Non-Maleficence

Closely related to beneficence, the principle of non-maleficence requires researchers to avoid causing harm. Synthetic biology carries inherent risks, including accidental release of genetically modified organisms, unforeseen ecological interactions, or misuse of technology for harmful purposes. Ethical guidelines advocate for rigorous risk assessment, containment measures, and safety protocols to prevent environmental, health, or societal damage. Non-maleficence also emphasizes the need to anticipate indirect consequences, such as economic disruption or exacerbation of social inequalities, ensuring that research does not unintentionally harm vulnerable populations. By prioritizing safety and harm reduction, synthetic biology research can proceed in a manner that respects both human and ecological well-being.

10.3.3 Transparency and Open Communication

Transparency is essential for maintaining public trust and accountability in synthetic biology research. Ethical guidelines recommend open communication about research goals, methodologies, potential risks, and expected benefits. Researchers should engage with stakeholders, including policymakers, ethicists, community representatives, and the general public, to foster informed dialogue and understanding. Open communication also supports reproducibility and peer evaluation, allowing the scientific community to validate findings and improve methodologies. Transparency ensures that research decisions are not made in isolation but reflect societal values and expectations. It also helps prevent misinformation, speculation, or public fear, which can hinder the responsible development and deployment of synthetic biology technologies.

Transparent Communication and Open Exchange

10.3.4 Dual-Use and Security Considerations

Synthetic biology has dual-use potential, meaning the same techniques that create beneficial applications can also be misused for harmful purposes. Engineered pathogens, toxins, or environmental modifications could pose significant security risks if applied maliciously. Ethical guidelines emphasize the responsibility of researchers to anticipate dual-use implications and implement preventive measures. This includes controlled access to sensitive materials, adherence to biosecurity protocols, and reporting mechanisms for potentially dangerous research. International collaboration and regulatory oversight are also critical to mitigate risks, as synthetic biology is a global field with cross-border implications. Addressing dual-use concerns ensures that innovation does not inadvertently become a threat to public health, national security, or global stability.

10.3.5 Equity and Access

Ethical guidelines for synthetic biology research also consider issues of equity and access. Technologies developed through synthetic biology may disproportionately benefit certain populations while leaving others underserved. Researchers are encouraged to consider the distribution of benefits and risks, promoting inclusive access to innovations such as medical treatments, bio-based materials, or environmental remediation technologies. Addressing equity involves designing research programs that are socially responsible, prioritizing global health challenges, and avoiding applications that exacerbate existing disparities. Ethical oversight mechanisms can help ensure that synthetic biology contributes to societal well-being in a manner that is just, fair, and broadly accessible.

10.3.6 Environmental Responsibility

The release of genetically engineered organisms into natural ecosystems carries potential environmental risks. Synthetic biology research must adhere to ethical principles of environmental stewardship, ensuring that experimental interventions do not disrupt ecological balance or biodiversity. Ethical guidelines recommend the use of containment strategies, controlled trials, and post-release monitoring when deploying engineered organisms. Researchers should also consider long-term ecological impacts, cumulative effects, and potential interactions with native species. By prioritizing environmental responsibility, synthetic biology research can contribute to sustainability and ecosystem health, aligning technological progress with ecological ethics.

10.3.7 Informed Consent and Human Subject Research

When synthetic biology research involves human participants, ethical guidelines emphasize the importance of informed consent, autonomy, and protection from harm. Participants must be provided with clear, comprehensive information about the purpose of the research, potential risks, expected benefits, and alternative options. Researchers are obligated to respect the autonomy of participants, ensuring voluntary participation without coercion or undue influence. Ethical oversight, typically through institutional review boards or ethics committees, is essential to safeguard human subjects and maintain compliance with international standards for biomedical research. Informed consent not only protects participants but also strengthens public confidence in synthetic biology applications in medicine and healthcare.

10.3.8 Accountability and Regulatory Compliance

Accountability is a cornerstone of ethical synthetic biology research. Researchers, institutions, and companies must adhere to national and international regulations, biosafety standards, and professional codes of conduct. Ethical guidelines emphasize the importance of documenting research activities, maintaining traceability of materials, and reporting results accurately. Institutional oversight, peer review, and audits help ensure compliance and foster a culture of responsibility. Accountability mechanisms also provide mechanisms for addressing adverse events, ethical violations, or unforeseen consequences, allowing corrective action and continuous improvement in research practices.

10.3.9 Education and Ethical Training

Ethical awareness is enhanced through education and training in responsible research practices. Researchers in synthetic biology should receive instruction on bioethics, biosafety, regulatory frameworks, and societal implications of their work. Educational programs encourage critical thinking about the ethical, social, and environmental dimensions of scientific innovation. By fostering a culture of ethical literacy, researchers are better equipped to anticipate dilemmas, make informed decisions, and engage constructively with stakeholders. Ethical training complements technical expertise, ensuring that synthetic biology advances responsibly and with societal benefit in mind.

Table: Ethical Principles for Synthetic Biology Research

Principle	Application
Beneficence	Maximize benefits while minimizing harm to society and the environment
Non-Maleficence	Prevent accidental or intentional harm from research activities
Transparency	Promote open communication, reproducibility, and stakeholder engagement
Dual-Use Awareness	Address potential misuse of synthetic biology technologies
Equity	Ensure fair distribution of benefits and access to innovations
Environmental Responsibility	Protect ecosystems and biodiversity from unintended impacts
Informed Consent	Safeguard autonomy and welfare of human research participants

10.3.10 Challenges in Implementing Ethical Guidelines

Despite clear ethical principles, implementing guidelines in practice can be challenging. Rapid technological advancement often outpaces regulatory frameworks, creating gaps in oversight. Global disparities in research infrastructure, funding, and expertise may result in inconsistent application of ethical standards. Complexities arise in balancing innovation with precaution, especially when predicting long-term ecological or societal impacts. Conflicts of interest, commercial pressures, and intellectual property considerations can further complicate adherence to ethical norms. Addressing these challenges requires ongoing dialogue between scientists, ethicists, policymakers, and the public, as well as adaptive governance structures that can respond to emerging technologies and evolving societal expectations.

10.4 SAFETY PROTOCOLS AND CONTAINMENT STRATEGIES

Safety protocols and containment strategies are essential enablers of synthetic biology's promise, balancing innovation with protection of health, environment, and society. Combining physical containment with genetic biocontainment, risk-based management, regulatory oversight, and fostering a robust safety culture establishes comprehensive safeguards. Emerging technologies enhance these frameworks, addressing novel challenges posed by increasingly complex synthetic biological systems. Continued interdisciplinary collaboration, innovation in safety science, and transparent governance will ensure synthetic biology advances responsibly, maintaining societal trust and maximizing benefits while minimizing risks. As synthetic biology matures, embedding rigorous safety and containment principles remains pivotal for sustainable and ethical biotechnology development.

10.4.1 Historical Context and Evolution of Safety Practices

The development of safety protocols in biotechnology initially drew upon experiences in microbiology and infectious disease research, where biosafety levels were codified to delineate appropriate laboratory containment based on organism pathogenicity and risk. The introduction of recombinant DNA technology in the 1970s prompted the establishment of guidelines by organizations such as the NIH Recombinant DNA Advisory Committee and the Asilomar Conference, which defined practices for safe genetic manipulation. As synthetic biology emerged with greater emphasis on designing novel genetic circuits and synthetic organisms, the range of potential risks expanded, necessitating new layers of control. Modern safety paradigms integrate traditional biosafety frameworks with innovations such as genetic biocontainment, which embeds safety features directly into engineered organisms. The iterative nature of safety protocol development mirrors technological progress and regulatory maturation.

10.4.2 Physical Containment Measures

Physical containment remains a frontline defense against unintended exposure and environmental release of engineered or pathogenic organisms. Facilities are classified by biosafety levels (BSL 1 to 4) based on hazard potential, dictating structural design, access control, ventilation systems, waste decontamination, and personal protective equipment requirements. Advanced infrastructure includes sealed negative-pressure laboratories, air filtration, and sterilization devices to safeguard operators and surroundings.

Standard operating procedures encompass practices like controlled material transfer, spill response, and decontamination protocols. In industrial biomanufacturing, closed bioreactors and secure waste handling procedures prevent escape of genetically modified strains. While effective, physical containment alone is insufficient for autonomous containment, especially for outdoor or therapeutic deployment, highlighting the need for complementary genetic safeguards.

10.4.3 Genetic Biocontainment and Safety-by-Design

Genetic biocontainment introduces fail-safe mechanisms within organisms themselves, enhancing containment "from within." These strategies include kill switches, which are synthetic genetic circuits triggering programmed cell death upon sensing environmental signals or mutation. Auxotrophy-based biocontainment requires microbes to depend on synthetic nutrients unavailable in natural environments, limiting survival outside controlled settings.

Another approach involves genome recoding to incorporate synthetic amino acids or nucleotides essential for viability, creating genetic firewalls against horizontal gene transfer and ecological establishment. These "safety by design" concepts aim to render synthetic organism's incapable of persisting or evolving outside intended contexts. Successful implementation necessitates balancing containment efficacy with organismal fitness and stability, often requiring iterative design and safety testing.

10.4.4 Risk Assessment and Management Frameworks

Effective safety protocols derive from comprehensive risk assessments identifying and quantifying potential hazards associated with engineered biological systems. Risk evaluations consider factors such as pathogenicity, transmission potential, environmental persistence, gene transfer risk, and human exposure scenarios. This informs risk management strategies encompassing containment levels, monitoring, emergency preparedness, and regulatory compliance. Dynamic risk assessment integrates new scientific data and operational feedback to update safety measures adaptively. Institutional biosafety committees and regulatory agencies oversee adherence to risk-based containment, ensuring proportional safeguards.

Integrating predictive modeling, laboratory diagnostics, and surveillance systems strengthens proactive risk mitigation and incident response capabilities.

10.4.5 Regulatory Oversight and International Standards

International and national regulatory bodies provide essential frameworks and guidelines that shape safety protocols and containment practices in synthetic biology and biotechnology. Agencies such as the US FDA, EPA, NIH, EFSA, WHO, and OECD establish standards governing laboratory safety, environmental release, clinical testing, and commercialization of genetically engineered organisms. Harmonizing regulations across jurisdictions remains a challenge due to differing risk perceptions, legal structures, and technological readiness. Cooperative efforts strive to develop common standards, transparent data-sharing, and rigorous approval mechanisms. These regulatory frameworks incentivize adherence to biosafety norms and hold stakeholders accountable. Furthermore, oversight extends to ethical governance, dual-use research policies, and public engagement to foster responsible innovation.

10.4.6 Training and Organizational Safety Culture

Personnel training and fostering a culture of safety are indispensable components of containment strategies. Comprehensive education in biosafety principles, risk awareness, emergency procedures, and ethical responsibilities empower researchers, technicians, and managers to maintain safe operations. Regular drills, safety audits, and incident reporting systems reinforce vigilant oversight and continuous improvement. Organizational commitment to safety, supported by leadership and resources, cultivates environments where adherence to best practices is standard. Safety culture incorporates openness to learn from near-misses and accidents, integrating lessons learned to prevent recurrence. Cross-disciplinary teams, including biosafety officers, ethicists, and engineers, collaborate to embed safety across research and production pipelines.

10.4.7 Emerging Technologies Enhancing Safety and Containment

Technological innovations increasingly support safety protocols by enabling precise control, monitoring, and risk reduction. Synthetic biology itself provides new tools for genetic containment and safety switches engineered into genomes. Real-time biosensors can detect pathogen leaks or environmental contamination, triggering automated responses. Advances in machine learning and data analytics enhance pattern recognition for anomaly detection in bioprocesses. Robotics and automation minimize human exposure to hazardous agents while improving reproducibility and containment fidelity. Digital twins and in silico modeling simulate containment scenarios and organism behavior, guiding proactive protocol design. These technologies collectively uphold rigorous safety standards amid expanding synthetic biology applications.

10.5 BEST PRACTICES IN INTELLECTUAL PROPERTY AND COMMERCIALIZATION

Intellectual property and commercialization are inextricably linked to business success. By adopting best practices—proactively protecting IP, strategically leveraging it for revenue, and navigating risks—companies can transform creative ideas into sustainable competitive advantages. As technology advances and markets globalize, the importance of a robust IP strategy will only grow.

Organizations that invest in understanding and implementing these practices will not only safeguard their innovations but also thrive in the innovation economy. Ultimately, effective IP management is about creating value, fostering growth, and securing a prosperous future in an increasingly knowledge-based world.

10.5.1 Best Practices for Protecting Intellectual Property

Protecting IP begins with a proactive and comprehensive strategy. The first step is identifying and documenting all valuable IP assets. This includes conducting regular IP audits to catalog patents, trademarks, copyrights, trade secrets, and know-how. Once identified, businesses must secure appropriate legal protections:

- **Patents:** For inventions, filing patents is crucial. A well-drafted patent application requires thorough prior art research and precise claims to ensure broad protection. Startups and established companies alike should file early to prevent competitors from encroaching on their innovations. International patent protection (e.g., via the PCT system) is vital for global markets.
- **Trademarks:** Registering trademarks protects brand identity. This includes names, logos, and slogans. Businesses should conduct clearance searches to avoid infringement and register trademarks in all relevant jurisdictions. Consistent brand enforcement prevents dilution and maintains market position.
- **Copyrights:** Automatically arising upon creation, copyrights protect original works like software, literature, and media. Registering copyrights provides stronger legal recourse. Companies should implement clear copyright notices and agreements with employees and contractors to ensure ownership.
- **Trade Secrets:** Confidential information (e.g., formulas, processes, customer lists) can be protected as trade secrets indefinitely, provided reasonable secrecy measures are in place. Non-disclosure agreements (NDAs), restricted access, and employee training are essential to maintain confidentiality.

Beyond formal protections, fostering a culture of IP awareness within the organization is critical. Employees should be educated on the importance of IP and their role in safeguarding it. Additionally, businesses must monitor the competitive landscape for potential infringements and respond swiftly to protect their rights.

10.5.2 Strategies for Leveraging IP for Commercial Success

Protecting IP is only the first step; leveraging it for commercialization is where value is realized. Businesses can monetize IP through various strategies:

- **Licensing:** Licensing IP to third parties allows companies to generate revenue without direct involvement in manufacturing or distribution. This can include exclusive or non-exclusive licenses, royalties, or upfront fees. Licensing agreements must be carefully negotiated to protect IP rights and ensure fair compensation.
- **Strategic Partnerships and Joint Ventures:** Collaborating with other companies can accelerate commercialization. Partnerships enable shared resources, expertise, and market access. IP assets can be pooled or cross-licensed to create innovative products or services.
- **IP-Backed Financing:** IP assets can be used as collateral for loans or to attract venture capital. Investors often view strong IP portfolios as indicators of a company's potential, making it easier to secure funding.
- **Direct Commercialization:** Developing and selling products or services based on proprietary IP. This requires effective go-to-market strategies, including marketing, sales, and distribution channels.
- **IP Portfolio Management:** Actively managing and optimizing the IP portfolio ensures that resources are focused on high-value assets. This may involve pruning unused patents, expanding protection for core innovations, and aligning IP strategy with business goals.

By integrating IP into overall business strategy, companies can maximize returns on innovation and sustain competitive advantage.

10.5.3 Navigating IP Challenges and Risks

Despite best efforts, IP holders face numerous challenges. Infringement is a pervasive risk; competitors may copy, reverse-engineer, or otherwise misuse protected IP. Vigilance is key:

- **IP Infringement Detection:** Using monitoring tools and services to identify potential infringements early. This includes watching patent databases, trademark registries, and market activities.

- **Enforcement:** Taking legal action against infringers is sometimes necessary. Cease-and-desist letters, litigation, or alternative dispute resolution (ADR) can stop unauthorized use. However, enforcement must be strategic, weighing costs against benefits.
- **Due Diligence:** Conducting thorough IP due diligence before mergers, acquisitions, or partnerships is essential. This involves verifying ownership, assessing the strength of IP assets, and identifying potential liabilities.
- **Global IP Risks:** In international markets, IP laws vary significantly. Companies must navigate differing regulations, enforceability issues, and cultural attitudes toward IP. Utilizing international treaties (e.g., Berne Convention, TRIPS Agreement) helps, but local legal counsel is often necessary.

Additionally, businesses must guard against internal risks, such as employee theft of IP or inadvertent disclosures. Robust internal controls, including confidentiality agreements and access restrictions, mitigate these threats.

CHAPTER 11

Appendices

11.1 GLOSSARY OF KEY TERMS IN SYNTHETIC BIOLOGY

BioBricks: Standardized DNA sequences with defined functions that can be combined to build complex genetic systems. BioBricks simplify the design, assembly, and testing of synthetic biology constructs.

Biocontainment: Safety strategies to prevent engineered organisms from surviving or spreading outside controlled laboratory or industrial settings. Examples include kill switches and nutrient dependency systems.

Bioproduction: Industrial-scale production of molecules like biofuels, enzymes, or pharmaceuticals using engineered organisms.

Biosensor: A genetically engineered device capable of detecting specific environmental signals or molecules and producing a measurable response, commonly used in diagnostics, environmental monitoring, and research.

Cell-Free Systems: Biological reactions performed outside living cells using extracts that contain the machinery for transcription and translation. These systems enable rapid prototyping of genetic circuits and protein production.

Chassis Organism: A host organism, often bacteria, yeast, or mammalian cells, used as a platform to implement synthetic biology designs. The choice depends on genetic manipulability, growth characteristics, and application requirements.

CRISPR-Cas9: A revolutionary gene-editing tool derived from bacterial immune systems, enabling precise targeting and modification of DNA in living cells. It is used for gene knockouts, insertions, and functional studies in synthetic biology.

Directed Evolution: A laboratory technique that mimics natural selection to evolve proteins, enzymes, or genetic pathways with enhanced or novel functions.

Ethical, Legal, and Social Implications (ELSI): Considerations of the societal, regulatory, and ethical issues associated with synthetic biology research, including biosafety, biosecurity, and environmental impact.

Genetic Circuit: An engineered network of genes, regulatory elements, and proteins designed to perform a specific function within a cell. Examples include switches, oscillators, and sensors that mimic electronic circuit behavior.

Genome Engineering: The large-scale modification of an organism's genome to add, remove, or alter genes. Techniques include CRISPR-Cas9, TALENs, and recombineering, allowing precise and controlled changes to DNA sequences.

Metabolic Engineering: The manipulation of cellular pathways to optimize the production of specific molecules, such as pharmaceuticals, biofuels, or industrial enzymes.

Modular Design: The concept of creating biological systems using interchangeable and standardized parts, allowing for efficient assembly, testing, and optimization of genetic circuits or pathways.

Orthogonal System: A synthetic system designed to operate independently of the host's natural biological machinery, minimizing cross-talk and unintended interactions.

Oscillator: A synthetic circuit designed to produce periodic fluctuations in gene expression or protein activity, analogous to a biological clock.

Protein Engineering: The design or modification of proteins to have novel functions, improved stability, or specificity, using rational design, directed evolution, or computational modeling.

Recombinant DNA Technology: Techniques for joining DNA from different sources to create new genetic combinations. It is foundational to synthetic biology applications.

Synthetic Minimal Genome: A genome engineered to contain only the essential genes required for life, providing a simplified chassis for research and applications.

Toggle Switch: A genetic circuit capable of existing in one of two stable states, switching between them in response to a specific input, functioning as a cellular memory element.

Xenobiology: A subfield exploring organisms with alternative biochemistries, such as synthetic nucleotides or amino acids not found in nature, to expand the capabilities of life.

11.2 RESOURCES AND RECOMMENDED READING

- **"Engineering Life: A Guide to Synthetic Biology"**–Mark A. Bedau, MIT Press, 2010. Examines both the scientific and philosophical aspects of designing new biological systems.

- **"Foundations of Synthetic Biology"**–Christophe d'Herouel & Markus Schmidt, MIT Press, 2018. Explores the theoretical frameworks, engineering principles, and ethical implications of synthetic biology.
- **"Regenesis: How Synthetic Biology Will Reinvent Nature and Ourselves"**–George M. Church & Ed Regis, Basic Books, 2012. Discusses the potential of synthetic biology to transform medicine, industry, and the environment.
- **"Synthetic Biology Handbook"**–Darren N. Nesbeth, Wiley, 2012. A practical reference covering methods, design strategies, and emerging applications.
- **"Synthetic Biology: A Primer"**–Paul S. Freemont & Richard I. Kitney, CRC Press, 2012. Provides a clear introduction to the principles, tools, and applications of synthetic biology, ideal for beginners and professionals.
- **"Synthetic Biology: A Very Short Introduction"**–Jamie A. Davies, Oxford University Press, 2018. Provides an accessible overview of the field for newcomers and general readers.
- **"Synthetic Biology: Industrial and Environmental Applications"**–F. G. Priest, Elsevier, 2016. Highlights applications in biofuels, bioremediation, and sustainable industrial biotechnology.
- **"Synthetic Biology: Tools for Engineering Biological Systems"**–Huimin Zhao, Springer, 2013. Focuses on laboratory techniques, genome engineering, metabolic engineering, and synthetic circuit design.
- **"Synthetic Genomics: Concepts, Methods, and Applications"**–Willem P.C. Stemmer, Springer, 2014. Covers genome synthesis, large-scale DNA assembly, and applications in therapeutics and bioengineering.
- **"The Biotech Primer: An Insider's Guide to the Science Driving the Biopharma Industry"**–BioTech Primer Inc., 2018. Connects synthetic biology concepts to biotechnology and pharmaceutical industry practices.

Journals:

- **Nature Biotechnology**: High-impact research in synthetic biology, metabolic engineering, and biomanufacturing.
- **ACS Synthetic Biology**: Focuses on experimental and computational approaches for engineering biological systems.
- **Synthetic Biology (Oxford Academic)**: Dedicated journal covering foundational and applied synthetic biology research.

- **Trends in Biotechnology**: Publishes reviews and perspectives on emerging trends and applications.
- **Journal of Biological Engineering**: Covers engineering approaches to synthetic biology, metabolic design, and microbial systems.

11.3 SYNTHETIC BIOLOGY ORGANIZATIONS AND RESEARCH INSTITUTES

11.3.1 International and Global Organizations

- **iGEM Foundation (International Genetically Engineered Machine Foundation):** A global non-profit promoting education and innovation in synthetic biology. It organizes an annual competition where students design projects using standardized genetic parts, and it maintains an open-access registry of BioBricks.
- **SynBioBeta:** A professional network connecting startups, investors, and researchers in synthetic biology. It provides industry news, investment opportunities, and insights into biomanufacturing, therapeutics, and innovation.
- **BioBricks Foundation:** Promotes open-source biological engineering, encouraging the development and sharing of standardized biological parts and ethical synthetic biology practices.
- **International Society for Synthetic Biology (ISSB):** A professional society fostering global collaboration through conferences, workshops, and networking, advancing both foundational and applied synthetic biology research.
- **Synthetic Biology Leadership Council (UK):** Advises the UK government on funding, strategy, and regulation of synthetic biology research, promoting academic-industry collaboration and responsible commercialization.

11.3.2 Major Research Institutes and Academic Centers

- **MIT Synthetic Biology Center (USA):** A leading hub for genome engineering, metabolic pathway design, and synthetic circuit development. It integrates engineering, biology, and computational modeling to drive innovation.
- **Wyss Institute for Biologically Inspired Engineering (Harvard University, USA):** Develops synthetic biology solutions for medicine, biofabrication, and environmental applications, focusing on translating research into real-world technologies.

- **Stanford Bioengineering and Synthetic Biology Laboratory (USA):** Focuses on genetic circuit design, microbial engineering, and computational approaches. Its SynBioHub platform shares standardized biological parts for research.
- **European Molecular Biology Laboratory (EMBL, Europe):** Conducts synthetic biology research in genome engineering, systems biology, and cellular design, while supporting training and collaborative projects.
- **Max Planck Institute for Molecular Plant Physiology (Germany):** Applies synthetic biology in plant sciences, metabolic engineering, and bio-based agriculture for sustainable solutions.
- **Imperial College London–Centre for Synthetic Biology (UK):** Integrates microbial engineering, metabolic pathway design, and computational modeling for biomanufacturing and industrial applications.
- **CSIRO Synthetic Biology Future Science Platform (Australia):** Leads research in bio-based manufacturing, environmental solutions, and agricultural biotechnology.
- **ETH Zurich–Synthetic Biology Center (Switzerland):** Specializes in systems biology, synthetic gene circuits, and metabolic engineering for biomedical and industrial applications.
- **Beijing Institute of Synthetic Biology (China):** A national hub for microbial engineering, genome editing, and bio-manufacturing, emphasizing applications in medicine, agriculture, and environmental biotechnology.
- **National Institute for Biological Standards and Control (NIBSC, UK):** Supports synthetic biology research for therapeutic development, vaccine design, and regulatory science to ensure biosafety and standardization.

11.3.3 Industry-Focused and Startup Incubators

- **Ginkgo Bioworks (USA):** Designs custom microbes for industrial, pharmaceutical, and agricultural applications.
- **Twist Bioscience (USA):** Specializes in DNA synthesis for rapid gene assembly and engineering.
- **Synlogic (USA):** Develops engineered microbial therapeutics for disease treatment.
- **Evonetix (UK):** Creates scalable DNA synthesis platforms using microfluidic technologies.

- **Cambridge Synthetic Biology Incubator (UK):** Supports startups and academic projects with mentorship, funding, and lab facilities.

11.3.4 Global Networks and Collaborations

- **European Synthetic Biology Network (ESynBioNet):** Promotes collaboration among European research institutions.
- **Synthetic Biology Open Language (SBOL) Community:** Provides standards for describing and sharing synthetic biology designs.
- **iGEM Regional Hubs:** Support local education, research collaboration, and innovation in synthetic biology.

11.4 FREQUENTLY ASKED QUESTIONS (FAQS)

- **What is synthetic biology?**
- **How is synthetic biology different from genetic engineering?**
- **What are the main applications of synthetic biology?**
- **What are BioBricks?**
- **What is a genetic circuit?**
- **What is a chassis organism?**
- **How is CRISPR-Cas9 used in synthetic biology?**
- **Are synthetic organisms safe?**
- **What is the difference between natural and synthetic genes?**
- **What is directed evolution?**
- **Can synthetic biology help the environment?**
- **How is synthetic biology regulated?**
- **What educational resources are available?**
- **What are the ethical concerns of synthetic biology?**
- **What is the future of synthetic biology?**

Reference

A Computational and Experimental Approach, Author: Marchisio, M.A. & Stelling, J., Publisher: World Scientific, 2019

A Critical Introduction, Author: Calvert, J., Publisher: The MIT Press, 2020

A Handbook for Educators, Author: Kuldell, N., Bernstein, R., Ingram, K., & Hart, K.M., Publisher: CreateSpace Independent Publishing, 2015

A Lab Manual, Author: Forster, A.C. & Church, G.M., Publisher: World Scientific Publishing, 2012

A Lab-based Introduction, Author: Baldwin, G. (Ed.), Publisher: Garland Science, 2016

A Multi-disciplinary Approach, Author: Deplazes-Zemp, A. & Biller-Andorno, N. (Eds.), Publisher: Springer, 2016

A New Era for Chemical Engineering, Author: Brück, T. (Ed.), Publisher: De Gruyter, 2018

A Practical Introduction, Author: Rudge, T.J. & Haseloff, J., Publisher: Royal Society of Chemistry, 2016

A Primer, Author: Baldwin, G., Bayer, T., Dickinson, R., Ellis, T., Freemont, P.S., Kitney, R.I., Polizzi, K., & Stan, G.-B., Publisher: Imperial College Press, 2012

A Synthesis, Author: Heinemann, M. & Panke, S., Publisher: Wiley-Blackwell, 2012

A Systems Biology Approach to Metabolic Engineering, Author: Lee, S.Y. & Kim, H.U., Publisher: Wiley-VCH, 2015

A User's Guide, Author: Cheng, A.A. & Lu, T.K., Publisher: Cold Spring Harbor Laboratory Press, 2015

A Very Short Introduction, Author: Davies, J.A., Publisher: Oxford University Press, 2018

Biosensing and Biocomputing, Author: Smolke, C.D. (Ed.), Publisher: Academic Press, 2018

Computer-Based Tools for Pathway Design, Author: Rodriguez, E. (Ed.), Publisher: Humana Press, 2022

Design and Analysis of Biological Systems, Author: Koeppl, H., Setti, G., di Bernardo, M., & Densmore, D. (Eds.), Publisher: Springer, 2013

Design and Build, Author: Purnick, P.E.M. & Weiss, R., Publisher: McGraw-Hill Education, 2014

Engineering Complexity and Refactoring Cell Systems, Author: Goni-Moreno, A. & Amos, M. (Eds.), Publisher: CRC Press, 2017

Engineering Evolution and Design of Biological Systems, Author: You, L. (Ed.), Publisher: Elsevier, 2019

Engineering Life for a Better Future, Author: Church, G.M. & Regis, E., Publisher: Yale University Press, 2014

Engineering Principles for Bio-based Products, Author: Kircher, M. (Ed.), Publisher: Wiley-Blackwell, 2019

Engineering Principles for Genetic Circuits, Author: Del Vecchio, D. & Murray, R.M., Publisher: Cambridge University Press, 2015

Engineering, Evolution and Design, Author: Strych, U. & Davison, M. (Eds.), Publisher: Wiley, 2020

Ethics and Policy, Author: Kaebnick, G.E. & Murray, T.H. (Eds.), Publisher: MIT Press, 2013

Foundations of Synthetic Biology, Author: Deplazes, A. & Huppenbauer, M. (Eds.), Publisher: Springer, 2015

From Bio-bricks to Whole Genomes, Author: Haseloff, J. & Ajioka, J. (Eds.), Publisher: Academic Press, 2017

From Genetic Engineering to Genome Editing, Author: Trefzer, A. & Lang, C. (Eds.), Publisher: Springer, 2020

From Genetic Switches to Oscillators, Author: Stricker, J. & Cookson, S., Publisher: Springer, 2015

From iGEM to the Artificial Cell, Author: Schmidt, M., Publisher: Springer, 2012

Genome Engineering, Author: Zhao, H. & Zeng, A.-P. (Eds.), Publisher: Springer, 2018

Governance and Regulation, Author: Oldham, P., Hall, S., & Burton, G., Publisher: Edward Elgar Publishing, 2016

Industrial and Environmental Applications, Author: Dietmar, H. (Ed.), Publisher: Wiley-Blackwell, 2012

Methods and Protocols, Author: Jensen, M.K. & Keasling, J.D. (Eds.), Publisher: Humana Press, 2022

Parts, Devices and Applications, Author: Nielsen, A.A.K. & Der, B.S. (Eds.), Publisher: Wiley-VCH, 2018

Philosophical and Scientific Perspectives, Author: Malaterre, C. & Boulanger, J. (Eds.), Publisher: Springer, 2017

Protocells and Non-Natural Life, Author: Rasmussen, S., Bedau, M.A., Chen, L., Deamer, D., Krakauer, D.C., Packard, N.H., & Stadler, P.F. (Eds.), Publisher: MIT Press, 2009

Standardization and Characterization, Author: Strych, U. (Ed.), Publisher: Springer, 2021

Synthetic Gene Circuits: Design and Implementation, Author: Wu, M. & Su, R., Publisher: Humana Press, 2021

The Design and Construction of Biological Systems, Author: Andrianantoandro, E., Basu, S., Karig, D.K., & Weiss, R., Publisher: Nature Publishing Group, 2006

The Engineering of a New Biology Era, Author: Collins, J.J., Publisher: HarperCollins, 2018

The Engineering of Biodegradation Pathways, Author: de Lorenzo, V. (Ed.), Publisher: Caister Academic Press, 2017

The Engineering of Biofuels, Author: Stephanopoulos, G. (Ed.), Publisher: Springer, 2014

The Engineering of Biological Circuits, Author: Alon, U., Publisher: CRC Press, 2020

The Engineering of Biological Security, Author: Lentzos, F. (Ed.), Publisher: Routledge, 2018

The Engineering of Biological Systems for Biotechnology, Author: Weber, W. & Fussenegger, M. (Eds.), Publisher: Wiley-VCH, 2012

The Engineering of Biology and Society, Author: Hilgartner, S., Miller, C., & Smith-Doerr, L. (Eds.), Publisher: The MIT Press, 2017

The Engineering of Biopharmaceuticals, Author: Mattanovich, D. & Gasser, B. (Eds.), Publisher: Humana Press, 2021

The Engineering of Biosecurity, Author: Trump, B.D. & Linkov, I. (Eds.), Publisher: Springer, 2020

The Engineering of Biosensors, Author: van der Meer, J.R. & Belkin, S., Publisher: Springer, 2016

The Engineering of Cell-Free Systems, Author: Noireaux, V. & Shin, J., Publisher: Springer, 2021

The Engineering of Nitrogen Fixation, Author: Dixon, R. & Geddes, B., Publisher: Springer, 2019

The Engineering of Open Source Biology, Author: Maurer, S.M. & Rai, A., Publisher: The MIT Press, 2013

The Engineering of Pattern Formation, Author: Isalan, M., Publisher: CRC Press, 2019

The Engineering of Photosynthesis, Author: Liu, W. & Blankenship, R.E., Publisher: Elsevier, 2020

The Engineering of Plant Systems, Author: Liu, W. & Stewart, C.N., Publisher: Elsevier, 2020

The Engineering of Population Dynamics, Author: You, L. & Hasty, J., Publisher: Springer, 2015

The Engineering of Programmable Materials, Author: Tang, T.-C. & Zhong, C., Publisher: Wiley-VCH, 2021

The Engineering of Protein-Based Devices, Author: Dueber, J.E. & Lim, W.A., Publisher: Springer, 2015

The Engineering of Public Engagement, Author: Balmer, A.S. & Calvert, J., Publisher: Palgrave Macmillan, 2016

The Engineering of Responsible Innovation, Author: Stilgoe, J., Guston, D.H., & Sarewitz, D., Publisher: Routledge, 2020

The Engineering of RNA Systems, Author: Chappell, J. & Lucks, J.B., Publisher: Wiley, 2019

The Engineering of Signaling Networks, Author: Lim, W.A. & Mayer, B., Publisher: Wiley, 2017

The Engineering of Space Biology, Author: Menezes, A.A. & Cumbers, J., Publisher: Elsevier, 2021

The Engineering of Sustainable Systems, Author: Arkin, A.P. & Fletcher, D.A., Publisher: MIT Press, 2016

The Engineering of the Bioeconomy, Author: Carlson, R., Publisher: Princeton University Press, 2016

The Engineering of the Future of Medicine, Author: Hood, L. & Price, N.D., Publisher: Basic Books, 2014

The Engineering of Therapeutic Systems, Author: Lu, T.K. & Collins, J.J., Publisher: Academic Press, 2018

The Engineering of Xenobiology, Author: Herdewijn, P. & Marlière, P. (Eds.), Publisher: Wiley-VCH, 2019

The Engineering of Yeast for Biotechnology, Author: Nevoigt, E. (Ed.), Publisher: Springer, 2017

The Intellectual Property Landscape, Author: Sherkow, J.S. & Abbott, R., Publisher: Oxford University Press, 2020

The Next Phase of Biotechnology, Author: Frow, E. & Calvert, J. (Eds.), Publisher: Palgrave Macmillan, 2018

The Quest for New Biological Functions, Author: Danchin, A., Publisher: World Scientific, 2013

The Role of Modeling and Simulation, Author: Marchisio, M.A., Publisher: Springer, 2021

The Science of Biobricks, Author: Torrance, A.W. & Kahl, L.J. (Eds.), Publisher: MIT Press, 2015

The Technoscience and its Societal Consequences, Author: Boldt, J. (Ed.), Publisher: Palgrave Macmillan, 2016

The Technoscience of Artificial Life, Author: Bensaude-Vincent, B. & Newman, S.A. (Eds.), Publisher: Routledge, 2020

The Technoscience of the 21st Century, Author: Boudry, M. & Pigliucci, M. (Eds.), Publisher: Harvard University Press, 2018

Tools and Applications, Author: Zhao, H. (Ed.), Publisher: Academic Press, 2013

Tools for Engineering Biological Systems, Author: Valla, S. & Lale, R. (Eds.), Publisher: Caister Academic Press, 2017

Index

U

V

W

X

Z

www.ingramcontent.com/pod-product-compliance
Lightning Source LLC
LaVergne TN
LVHW071245210926
848486LV00008B/352

* 9 7 8 1 7 8 7 1 5 6 5 8 6 *